Physicochemical Groundwater Remediation

Physicochemical Groundwater Remediation

Edited by

James A. Smith

and

Susan E. Burns

Department of Civil Engineering
University of Virginia
Charlottesville, Virginia

Springer Science+Business Media, LLC

Library of Congress Cataloging-in-Publication Data

Physicochemical groundwater remediation/edited by James A. Smith and Susan E. Burns.
p. cm.
Includes bibliographical references and index.

DOI 10.1007/978-0-306-46928-2
1. Groundwater—Purification. I. Smith, James A. II. Burns, Susan E.

TD426 .P48 2001
628.1'68—dc21

2001023460

Originally published by Kluwer Academic/Plenum Publishers, New York in 2001
MyCopy version of the original edition 2001

http://www.wkap.nl/

10 9 8 7 6 5 4 3 2 1

A C.I.P. record for this book is available from the Library of Congress

Preface

Timely and cost-effective restoration of contaminated soil and groundwater continues to be one of the most challenging problems facing environmental scientists and engineers. Remediation of contaminated subsurface environments is a multi-billion dollar business owing to years of improper waste management and disposal practices at tens of thousands of sites around the globe, and the current absence of technologies to effectively clean many of these sites.

Several factors contribute to our inability to remediate contaminated subsurface environments. First, subsurface heterogeneities often create regions of low water permeability. Over long periods of contaminant release to the subsurface, pollutants can diffuse into these low-permeability zones. When a conventional remediation technology such as pump-and-treat is applied, pollutants must diffuse back out of these low-permeability zones before they can be captured by the pump-and-treat system. Second, pollutants may desorb slowly from natural soil, or, in cases of gross subsurface contamination, pollutants may be present as nonaqueous phase organic liquids (NAPLs). For these cases, remediation efforts may be further limited by pollutant desorption and/or dissolution rates. Many organic pollutants are highly resistant to biodegradation or chemical transformation under natural conditions, and usually cannot be transformed unless present in the aqueous phase. Oftentimes, groundwater is contaminated by a variety of organic and inorganic pollutants, and a remediation technology that works well for one pollutant may not be appropriate for another. When pump-and-treat remediation systems are used, they are rarely designed for optimum performance.

Most of us would agree that much work remains to be done. As we transition into the 21st century, it is also apparent that this is an exciting time for engineers and scientists studying remediation technologies. Our research activities of the past and next 10 years will have a dramatic impact on the quality of the subsurface environment for the next century. In 20, or even 10 years from now, our approach to subsurface remediation will probably be vastly different than it is today. Many of the emerging technologies presented here will form the basis of standard remediation practices.

The first two chapters of this book address the design of pump-and-treat systems for remediation and/or containment of contaminated groundwater. Using powerful optimization algorithms built on top of conventional flow and solute-transport models, groups from the University of Massachusetts and the University of Virginia demonstrate the optimal design of pump-and-treat remediation systems. Using appropriate constraints and penalty functions, the optimization algorithms efficiently search for the optimal system design with regard to variables such as well pumping rate, well location, and management periods. These relatively inexpensive simulations can significantly reduce the cost and improve the effectiveness of pump-and-treat remediation technologies.

Chapters 3 through 5 present emerging technologies for chemical, electrochemical, and biochemical remediation processes. Munakata and Reinhard from Stanford University discuss the advantages and limitations of palladium catalysts for the reductive dehalogenation and hydrogenation of a variety of contaminants in wastewater. Researchers from Northeastern University and the U.S. Environmental Protection Agency discuss the transport and degradation of groundwater pollutants subjected to electric fields and include a discussion of the mobilization of pollutants from low-permeability media.

Chapters 6 through 8 focus on the use of sorptive and reactive in situ treatment walls. Alan Rabideau and his co-workers begin this series with an overview of sorptive vertical barrier technologies. They explore both low- and high-permeability systems and present two case studies of barrier performance. Baolin Deng and Shaodong Hu from New Mexico Tech discuss the reaction rates of chlorinated solvents on zero-valent iron surfaces. Their analyses will help guide future designs of these reactive treatment walls. Robert Bowman and his co-workers describe a pilot-scale test of a permeable, sorbing treatment wall composed of a surfactant-modified zeolite at the Large Experimental Aquifer Facility of the Oregon Graduate Institute. In this study, they observe and analyze problems associated with permeability reductions in the treatment zone.

Surfactant-enhanced aquifer remediation for both sorbed pollutants and nonaqueous phase liquids is addressed in Chapters 9 through 13. Chapters 9 and 10 discuss the effects of surfactants on the sorption of organic contaminants to natural soil. Seok-Oh Ko and co-workers present data on the equilibrium distribution of pollutants in the presence of surfactants, whereas James Deitsch and Elizabeth Rockaway discuss how surfactants can increase a pollutant's desorption rate by increasing the desorption mass-transfer coefficient and the desorption concentration gradient. In Chapter 11, Wu and co-workers from the University of Oklahoma discuss surfactant selection for separate-phase oil removal from the subsurface. They demonstrate that surfactant hydrophobicity should relate to oil hydrophobicity, and they describe a method to quantify the hydrophobicity of multicomponent nonaqueous phase liquids. In the next chapter, Kibbey and co-workers examine the use of surfactant/alcohol mixtures to reduce the density of nonaqueous phase liquids prior to their mobilization. This approach prevents downward migration of the nonaqueous phase liquid and possible complications with its extraction from the subsurface. In Chapter 13, Taylor and Pennell from Georgia Tech report on the use of surfactant/ethanol formulations to increase the solubilization of nonaqueous phase tetrachloroethene.

Finally, Chapters 14 and 15 address the remediation of the unsaturated zone. In Chapter 14, researchers from the University of Virginia study the effects of natural atmospheric pressure variations on the flow of air into and out of the unsaturated zone at the Picatinny Arsenal in northern New Jersey. This "barometric pumping" contributes to the natural remediation of the shallow, trichloroethylene-contaminated groundwater. In the last chapter of the book, Richard Meixner and co-workers present a detailed field study documenting the effectiveness of soil-vapor extraction to remediate gasoline hydrocarbons in the unsaturated zone.

Acknowledgements

The "seed" for this book was planted at a meeting of the Water Quality committee of the Hydrology Section of the American Geophysical Union (AGU) sometime in late 1997. At that meeting, we proposed to the committee a special session on "Physicochemical Remediation of the Subsurface Environment". Based on the favorable response of the committee, we moved forward with this special session for the Fall '98 AGU meeting in San Francisco, and our seed had taken root. The session was highly successful, and the discussions were lively and interesting. It was apparent to us that ideas for many new and exciting physicochemical remediation technologies were developing and growing. Following the meeting, we contacted participants in the session and other experts in this subject area and invited them to contribute a chapter to the book you are reading today. We think you will find an impressive array of research results and analyses herein. All of the chapters submitted to us were critically reviewed by two or more anonymous peer reviewers, and only the most favorably reviewed submissions were included in this book. As a result of this multi-stage review process (beginning back at that first AGU Water Quality committee meeting), we think that our early efforts have grown into a valuable book for practicing environmental scientists and engineers, environmental decision makers, and environmental researchers in academia and government.

To this end, we thank the members of the AGU Water Quality committee for encouraging us to proceed with our original idea. We also thank the many participants of the Fall '98 session for their presentations and lively discussions, and for their enthusiasm for extending the session topic into a refereeed book. Certainly, this book would not have been possible without the high-quality contributions of each of the author groups and the careful technical evaluations of the anonymous reviewers. We offer all of you a sincere "thank you" for your hard work.

James A. Smith
Susan E. Burns

Contents

Chapter One

Dynamic Optimal Design of Groundwater Remediation Using Genetic Algorithms

AMY CHAN HILTON[1], AYSEGUL AKSOY[2], and TERESA B. CULVER[2]

[1] *Department of Civil Engineering, FAMU - FSU College of Engineering, Florida State University, Tallahassee, FL 32310*
[2] *Department of Civil Engineering, University of Virginia, Charlottesville, VA 22904*

Key words: genetic algorithm, optimal design, dynamic optimization

Abstract: The use of genetic algorithms for the dynamic optimal design of pump-and-treat groundwater remediation systems is demonstrated through two new dynamic formulations. In the first formulation in which the contaminant sorption was assumed to be in equilibrium, the lengths of management periods were decision variables. The second formulation assumed a pulsed pumping approach to remove a contaminant with mass-transfer-limited sorption. While the genetic algorithm could successfully solve these dynamic problems, only small percentage reductions in the overall remediation costs were achieved. However, the savings in the operational costs were more significant with the mass transfer-limited pulsed pumping example saving up to 10% compared to continuous pumping and the flexible-length management periods saving up to 3% compared to fixed-length periods. With the high costs of remediation, even a small percentage of savings in operational costs could be significant. For instance, the 3% savings with flexible management periods corresponded to a cost reduction of more than $26,000 that was achieved by allowing for variable length management periods, a relatively simple change within the GA algorithm. Dynamic pumping that is adapted over time to the unique site conditions may be an option to improve the cost-effectiveness of a remediation design, especially for mass-transfer limited sites.

1. INTRODUCTION

The application of optimization to field-scale groundwater quality problems is increasing rapidly (Ahlfeld *et al.*, 1995; Gailey, 1999; Gailey and Gorelick, 1993; Marryott *et al.*, 1993; Nishikawa, 1998; Pinder *et al.*, 1995; Wagner, 1995). Not only can optimal design provide policies superior to those that have been implemented (Wagner, 1995), but it also can be a powerful and efficient design approach to explore varying conditions or constraint sets (Gailey, 1999; Nishikawa, 1998; Pinder *et al.*, 1995). Optimal design can also demonstrate feasibility or infeasibility (Ahlfeld *et al.*, 1995). With the development of optimal design utilities (Ahlfeld and Reifler, 1998; Greenwald, 1993; Sun and Zheng, 1999; Zheng, 1999) for the popular groundwater simulation model, MODFLOW (Harbaugh and McDonald, 1996), the use of optimal remedial design is expected to continue to expand.

Groundwater remediation systems have been designed using a wide variety of optimization approaches, including linear programming (Atwood and Gorelick, 1985), nonlinear programming (McKinney and Lin, 1996; McKinney and Lin, 1995), dynamic programming (Culver and Shoemaker, 1992; Culver and Shoemaker, 1993), and simulated annealing (Marryott et al., 1993; Rizzo and Dougherty, 1996). Recently, genetic algorithms (GAs) have also been utilized for optimal groundwater remediation design. A GA is a random search procedure that uses probabilistic search rules. Previous works applying GAs to optimal groundwater remediation found that this probabilistic search technique successfully identified optimal or near-optimal solutions (McKinney and Lin, 1994; Rogers et al., 1995; Wang and Zheng, 1997; Wang and Zheng, 1998). GAs also have been employed to solve multi-objective groundwater problems, such as groundwater monitoring (Cieniawski et al., 1995) and containment (Ritzel et al., 1994).

There are several advantages to applying GAs to groundwater remediation. GAs, which do not require derivatives, are usually straightforward to implement even for challenging complex problems (Goldberg, 1989), such as groundwater remedial design in which cost functions may be discontinuous and non-differentiable (Culver and Shoemaker, 1997; McKinney and Lin, 1995) and for which the system is usually non-convex (Ahlfeld and Sprong, 1998; Culver and Shenk, 1998). Flexibility is especially important in remedial design given the wide range of remediation alternatives and diversity of field conditions. While a GA approach can readily be adapted to new simulation models, control theory approaches, such as SALQR, may require substantial effort to adapt to new models (Yoon and Shoemaker, 1999).

For groundwater reclamation, dynamic policies, in which the policies change between management periods, have been shown to be more cost-effective than the best steady-state policies (Ahlfeld, 1990; Chang et al., 1992; Culver and Shoemaker, 1992; Karatzas et al., 1996; Minsker and Shoemaker, 1998). Control theory approaches, such as the Successive Approximations Linear Quadratic Regulator method (SALQR), are specifically formulated to address differentiable dynamic systems and have been applied to dynamic groundwater remedial design (Culver and Shoemaker, 1992; Culver and Shoemaker, 1993; Minsker and Shoemaker, 1998). Time-varying remedial design policies also have been determined by other optimization algorithms, such as nonlinear programming (Ahlfeld, 1990), the outer approximation method (Karatzas et al., 1996), genetic algorithms (Huang and Mayer, 1997; Wang and Zheng, 1997; Yoon and Shoemaker, 1999), simulated annealing (Rizzo and Dougherty, 1996), and search techniques (Sun and Yeh, 1998; Yoon and Shoemaker, 1999). While the majority of these alternative approaches have solved problems with relatively few management periods (Sun and Zheng, 1999), problems with as many as twenty-four management periods have been solved (Yoon and Shoemaker, 1999).

This work will demonstrate the utility of GAs with two new applications of dynamic remedial design. In the following section, the genetic algorithm process for remedial design is briefly described. Then to demonstrate the use of GAs, two dynamic remedial design problems are developed and solved with GAs. One optimal design problem demonstrates the use of variable-length management periods and the other solves for pulsed pumping policies.

2. THE GENETIC ALGORITHM FOR REMEDIAL DESIGN

2.1 Genetic Algorithms

GAs are probabilistic search methods based on the mechanics of natural selection and genetics. The basic idea in using a GA as an optimization method is to represent a population of possible solutions in a chromosome-type encoding, called strings, and evaluate these encoded solutions through simulated reproduction, crossover, and mutation to reach an optimal or near-optimal solution. The GA starts with the creation of an initial population of strings whose values are selected randomly. Binary encoding is widely used for encoding of the decision variables. Next the strings are evaluated using an objective function or fitness measure. A mating pool of strings is then generated using a stochastic selection process, such as tournament selection. Tournament selection begins by randomly picking two strings from the population. These two strings are then pitted against each other in a tournament, and the one with the better fitness wins. A copy of the winner is placed in the mating pool. When the number of strings in the mating pool reaches the specified replacement population size, crossover begins. Two strings are selected from the mating pool randomly. A crossover site is selected, and the genetic material at the crossover site is exchanged between the two selected strings. Crossover is performed on each mated pair with a specified probability. The last step is gene mutation, which also is controlled by a specified probability. After mutation, the new population, or generation, is complete. The above procedure is repeated for many generations until a stopping criterion is met. More detailed descriptions of the GA process can be found in Goldberg (1989) and Wang and Zheng (1997).

2.2 Simulation and Cost Models

The optimization model combines the GA with a two-dimensional groundwater flow and transport simulation model, BIO2D-KE (Culver et al., 1996). The GA model was coded using the PGAPack library (Levine, 1996). BIO2D-KE is used in order to evaluate the objective function, and thus the fitness of each string. It implements the finite element method to simulate the advective and dispersive transport of a contaminant that may undergo abiotic and/or biotic degradation and equilibrium and/or kinetic sorption. Contaminant biodegradation is not considered in this work. Within a management period, during which the pumping remains constant, steady-state, saturated flow is assumed, which is described by:

$$\nabla \cdot (T\nabla h) + Q_w = 0 \tag{1}$$

where T is the transmissivity (L^2/T), h is the hydraulic head (L), and Q_w is the well flow rate (L^3/L^2T). The contaminant transport is given by the following relationships:

$$b(n + \rho_b f K_D)\frac{\partial C}{\partial t} = \nabla \cdot (bD\nabla C) - bv \cdot \nabla C + (C_w - C)Q_w - \alpha b\rho_b[(1-f)K_D C - S_K] \tag{2}$$

$$\frac{\partial S_k}{\partial t} = \alpha\left[(1-f)K_D C - S_K\right] \tag{3}$$

where b is the saturated aquifer depth (L), n is the porosity, ρ_b is the dry bulk density of the porous media (M/L^3), f is the fraction of sorption sites in equilibrium, K_D is the equilibrium partitioning coefficient (L^3/M), C is the aqueous concentration of the contaminant (M/L^3), t is time (T), D is the hydrodynamic dispersion coefficient (L^2/T), v is the Darcy velocity (L/T), C_w is the contaminant concentration in the well water (M/L^3), α is the mass transfer rate coefficient (1/T), and S_K is the mass of contaminant sorbed kinetically per unit mass of soil (M/M).

In general, the design objective is to determine the most cost-effective remedial design. The cost functions for pump-and-treat remediation using granular activated carbon (GAC) derived by Culver and Shenk (1998) were adapted for this analysis. In addition to the operating costs and the treatment capital costs considered by Culver and Shenk (1998), the capital costs of well installation have also been included, and the carbon utilization costs are calculated at every simulation time step. Thus the objective function can be described as follows:

$$\min Cost = \sum_{m=1}^{M} G_{pump,m} + \sum_{t=1}^{T} G_{carbon,t} + \sum_{w=1}^{W} G_{cap\ \mathrm{well},w} + G_{cap\ treat} \tag{4}$$

where:

$$G_{pump,m} = \frac{.002725\, S_p t_m}{\varepsilon} \sum_{e=1}^{E} \left[E_{e,m}\left(H_z - h_{e,m} + 0.7P_a\right)\right] \tag{5}$$

$$G_{carbon,t} = \frac{S_c t_s}{d} \beta (C_t)^{\eta} E_{tot,m} \tag{6}$$

$$G_{cap\ \mathrm{well},w} = 5543\, I_w (H_z)^{0.299} \tag{7}$$

$$G_{cap\ \mathrm{treat}} = A\, n_{\mathrm{ads}} \tag{8}$$

$$n_{ads} = \mathrm{ceiling}\left[\frac{t_a}{V_a} \max_m \left(E_{tot,m}\right)\right] \tag{9}$$

$$E_{tot,m} = \sum_{e=1}^{E} E_{e,m} \tag{10}$$

$$C_t = \frac{1}{E_{tot,m}} \sum_{e=1}^{E} E_{e,m} \left(\frac{C_{e,t0} + C_{e,t1}}{2} \right) \tag{11}$$

where $G_{pump,m}$ is the pumping operational costs for management period m; $G_{carbon,t}$ is the operational cost of the GAC treatment facility during simulation time step t; $G_{cap\ well,\ w}$ is the capital and installation cost for well w; $G_{cap\ treat}$ is the capital cost of GAC adsorbers; and M, T and W are the total number of management periods, the total number of simulation time steps and the total number of potential wells (injection or extraction), respectively. Furthermore, t_m is the length of management period m (s); E is the total number of potential extraction wells; $E_{e,m}$ is the volumetric extraction rate at potential extraction well e during management period m (m^3/s); $h_{e,m}$ is the hydraulic head, relative to the well depth, at the end of the management period m at potential extraction well e (m); t_s is the length of a simulation time step (s); β and η are the coefficient and exponent for the carbon adsorption isotherm, respectively; C_t is the weighted average influent concentration to the adsorbers for simulation time step t; $E_{tot,m}$ is the total extraction rate during management period m (m^3/s); and I_w is a flag indicating whether well w is active during any period m, where $I_w = 0$ if well w is never used and $I_w = 1$ if it is ever pumped; n_{ads} is the total number of adsorbers required for the treatment system; and, $C_{e,t0}$ and $C_{e,t1}$ are the aqueous concentrations of the contaminant at extraction well e at the beginning and end of simulation time step t, respectively. The remainder of the model parameters are described in Table 1. Equation 9 ensures that sufficient absorbers are utilized to maintain sufficient contact time. Details on the derivation of the above equations may be found in Culver and Shenk (1998).

Two examples follow which explore different and potentially useful formulations of dynamic design. The first is a flexible-length management period formulation and the second includes pulsed pumping. As described in the following sections, each of the example problems will require constraints, and the decision variables will also vary between problems. When necessary, a multiplicative penalty method (Chan Hilton and Culver, 2000) was used to manage constraints. In applying this penalty method, the cost function (Equation 4) is multiplied by a factor proportional to any constraint violations that a design may have.

Table 1. GAC cost coefficients

Coefficient	Value
Energy cost, S_p (\$/kWh)	0.07
Wire to water pump efficiency, ε	0.5
Pressure required for the adsorber, Pa (psi)	15
Carbon cost coefficient, S_c (\$/kg)	2.20
Fraction of carbon used at the time of removal, d	0.9
Cost per adsorber unit, A (\$/unit)	100,000
Required contact time in adsorber, t_a (s)	900
Pore volume of an adsorber, $V_a(m^3)$	20

3. EXAMPLE 1: FLEXIBLE-LENGTH MANAGEMENT PERIODS

Most previous dynamic optimal design problems have assumed that the total time to remediate the site is predetermined and that all management periods have equal lengths. However, it may be possible to have a more cost-effective remediation strategy by allowing the optimization model to choose appropriate values for these time variables. There has been little published work in which the length of remediation is incorporated into the optimization model. Greenwald and Gorelick (1989) solved their optimization problem with a range of total remediation periods to evaluate the sensitivity of the policies to remediation length.

The objective of this work is to determine whether allowing the length of each management period to be a decision variable will result in significant cost savings. In this example, while the total length of the remediation periods is held fixed, the lengths of individual management periods within the total remediation period will be decision variables. The decision variables for this example are the management period lengths, t_m, and the extraction rates, $E_{e,m}$, at eighteen potential extraction wells.

The results are compared with solutions determined using fixed-length management periods. The hypothetical homogeneous, isotropic, confined aquifer is comprised of 60 finite elements and 77 nodes, with dimensions 1500 m by 900 m (Culver and Shenk, 1998). The initial contaminant plume, which has a maximum toluene concentration of 40 mg/L, is shown in Figure 1. An easterly steady flow was maintained with a constant hydraulic head of 12.0 m and contaminant concentration of 0.0 mg/L on the left side, a constant hydraulic head of 0.0 m and contaminant concentration of 0.0 mg/L on the right side, and no flow at the top and bottom boundaries. In this example, the sorbed phase is assumed to remain in equilibrium with the aqueous phase. Additional aquifer and GAC isotherm parameters are listed in Table 2.

It is assumed that treated water cannot be reinjected. The aqueous concentration at thirteen observation wells must be less than or equal to the water quality goal (0.5 mg/L) by the end of a 5-year remediation period. The locations of these potential extraction wells and observation wells are indicated in Figure 1. Thus, to complete the formulation of this problem, the following three constraints must be added to the management model described in Equations 4 to 11:

$$\sum_{m=1}^{M} t_m = t_R \tag{12}$$

$$C_j \leq C_{goal}, \qquad \forall j \tag{13}$$

$$0 \leq E_{e,m} \leq E_{max,e} \tag{14}$$

where t_R is the total remediation time (5 years); C_j is the aqueous concentration at the jth observation well at the end of the remediation period; C_{goal} is the water quality goal; and $E_{max,e}$ is the upper bound on $E_{e,m}$ (15.5 L/s).

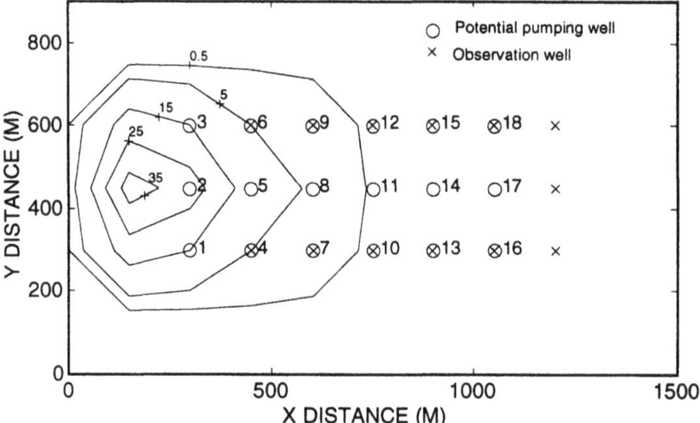

Figure 1. Initial toluene plume (mg/L) and observation and potential extraction wells for Example 1.

Table 2. Aquifer and GAC parameters for Example 1.

Parameter	Value
Hydraulic conductivity, K (m/s)	4.31×10^{-4}
Longitudinal dispersivity, α_L (m)	70
Transverse dispersivity, α_T (m)	3
Porosity, n	0.2
Media bulk density, ρ_b (g/cm^3)	2.12
Aquifer thickness, b (m)	10
Well depth, H_Z (m)	100
Equilibrium partitioning coefficient, K_D (cm^3/g)	0.245
Fraction in sorption equilibrium, f	1
GAC isotherm coefficient, β	0.0462
GAC isotherm exponent, η	0.136

This example was first solved using one, two, and three equal-length management periods (with management period lengths of 60 months, 30 months, and 20 months, respectively). Thus the extraction rates, $E_{e,m}$, were the only decision variables. The problem then was re-solved with flexible-length management periods, in which the GA determined both the extraction rates and the optimal length of each management period for the two and three period cases.

For the formulation in which the management period lengths are not included as decision variables, the string, which represented the extraction well rates, had a length of 60 bits (*len* =60) for each management period. The well rates for 12 of the 18 potential wells (those wells located at the center and bottom rows in Figure 1) were determined by the GA. Assuming symmetry, well rates of the top row of wells are set to the rates of the corresponding wells on the bottom row. The population size was 2 *len* to 3 *len*. 1.5 *len* to 2 *len* strings were replaced each generation and the GA was run for 2.5 *len* to 3 *len* generations. For dynamic runs, the initial population contained 2 or 4 copies of feasible solution strings found from static formulation trials. For the flexible length management periods formulation, the string length was again 60 bits for the extraction well rates for each management period and 4 to 6 bits for the length of each management period. The population size was 1 *len* to 2 *len* with 0.6 *pop* to *pop*-1 strings replaced each generation, where *pop* is the population size. The number of generations that the GA was run was 2.4 *len* to 10 *len*.

The optimal predicted remediation costs for all formulations are presented in Table 3. For both the equal- and flexible-length management period formulation, the optimal cost decreases as the number of management periods, M, increases (Table 3), with the three-management period policies saving 15% compared to the static policy. Each of the flexible-length policies is less expensive than their respective equal-length policies, although the savings are not large (Table 3). Although the static policy had 20% lower well capital cost than the other cases, which is due to the use of one less well, its optimal total cost was 9% to 15% higher than the costs of the dynamic cases. The treatment capital cost was the same for all formulations. The operating costs decrease as the number of management periods increases. Additionally the operational costs are slightly lower for the flexible-length period formulations than their respective equal-length formulations, with reductions of 3% (or $26,000) and 0.1% (or $800) for the two-and three-management period cases, respectively (Table 3).

Table 3. Results for Example 1: Remediation costs of optimal policies for equal and flexible length management periods.

	M=1	M=2		M=3	
Cost type	Static	Equal	Flexible	Equal	Flexible
$G_{cap\ well}$ (\$)	87,860	109,800	109,800	109,800	109,800
$G_{cap\ treat}$ (\$)	200,000	200,000	200,000	200,000	200,000
$\Sigma G_{pump,\ m}$ (\$)	244,900	202,300	197,500	183,000	182,700
$\Sigma G_{carbon,\ m}$ (\$)	769,900	673,100	651,700	612,300	611,800
Total Cost ($)	1,302,600	1,185,200	1,159,000	1,105,100	1,104,300

The same five wells are utilized in each of the dynamic cases (Figure 1 and Table 4). Although the two-period policies extract similar total volumes, the pumping policies, including the lengths of the management periods and the extraction rates (Table 4), are quite different. For example, during the initial pumping period, the flexible-length policy extracts at a rate that is 32% less than the extraction rate used in the equal-length policy, while in the second management period the flexible-length rate is 45% greater than the equal-length rate. The policies, including management period lengths, are quite similar for the three-management period cases (Table 4).

For this example, using a dynamic policy, as opposed to a static policy, is far more important than using flexible-length management periods instead of fixed-length periods. However, there were no time-dependent fate and transport mechanisms in this problem, such as mass-transfer limitations or biodegradation kinetics. Time-dependent mechanisms may increase the importance of flexibility in the time domain. Thus, the second dynamic example will include mass-transfer limitations.

Table 4. Results for Example 1: Summary of results for flexible management period experiment.

Case	Well Numbers of Selected Wells	Total Volume Extracted (10^6 m^3)	Total Extraction Rate, $E_{tot,m}$ (m^3/d)	Period Lengths (months)
M=1, static	5, 7, 8, 9	5.99	3280	60
M=2, equal	5, 7, 8, 9, 11	4.97	3500, 1940	30, 30
M=2, flexible	5, 7, 8, 9, 11	4.85	2380, 2810	21, 39
M=3, equal	5, 7, 8, 9, 11	4.49	2380, 3460, 1560	20, 20, 20
M=3, flexible	5, 7, 8, 9, 11	4.49	2380, 3280, 1600	21, 21, 18

4. EXAMPLE 2: PULSED PUMPING

Pulsed pumping, in which pumps are periodically turned off and on, has been suggested as a means of reducing the costs of groundwater remediation (Keely, 1989; Nowack, 1999; Sullivan, 1996). This may be useful for *in situ* bioremediation systems (Minsker and Shoemaker, 1998) and for sites that are mass-transfer limited, such as sites that have been contaminated for extensive periods (Culver *et al.*, 1997; Pavlostathis and Mathavan, 1992; Smith *et al.*, 1990; Steinberg *et al.*, 1987). Slow mass transfer can limit the rate at which a contaminant moves from the sorbed phase to the aqueous phase. Thus, while a pump-and-treat system may initially remove a significant mass of contaminant, the concentration of the extracted water will begin to drop as slow mass transfer limits the concentration in the aqueous phase. As a result, relatively clean water is extracted at the later stages of continuous pumping. Pulsed pumping may be more cost-effective in that the operating costs can be reduced by incorporating resting periods during which the concentration gradient between the sorbed and aqueous phases drives the sorbed contaminant into the aqueous phase. However, several previous studies of pulsed remediation strategies have found that there was little if any improvement in overall mass removal efficiency as compared to continuous pumping (Borden and Kao, 1992; Harvey *et al.*, 1994; Rabideau and Miller, 1994). Here GAs are used to explore the cost benefits that may be gained by pulsed pumping versus continuous pumping. Other pulsed pumping studies did not use remediation costs as the basis of comparison.

A hypothetical aquifer, which is depicted in Figure 2, was used for this example. A homogeneous, isotropic, and confined aquifer, which is similar to that of McKinney and Lin (1996), is 335 m by 244 m in dimensions and has no-flow conditions at the top and bottom boundaries. A constant head difference of 4.2 m between the right and left boundaries creates a flow in the east direction. Constant concentration boundaries on the right and left sides of the aquifer are set to 0 mg/L. The aquifer is discretized into 352 nodes with 391 grid points. The contaminant transport, including mass transfer limitations, was described by Equations 1 to 3. The initial contaminant plume was created by simulating the plume generated from 15 years of contamination from an area source of tetrachloroethane (PCE) with a maximum concentration of 30 mg/L. Sorption parameters were selected based on the study of Goltz and Roberts (1986) and reported mass transfer rates of contaminants for long-term contaminated soils (Culver *et al.*, 1997; Koller *et al.*, 1996). Aquifer and sorption parameters are given in Table 5. The initial PCE plume in Figure 2 is shown in terms of total concentration (aqueous and sorbed) per liter of aquifer. Details of the initial plume generation can be found in Aksoy and Culver (2000).

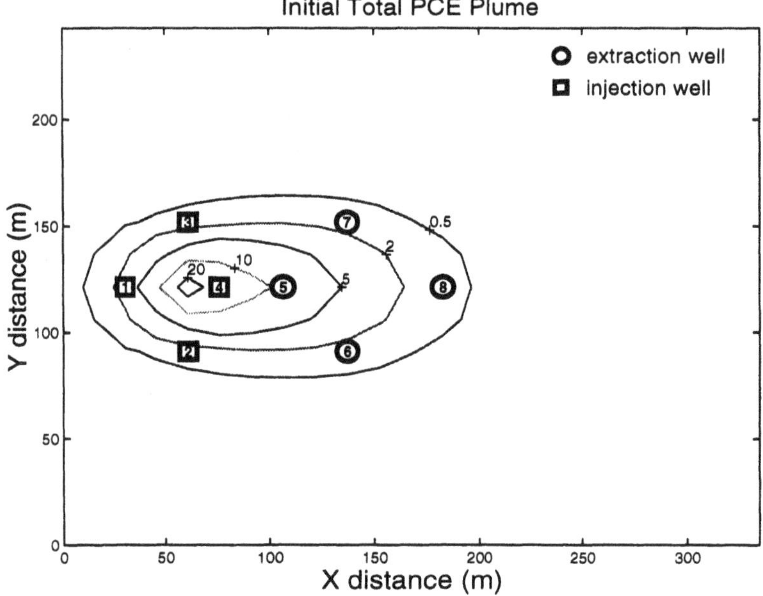

Figure 2. Initial PCE plume in total (aqueous plus solid) concentration (mg/L aquifer) and potential well locations for Example 2.

Table 5. Aquifer and GAC parameters for Example 2.

Parameter	Value
Hydraulic conductivity, K (m/s)	6.1×10^{-6}
Saturated thickness, b (m)	30.5
Longitudinal dispersivity, α_L (m)	21.3
Transverse dispersivity, α_T (m)	3.2
Porosity, n (-)	0.2
Bulk density, ρ_b (g/cm^3)	1.81
Fraction of equilibrium sites, f (-)	0.75
Equilibrium partitioning coefficient, K_d (cm^3/g)	0.36
Mass transfer rate, α (1/d)	0.0025
GAC isotherm coefficient, β	0.017
GAC isotherm exponent, η	0.52

Pumping scenarios with 1, 3, or 5 management periods were considered (M=1, 3, or 5). Odd-numbered periods were the pumping periods, whereas even-numbered periods were the resting periods during which the pumps were turned off. The case with a single management period represents the static, continuous pumping scenario. Decision variables were the extraction and injection rates at prescribed potential well locations for each pumping period, and the lengths of the pumping and resting periods. For the scenarios in which M=3 or 5, one decision variable was utilized for the length of the resting periods, since all resting periods within the total management time were assumed to have equal lengths. Because the contaminant removal efficiency may be greatest for the initial pumping period, this period was allowed to have a unique length. All subsequent pumping periods were assumed to be of equal lengths. Assuming the total amount of pumping time (excluding rest periods) is fixed at 1260 days, a single decision variable can be used for the length of subsequent pumping periods. The length of the first pumping period, t_1, is then calculated using the following constraint:

$$t_1 = 1260 - \sum_{j=?}^{\left(\frac{M-1}{2}+1\right)} t_{2j-1} \tag{15}$$

In addition, it is assumed that extracted water will be re-injected after treatment, so that:

$$\sum_{e=1}^{E} E_{e,m} = \sum_{i=1}^{I} E_{i,m} \tag{16}$$

where $Q_{i,m}$ is the injection rate at potential site i during management period m, and I is the number of injection wells. Upper and lower head levels were enforced such that:

$$h_{\min} \geq h^*_{\min} \tag{17}$$

$$h_{\max} \leq h^*_{\max} \tag{18}$$

Upper and lower bounds on extraction and injection rates were enforced by the binary encoding of the decision variables. These bounds were set as:

$$0.0 \leq E_{e,m} \text{ or } E_{i,m} \leq 7.75\ L/s \tag{19}$$

where h_{min} and h_{max} are the minimum and maximum heads anywhere within the aquifer at anytime during the remediation, respectively (m); and

h^*_{min} and h^*_{max} are the minimum and maximum head bounds, respectively.

The goal of this remediation example is to determine the most cost-effective management policy using a specified number of active pumping days to remediate a site. For a mass transfer limited-site, both the aqueous and total concentrations must be considered to prevent aqueous concentration rebound (Aksoy and Culver, 2000; Harvey *et al.*, 1994). Instead of using observation wells, the water quality goal constrains the maximum aqueous and total concentrations anywhere within the aquifer (Aksoy and Culver, 2000; Aly and Peralta, 1999a; Aly and Peralta, 1999b), as follows:

$$C_{max} \leq C_{goal} = 1 \text{ mg/L} \tag{20}$$

$$C_{Tot\ max} \leq C_{Tot\ goal} = C_{goal}\,(K_D\rho_b + n) = 0.852 \text{ mg/(L aquifer)} \tag{21}$$

where C_{max} and $C_{Tot\ max}$ are the maximum aqueous and total concentrations anywhere within the aquifer at the end of the remediation period (t_R), and the total concentration goal, $C_{Tot\ goal}$, is equal to the sum of the aqueous and sorbed concentration at sorption equilibrium, which is expressed in terms of the mass of contaminant per liter of aquifer. Thus, Equations 4 to 11 and 15 to 21 define the optimal design formulation for the pulsed pumping problem.

Results obtained from the dynamic pulsed pumping problems will be compared to the optimum continuous pumping (M=1) policy, given the same period of active pumping (1260 days). This formulation does not, however, constrain the total time period of remediation, t_R. Note that from a strict engineering economics approach, the costs in Equation 4, should include discount factors for costs incurred at later times. However, polices that require longer periods of remediation are generally less preferred than faster approaches. The relative importance of costs through time versus remediation speed is a qualitative decision. In this work, we have assumed that the preference for early remediation exactly counter-balances the discount factor for costs that occur at later time periods.

Values of the total extraction rates (Q) and lengths of management periods (t_m) obtained through the optimization process and resulting extraction volumes (V) are given in Table 6. Only one extraction well (well 5) and one injection well (well 1) were selected for all pulsed pumping cases (see Figure 2). Optimization results show that the required total volume of water extracted decreases with the application of the dynamic pulsed pumping as compared to the continuous pumping (Table 6). With pulsed pumping, less water is extracted at the later pumping periods. As a result, compared to continuous pumping, reductions of 11% and 14% were obtained in the total extraction volume for pulsed pumping cases with 3 and 5 management periods, respectively. Results on water extraction were also analyzed based on the volume of water extracted per mass of PCE removed. As seen in Figure 3, together with the reduction in total water extraction, water extracted per mass of contaminant removed is also reduced with dynamic pulsed pumping. These reductions correspond to 8% and 11% for M=3 and M=5, respectively.

Table 6. Results for Example 2: Design summaries for continuous (M=1) and pulsed pumping cases with 3 and 5 management periods; Q = pumping rates (m^3/d), t_m= management period length (d), and V = resulting extraction volume (m^3).

Period	M=1			M=3			M=5		
m	Q	t_m	V	Q	t_m	V	Q	t_m	V
1	345.6	1260	435500	302.4	1230	372000	302.4	1200	362900
2				0.0	330	0	0.0	384	0
3				540.0	30	16200	345.6	30	10400
4							0.0	384	0
5							21.6	30	600
Total		1260	435500		1590	388200		2028	373900

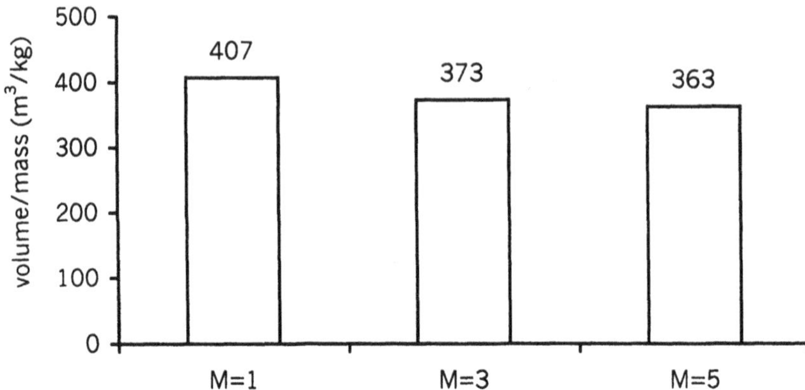

Figure 3. Volume of water extracted per mass of PCE removed from aquifer given different numbers of management periods for Example 2.

Remediation costs for optimal policies obtained for the continuous and pulsed pumping cases are given in Table 7. Pulsed pumping resulted in significant savings in total operating costs. Compared to continuous pumping, approximately 6% and 8% savings were achieved in treatment operating costs for pulsed pumping with 3 and 5 periods, respectively. Savings in pumping operating costs were 14% and 17% for 3 and 5 management periods, respectively. The capital costs were unchanged. However, since the operating costs represent only 21% of the continuous pumping costs for this example, the overall costs savings were minor (up to 2%). As can be seen in Table 6, the pulsed pumping scenarios required significantly longer remediation periods (26% and 61% for the M=3 and M=5 cases, respectively).

Table 7. Results for Example 2: Remediation costs of the optimal policies for continuous (M=1) and pulsed pumping cases with 3 and 5 management periods.

cost type	M=1	M=3	M=5
$G_{cap\ well}$ ($)	38000	38000	38000
$G_{cap\ treat}$ ($)	100000	100000	100000
$\Sigma G_{pump,\ m}$ ($)	8800	7600	7300
$\Sigma G_{carbon,\ m}$ ($)	27500	25900	25400
Total Cost ($)	174300	171500	170700

Although optimized pulsed pumping does not remove more contaminant mass than optimized continuous pumping, when properly designed it can remove as much contaminant as continuous pumping (Harvey, et al., 1994). One of the major advantages of pulsed pumping may be reduction of the total volume of water extracted as shown in the dynamic pulsed pumping example. With pulsed pumping, it was possible to meet the water quality standards with less water extracted per mass of contaminant removed within the same pumping time as continuous pumping.

5. CONCLUSIONS

This work has demonstrated the use of genetic algorithms for the optimal dynamic design of remediation systems. In this study, small percentage reductions in the overall remediation costs were achieved through flexible-length management periods and pulsed pumping. However, the savings in the operational costs were more significant with the mass transfer-limited pulsed pumping example saving up to 10% compared to continuous pumping and the flexible-length management periods saving up to 3% compared to fixed-length periods. Reported annual operation and maintenance costs for pump-and-treat remediation of Superfund sites reach up to 35% of the total capital costs (Du Teaux, 1997). Considering the estimated average time of cleanup for 1249 Superfund sites is at least 12 years (Reisch and Bearden, 1997), even a small percentage of savings in operational costs could be significant. For instance, the 3% savings with flexible management periods corresponded to a cost reduction of more than $26,000 that was achieved by allowing for variable length management periods, a relatively simple change within the GA algorithm. With the high costs of remediation and large number of contaminated sites, dynamic pumping that is adapted over time to the unique site conditions may be an option to improve the cost-effectiveness of a remediation design, especially for mass-transfer limited sites.

There is potential for even greater savings in operational costs with the optimal dynamic designs. Flexible-length management periods may be more appropriate for systems where the transport mechanisms are kinetic. Furthermore, we made several assumptions about the lengths of management periods in the pulsed pumping case (equal-length rest periods and the length of the second and third pumping periods were assumed constant). By relaxing these assumptions for the pulsed pumping problem so that each period length is a variable and by allowing for a larger number of pumping periods, it is likely that the GA would find even more cost-effective policies. Also, we did not consider that labor costs may be reduced with pulsed pumping due to discontinuous operation. These issues will be explored in future research.

ACKNOWLEDGMENTS

This work was supported in part by a National Science Foundation (NSF) CAREER award to T.B. Culver, by an NSF traineeship to A.B. Chan Hilton through the University of Virginia's Program in Interdisciplinary Research in Contaminant Hydrogeology, by a doctoral fellowship award to A.B. Chan Hilton from the American Association of University Women, and by a graduate fellowship to A. Aksoy from TUBITAK, the Turkish Scientific and Technical Research Council. Computational resources were provided through the IBM Shared University Research program, an IBM Environmental Research Grant, NSF, and the Division of Information Technology and Communications at the University of Virginia.

BIBLIOGRAPHY

Ahlfeld, D. P. (1990). "Two-stage ground-water remediation design." J. Water Resour. Plng. and Mgmt Div., ASCE, 116(4), 517-529.

Ahlfeld, D. P., Page, R. H., and Pinder, G. F. (1995). "Optimal ground-water remediation methods applied to a superfund site: From formulation to implementation." Ground Water, 33(1), 58-70.

Ahlfeld, D. P., and Reifler, R. G. (1998). "Documentation for MODOFC: A Program for solving optimal flow control problems based on MODFLOW simulation, Version 2.1." Available from http://www.ecs.umass.edu/modofc.

Ahlfeld, D. P., and Sprong, M. P. (1998). "Presence of nonconvexity in groundwater concentration response functions." J. Water Resour. Plng. and Mgmt Div., ASCE, 124(1), 8-14.

Aksoy, A., and Culver, T. B. (2000). "Effect of sorption assumptions on the optimization of aquifer remediation designs." Ground Water, 38(2),200-208.

Aly, A. H., and Peralta, R. C. (1999a). "Comparison of a genetic algorithm and mathematical programming to the design of groundwater cleanup systems." Water Resour. Res., 35(8), 2415-2425.

Aly, A. H., and Peralta, R. C. (1999b). "Optimal design of aquifer cleanup systems under uncertainty using a neural network and a genetic algorithm." Water Resour. Res., 35(8), 2523-2532.

Atwood, D. F., and Gorelick, S. M. (1985). "Hydraulic gradient control for groundwater contaminant removal." J. Hydrol., 76, 85-106.

Borden, R. C., and Kao, C.-M. (1992). "Evaluation of groundwater extraction for remediation of petroleum-contaminated aquifers." Water Environ. Res., 64(1), 28-36.

Chan Hilton, A. B., and Culver, T. B. (2000). "Constraint-handling methods for genetic algorithms in optimal pump-and-treat design." J. Water Resour Plng. and Mgnt Div., ASCE, 126(3), 128-137.

Chang, L. C., Shoemaker, C. A., and Liu, P. L. F. (1992). "Optimal time-varying pumping rates for groundwater remediation: Application of a constrained optimal control algorithm." Water Resour. Res., 28(12), 3157-3171.

Cieniawski, S. E., Eheart, J. W., and Ranjithan, S. (1995). "Using genetic algorithms to solve a multiobjective groundwater monitoring problem." Water Resour. Res., 31(2), 399-409.

Culver, T. B., Earles, T. A., and Gray, J. P. (1996). "Numerical modeling of in situ bioremediation with sorption kinetics." Rep. No. DAAL03-91-C-0034; TCN95-066. U.S. Army Research Office, Research Triangle Park, NC.

Culver, T. B., Hallisey, S. P., Sahoo, D., Deitsch, J. J., and Smith, J. A. (1997). "Modeling the desorption of organic contaminants from long-term contaminated soil using distributed mass transfer rates." Environ. Sci. Technol., 31(6), 1581-1588.

Culver, T. B., and Shenk, G. W. (1998). "Dynamic optimal groundwater remediation by granular activated carbon." J. Water Resour. Plng. and Mgmt Div., ASCE, 124(1), 59-64.

Culver, T. B., and Shoemaker, C. A. (1992). "Dynamic optimal control for groundwater remediation with flexible management periods." Water Resour. Res., 28(3), 629-641.

Culver, T. B., and Shoemaker, C. A. (1993). "Optimal control for groundwater remediation by differential dynamic programming with quasi-Newton approximations." Water Resour. Res., 29(4), 823-831.

Culver, T. B., and Shoemaker, C. A. (1997). "Dynamic optimal ground-water reclamation with treatment capital costs." J. Water Resour. Plng. and Mgmt Div., ASCE, 123(1), 23-29.

Du Teaux, S. B. (1997). "A compendium of cost data for environmental remediation technologies." Rep. No. LA-UR-96-2205. Energy and Environmental Analysis Group, Los Alamos National Laboratory, Los Alamos, New Mexico. Available from http://www.lanl.gov/projects/etcap.

Gailey, R. (1999). "Application of Mixed-integer linear programming techniques for water supply wellfield management and plume containment at a California EPA site." Proc., 26th Annual Water Resources Planning and Management Conference., ASCE, Tempe, Arizona, 1-8.

Gailey, R., and Gorelick, S. M. (1993). "Design of optimal, reliable plume capture schemes: application to the Gloucester Landfill ground-water contamination problem." Ground Water, 31(1), 107-114.

Goldberg, D. E. (1989). "Genetic algorithms in search, optimization, and machine learning," Addison-Wesley Publishing, Reading, Massachusetts.

Goltz, M. N., and Roberts, P. V. (1986). "Interpreting organic solute transport data from a field experiment using physical nonequilibrium models." J. Contam. Hydrol., 1(1), 77-93.

Greenwald, R. M. (1993). "Documentation and User's Guide: MODMAN, An Optimization Module for MODFLOW, Version 3.02." Colorado School of Mines, International Ground Water Modeling Center.

Harbaugh, A. W., and McDonald, M. G. (1996). "User's documentation for MODFLOW-96, an update to the U.S. Geological Survey Modular finite-difference ground-water flow model." Open-File Report 96-485. U.S. Geological Survey, Reston, VA. Available at http://water.usgs.gov/software/modflow-96.html.

Harvey, C. F., Haggerty, R., and Gorelick, S. M. (1994). "Aquifer remediation: A method for estimating mass transfer rate coefficients and an evaluation of pulsed pumping." Water Resour. Res., 30(7), 1979-1991.

Huang, C., and Mayer, A. S. (1997). "Pump-and-treat optimization using well locations and pumping rates as decision variables." Water Resour. Res., 33(5), 1001-1012.

Karatzas, G. P., Spiliotopoulos, A. A., and Pinder, G. F. (1996). "A multiperiod approach for the solution of groundwater management problems using the outer approximation method." Proc., North American Water and Environment Congress, A.S.C.E., June 23-28, 1996., Anaheim, California,

Keely, J. F. (1989). "Performance evaluations of pump-and-treat remediations." Rep. No. EPA/540/4-89/005. U.S. Environmental Protection Agency, Washington, D.C.

Koller, D., Imbrigiotta, T. E., Baehr, A. L., and Smith, J. A. (1996). "Desorption of trichloroethylene from long-time contaminated aquifer sediments to ground water at Picatinny Arsenal, New Jersey." Proc., U.S. Geological Survey Toxics Substances Hydrology Program--Proceedings of the technical meeting September 20-24, 1993., U.S. Geological Survey Water-Resources Investigations Report 94-4015, Colorado Springs, Colorado, 329-338.

Levine, D. (1996). "User's guide to the PGAPack parallel genetic algorithm library." Rep. No. ANL-951/18. Available from ftp://ftp.mcs.anl.gov/pub/pgapack/pgapack.tar.Z.

Marryott, R. A., Dougherty, D. E., and Stollar, R. L. (1993). "Optimal groundwater management - 2. Application of simulated annealing to a field-scale contamination site." Water Resour. Res., 29(4), 847-860.

McKinney, D. C., and Lin, M.-D. (1994). "Genetic algorithm solution of groundwater management models." Water Resour. Res., 30(6), 1897-1906.

McKinney, D. C., and Lin, M.-D. (1996). "Pump-and-treat ground-water remediation system optimization." J. Water Resour. Plng. and Mgmt Div., ASCE, 122(2), 128-136.

McKinney, D. C., and Lin, M. D. (1995). "Approximate mixed-integer nonlinear programming methods for optimal aquifer remediation design." Water Resour. Res., 31(3), 731-740.

Minsker, B. S., and Shoemaker, C. A. (1998). "Dynamic optimal control of in-situ bioremediation of ground water." J. Water Resour. Plng. and Mgmt Div., ASCE, 124(3), 149-161.

Nishikawa, T. (1998). "Water-resources optimization model for Santa Barbara, California." J. Water Resour. Plng. and Mgmt Div., ASCE, 124(5), 252-263.

Nowack, K. W. (1999). "An analysis of pulse pumping and bioremediation and the effects of rate limited sorption". Master's Thesis, Department of Civil Engineering, University of Virginia, Charlottesville, Virginia.

Pavlostathis, S. G., and Mathavan, G. N. (1992). "Desorption kinetics of selected volatile organic compounds from field contaminated soils." Environ. Sci. Technol., 26, 532-538.

Pinder, G. F., Ahlfeld, D. P., and Page, R. H. (1995). "Cleanup Solution - Conflict Resolution." Civil Engineering, 65(3), 59-61.

Rabideau, A. J., and Miller, C. T. (1994). "Two-dimensional modeling of aquifer remediation influenced by sorption non-equilibrium and hydraulic conductivity heterogeneity." Water Resour. Res., 30(5), 1457-1470.

Reisch, M., and Bearden, D. M. (1997). "Superfund Fact Book." Rep. No. 97-312ENR. Congressional Research Service. Available from http://www.cnie.org/nle/waste-1.html.

Ritzel, B. J., Eheart, J. W., and Ranjithan, S. (1994). "Using genetic algorithms to solve a multiple-objective groundwater pollution containment-problem." Water Resour. Res., 30(5), 1589-1603.

Rizzo, D. M., and Dougherty, D. E. (1996). "Design optimization for multiple management period groundwater remediation." Water Resour. Res., 32(8), 2549-2561.

Rogers, L. L., Dowla, F. U., and Johnson, V. M. (1995). "Optimal field-scale groundwater remediation using neural networks and genetic algorithms." Environ. Sci. Technol., 29(5), 1145-1155.

Smith, J. A., Chiou, C. T., Kammer, J. A., and Kile, D. E. (1990). "Effect of soil moisture on the sorption of trichloroethene vapor to vadose-zone soil at Picatinny Arsenal, New Jersey." Environ. Sci. Technol., 24(5), 676-683.

Steinberg, S. M., Pignatello, J. J., and Sawhney, B. L. (1987). "Persistence of 1,2-dibromoethane in soils: entrapment in intraparticle micropores." Environ. Sci. Technol., 21, 1201-1208.

Sullivan, R. A. (1996). "Pump and treat and wait." Civil Engineering, 66(11), 8A-12A.

Sun, M., and Zheng, C. (1999). "Long-term groundwater management by a MODFLOW based dynamic optimization tool." J. AWRA, 35(1), 99-111.

Sun, Y.-H., and Yeh, W. W.-G. (1998). "Location and schedule optimization of soil vapor extraction system design." J. Water Resour. Plng. and Mgmt Div., ASCE, 124(1), 47-58.

Wagner, B. J. (1995). "Recent advances in simulation-optimization groundwater management modeling." Rev. Geophys., (Supplement), 1021-1028.

Wang, M., and Zheng, C. (1997). "Optimal remediation policy selection under general conditions." Ground Water, 35(5), 757-764.

Wang, M., and Zheng, C. (1998). "Application of genetic algorithms and simulated annealing in groundwater management: formulation and comparison." J. AWRA, 34(3), 519-530.

Yoon, J.-H., and Shoemaker, C. A. (1999). "Comparison of optimization methods for ground-water bioremediation." J. Water Resour. Plng. and Mgmt Div., ASCE, 125(1), 54-63.

Zheng, C. (1999). "ModGA- A Modular Genetic Algorithm Based Flow and Transport Optimization Model for MODFLOW and MT3D", Summary of the Technical Report to DuPont Company. Hydrogeology Group, University of Alabama, Tuscaloosa. Available from http://hydro.geo.ua.edu/mt3d/modga.htm.

Chapter Two

Optimal Plume Capture Design In Unconfined Aquifers

ANN E. MULLIGAN[1, 2] and DAVID P. AHLFELD[1]

[1] *Department of Civil and Environmental Engineering, University of Massachusetts, Amherst, MA 01003*
[2] *now at Department of Geology and Geophysics, Woods Hole Oceanographic Institution, Woods Hole, MA 02543*

Key words: groundwater remediation design, optimal plume control, groundwater quality management

Abstract: A combined simulation and optimization model for designing groundwater plume control systems in unconfined aquifers is presented and demonstrated on two- and three-dimensional aquifer problems. Unconfined aquifer simulation poses numerical challenges that are not present when confined conditions are assumed. In unconfined aquifers, excessive extraction from a pumping well may result in drying of the well and cessation of pumping. In the simulation model, excessive extraction rates result in dewatering portions of the numerical domain. To avoid numerical dewatering, constraints can be placed on hydraulic head within the well cell. However, head is a nonlinear function of pumping in unconfined aquifers. The nonlinearity of head and the potential for dewatering must both be considered when applying simulation-optimization models to design containment systems in unconfined aquifers. In this chapter, optimization search procedures for an advective-control model are developed for accommodating unconfined aquifer simulation. The simulation-optimization model represents advective contaminant transport using particle tracking techniques. The goal of the groundwater management strategy is to contain plume migration while minimizing total pumping of the containment system.

1. INTRODUCTION

Left unabated, groundwater contaminant plumes can migrate through the subsurface to downgradient receptors such as water supply wells or surface water bodies and can pose significant risks to human and ecological health. Over the past few decades, many technologies have been developed to control and remediate subsurface contamination. One of the oldest and most widely used remedial techniques is the pump-and-treat method in which a system of extraction and injection wells is used to control the subsurface flow regime such that contaminants are captured and removed by extraction wells. Pump-and-treat design, which requires specifying the location and magnitude of pumping stresses to be imposed on the aquifer system, involves a sequence of decisions that ultimately affects the success and cost of the project. However, the hydrogeologic complexities at many groundwater contamination sites can make it very difficult to use trial-and-error to identify a set of stresses that accomplishes remedial goals and is also cost effective. Combined optimization-simulation models are powerful tools that allow the designer to simultaneously address multiple project objectives, such as identifying the least cost design that satisfies all remedial goals. In this chapter, an optimization-simulation model for designing a set of pumping stresses to contain plume migration in unconfined aquifers is developed and demonstrated.

Optimization methods provide an alternative approach to trial-and-error design and require that the design problem be stated in a mathematical formulation as a set of project goals and constraints; the formulation is then solved numerically. By integrating mathematical optimization tools into the engineering design process, the space of feasible engineering designs can be searched in a systematic manner for the solution that best meets the stated management objective, such as minimizing cost.

Applications of optimization models to site-specific groundwater management problems are described in the literature and demonstrate that the optimization approach can indeed identify strategies that are less costly than the pumping schemes currently operating at the sites of interest (see the reviews by Ahlfeld and Heidari 1994 and Wagner 1995). In some cases, competing constraints render a problem infeasible, an outcome that can be conclusively identified by optimization models (e.g., Ahlfeld et al. 1995) but that can be virtually impossible to determine with the trial-and-error approach.

Three general classes of optimization models for groundwater plume containment design are described in the literature, including hydraulic control models, advective control models, and concentration control models. All three classes of optimization models require that the groundwater flow system be simulated. Hydraulic control models are constructed with the flow simulator only, advective control models append a particle tracking algorithm to simulate advective transport, and concentration control models require a concentration-based advective-dispersive contaminant transport model. In all cases, the design variables are the well locations and pumping rates of the pump-and-treat system and an optimization algorithm is used to find the design that best meets a specified management objective. Several reviews of optimization-simulation models in groundwater quality management are available, including those by Gorelick (1983, 1990), Ahlfeld and Heidari (1994), and Wagner (1995).

Hydraulic control formulations indirectly manage plume migration by constraining either hydraulic gradients (e.g. Lefkoff and Gorelick 1986) or velocity vectors (e.g. Colarullo et al. 1984) at specified locations within the model domain. The constraint values are chosen to force the flow direction along the edge of the plume inwards and toward extraction wells, thereby preventing plume migration beyond constraint locations. Because hydraulic constraints are used as surrogates for the true design objective of capturing the plume, hydraulic control solutions should be checked with post-optimization simulation to ensure that the plume is captured by the control system (Ahlfeld et al. 1995; Ahlfeld and Mulligan 2000).

Concentration control optimization models typically constrain the design stresses to meet specified concentration standards within the aquifer at specified times (Gorelick et al. 1984). These models have nonlinear properties that make them difficult to solve (Ahlfeld and Sprong 1998) and many different optimization methods have been investigated in an effort to identify a technique that is robust in the face of these numerical challenges. Chan Hilton et al. (this book) review some of the different techniques used to solve concentration control formulations and present results of their work in applying a genetic algorithm to designing time-varying pumping schemes subject to concentration constraints.

The third type of groundwater quality management model is based on advective control and uses particle tracking methodologies to determine either contaminant pathlines (Shafer and Vail 1987; Greenwald and Gorelick 1989; Mulligan and Ahlfeld 1999a) or capture zone location (Varljen and Shafer 1993; Mulligan and Ahlfeld 1999a). The optimization formulation seeks a pumping system that succeeds in either capturing the contaminant pathlines or in establishing a capture zone that fully encompasses the contaminant plume.

The advective control model presented in this chapter is an extension of the formulation presented by Mulligan and Ahlfeld (1999a) that uses both contaminant pathline and capture zone simulation to constrain plume capture designs. In this chapter, additional constraints and algorithmic procedures are introduced so that the optimization model can be applied to unconfined aquifer problems.

The formulation described by Mulligan and Ahlfeld (1999a) is insufficient for solving design problems under many unconfined aquifer conditions because there are no provisions to restrict the optimization search to extraction rates that avoid physical and numerical dewatering of portions of the aquifer. Under certain physical conditions, such as low permeability sediments or a thin saturated thickness, extraction from a well may be higher than the flow into the well from the aquifer. Eventually the well becomes dry and extraction ceases. This physical condition manifests itself in numerical models when the head within a grid cell falls below the elevation of the cell. When this occurs, the cell is said to have dewatered and the simulation model fails to converge to a meaningful solution. Grid cell dewatering results in a numerical discontinuity in the optimization model because both head and particle pathlines change abruptly when a cell dewaters. This discontinuity can result in failure of a gradient-based optimization search process. Appropriate steps must therefore be incorporated into the optimization algorithm to accommodate the potential numerical problems associated with simulating extraction pumping in unconfined aquifers.

To avoid dewatering the aquifer, constraints can be imposed on either the upper bounds of extraction rates (e.g., Rastogi 1989; Marryott et al. 1993) or the minimum saturated thickness at numerical grid cells (e.g., Ahlfeld et al. 1998). The later approach pushes the physical limits of the system by implicitly allowing the system behavior to define the upper pumping bounds. In this later approach, however, care must be taken to ensure that the optimization search procedure remains feasible with respect to the saturated thickness constraints.

In unconfined aquifers, head is a nonlinear function of pumping and minimum saturated thickness constraints, which are equivalent to lower head bound constraints, are therefore nonlinear. When solving an optimization problem with such constraints, the nonlinearity can be treated by assuming that the nonlinearity is weak and that head changes linearly with pumping. When drawdown is small relative to total saturated thickness, this approximation is likely adequate (Colarullo et al. 1984). However, if drawdown is significant, then the nonlinearity can be significant and linear approximations of the head response must be updated periodically to reflect the current state of the system (e.g., Willis and Finney 1985; Jones et al. 1987; Riefler and Ahlfeld 1996). An alternative approach to accommodating the nonlinear head response is to use nonlinear programming techniques (e.g., Tiedeman and Gorelick 1993; Barlow et al. 1996).

In this paper, the advective control model for groundwater plume capture design is described, algorithmic requirements to accommodate unconfined aquifer simulation are presented, and two- and three- dimensional example problems are used to demonstrate the optimization model capabilities and design implications. The model is applicable for designing long-term plume containment systems and as such assumes steady-state flow and time-invariant pumping.

2. GOVERNING EQUATIONS

In the optimization model presented in the next section, advective contaminant transport is simulated by particle tracking, which is accomplished by numerically tracking a hypothetical particle through the groundwater velocity field. In forward tracking, the particle represents a small volume of groundwater and the numerical procedure maps fluid pathlines. For steady flow systems, solute pathlines are coincident with groundwater pathlines and a set of particles originating within or on the contaminant plume boundary simulates advective transport. In reverse tracking, the direction of the groundwater velocity vectors is reversed and particles originating adjacent to an extraction well map the capture zone of the well.

2.1 Groundwater Flow

Prior to tracking particles, the groundwater flow equation must be solved in order to determine the hydraulic head distribution within the aquifer. The two-dimensional, steady-state groundwater flow equation is

$$\nabla \cdot (\mathbf{K} b \nabla h) + q = 0 \tag{1}$$

where $\mathbf{K}$ is the hydraulic conductivity (L/T), h is hydraulic head (L), b is the saturated thickness (L), and q represents sources ($q > 0$) and sinks ($L^3/T/L^2$). In unconfined aquifers, the saturated thickness is a function of hydraulic head and equation (1) is therefore a nonlinear partial differential equation.

A common approach in three-dimensional groundwater simulation (e.g., McDonald and Harbaugh 1984) is to represent the vertical dimension with several numerical grid layers and to determine the head within each layer using equation (1). Flow between the layers is accommodated with a vertical conductance term.

The groundwater velocity field is determined by Darcy's law after the head distribution from solution of (1) has been determined:

$$\mathbf{U} = -\frac{1}{\Theta}\mathbf{K}\,(\nabla h) \tag{2}$$

where U is the velocity vector (L/T) and Θ is porosity (dimensionless).

2.2 Advective Transport

Advective transport is simulated in the advective control model by particle tracking after the velocity field is determined. Following solution of equations (1) and (2), a particle can be tracked as it travels through any part of the domain (Shafer 1987). A particle pathline is the trace of particle position over time, which is determined as

$$\int_{s_o}^{s_f} ds = \int_{t_o}^{t} u_s\, dt \tag{3}$$

where s is the particle coordinate along the pathline, u_s is the particle velocity at location s, t is time, and s_o and s_f are initial and final particle locations, respectively. Forward tracking is used in the advective control model to simulate the advective transport of groundwater contaminants.

Reverse tracking is also used in the advective control model, where the particles are used to delineate the capture zone of all extraction wells. Reverse tracking is performed by reversing the direction of all velocity vectors and integrating from the final location backward to the initial location

$$\int_{s_f}^{s_o} ds = -\int_{t}^{t_o} u_s\, dt \tag{4}$$

3. THE ADVECTIVE CONTROL MODEL

3.1 Model Formulation

The advective control model is a combined simulation-optimization model that seeks to determine the minimum total pumping needed to capture a specified contaminant plume. Two reference frames are used to simulate the plume control problem: (1) reverse particle tracking is used to identify the capture zone of a specific groundwater extraction well, and (2) forward tracking is used to explicitly represent plume transport. The simulation-optimization model uses both approaches in order to take advantage of specific numerical properties of each approach (Mulligan and Ahlfeld 1999a). The formulation is solved in two stages where the goal of the first stage is to find a feasible solution and the goal of the second stage is to converge to an optimal solution. During the first stage, the plume is represented by a set of stationary control points that are constrained to lie within the capture zone of extracting wells. Forward tracking is used in the second stage of the solution process and all particles representing the contaminant plume are constrained to leave the simulation model domain through an extraction well.

The simulation-optimization model incorporates the USGS simulation models MODFLOW (McDonald and Harbaugh 1984) and MODPATH (Pollack 1994) to determine the system state under a set of imposed stresses. More detailed discussions of the simulation-optimization model, model parameterization, and solution characteristics are presented in Mulligan and Ahlfeld (1999a) and Mulligan (1999).

A typical optimization based approach to the pump-and-treat design problem is to specify a set of fixed locations where a well can be installed. The well locations in this set are called candidate wells and the management model must determine the appropriate pumping rates from this set (Gorelick et al. 1984; Ahlfeld and Heidari 1994). Another common approach for simplifying the optimization problem is to assume that well installation costs are small relative to operational costs for long-term plume control systems (Ahlfeld and Heidari 1994). With this assumption, the cost function to be minimized can be represented as a continuous function of the pumping rates.

During simulation of unconfined aquifers, care must be exercised to avoid dewatering numerical grid cells because of excessive pumping. If extraction well locations are assumed to be most prone to dewatering then either upper bounds on extraction rates or lower bounds on head at the well cell can be imposed on the pumping solution. Imposing explicit upper bounds on extraction may artificially eliminate the best solutions from consideration. With this in mind, the approach taken here is to impose lower head bounds at each candidate well location without imposing explicit restrictions on the extraction rates at those wells.

In the first stage of the solution process, the advective control model seeks a pumping scheme in which the capture zone fully encompasses all control points representing the contaminant plume. The capture zone is simulated by tracking particles from extraction wells backwards through the velocity field. To represent the plume capture constraints numerically, a distance measure is used in which the minimum distance between each plume control point and all particles (see Figure 1) is constrained. When the distance between a control point and particle pathline equals zero then the plume control point lies within the capture zone. To ensure capture of the entire plume, the constraint function must equal zero for all control points. The reverse tracking formulation is stated as

$$\text{Minimize } f = \sum_{j=1}^{n} \left(\alpha_j \left| q_j \right| \right) \tag{5a}$$

such that

$$p_k(\mathbf{q}) = \min \left[\frac{\left(\hat{r}_{k,m} - \delta \right)}{\left(\hat{r}_{o,k} - \delta \right)} \middle| \forall M, t \right] \leq 0 \qquad \forall\, k \in K$$

$$h_j \geq h_j^l \qquad j = 1, \ldots, n \tag{5b}$$

where n is the number of candidate wells, α_j is the unit cost of pumping from well j, $\mathbf{q}$ is the vector of pumping rates at the n candidate wells, $\hat{r}_{k,m}$ is the minimum distance between control point k and particle m, $\hat{r}_{o,k}$ is the minimum distance between control point k and all wells prior to remedial pumping, δ is a user specified maximum distance around the plume control point, K is the set of plume control points, t is time, M is the set of particles being tracked, h_j is the head at well j, and h^l_j is a specified lower bound on head at well j. The constraint function p_k determines the minimum normalized distance of closest approach between all particles m and control point k.

The groundwater flow and particle tracking simulation models are linked to the optimization model via the constraints in (5b). As stated previously, the goal when using the formulation in equations (5a) and (5b) during the first stage of the solution process is to find a set of pumping rates that satisfy all of the constraints in (5b). Although it is unlikely that the set $\mathbf{q}$

will also satisfy the objective in (5a) because of the numerical behavior of this formulation (Mulligan and Ahlfeld 1999a), the objective function is still used to guide the search for a feasible solution.

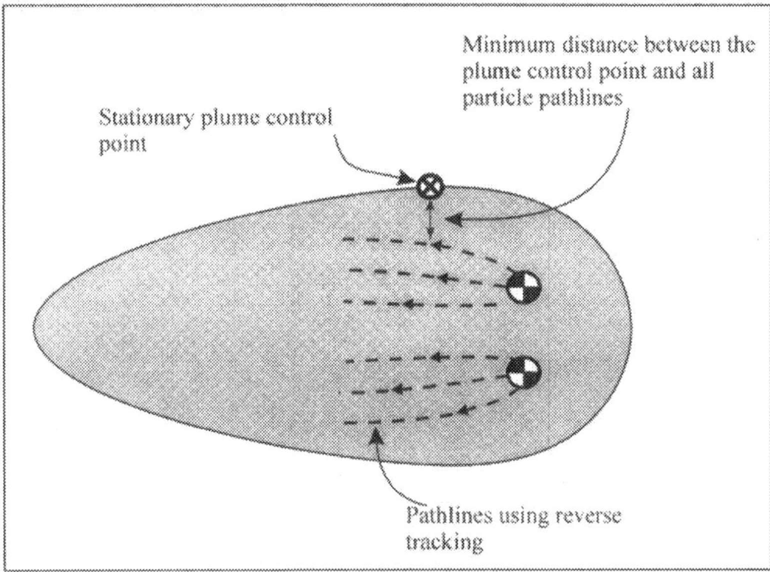

Figure 1. In reverse tracking, pathlines represent the capture zone and the constraint function measures the minimum distance between a plume control point and all pathlines. Pathlines are shown as dashed curves and the gray area represents the contaminant plume to be contained.

During the second stage of the solution process, the feasible region is searched for a set of pumping rates that minimizes the objective function. In this stage, the contaminant plume is represented by a set of mobile particles that are tracked forward through the velocity field. The qualitative constraint that all particles must be captured by an extraction well is represented numerically by requiring that the minimum distance between a particle (or contaminant) pathline and all extraction wells (see Figure 2) equals zero. The stage 2 formulation is stated as

$$\text{Minimize} \quad f = \sum_{j=1}^{n} \left(\alpha_j \left|q_j\right|\right) \tag{6a}$$

such that

$$g_i(\mathbf{q}) = \min\left\{ \frac{\hat{r}_{i,j}}{\hat{r}_{o,i}} \middle| j = 1, 2, ..., n; \forall t \right\} = 0 \qquad \forall\ i \in I$$

$$h_j \geq h_j^l \qquad j = 1, ..., n \tag{6b}$$

where $\hat{r}_{i,j}$ is the minimum distance between particle i and well j, $\hat{r}_{o,i}$ is the minimum distance between particle i and all wells prior to remedial pumping, t is time, and I is the set of particles used to represent the plume.

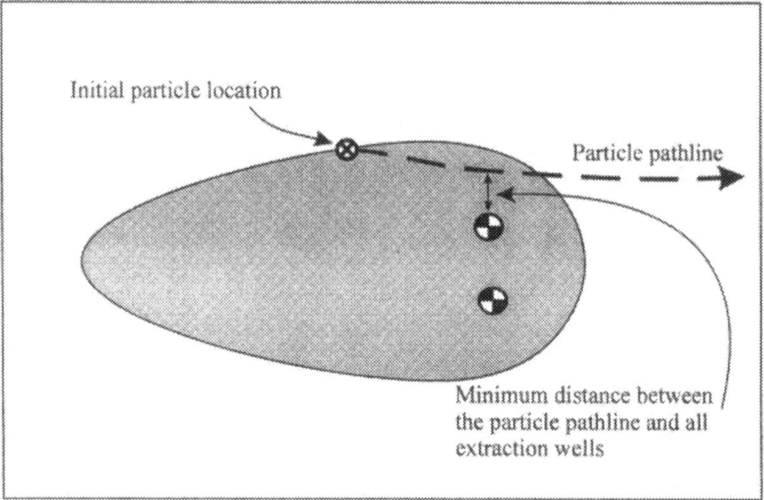

Figure 2. In forward tracking, pathlines represent advective transport and the constraint function measures the minimum distance between a pathline and all extraction wells. A pathline is shown as the dashed curves and the gray area represents the contaminant plume to be contained.

In both stages of the solution process, the plume capture constraints are handled using the penalty method (Mulligan and Ahlfeld 1999a). In this approach, constraint functions are added to the objective function and constraint violations are penalized so that infeasible solutions do not appear to be optimal. For the first stage search, the conjugate-gradient method (Luenberger 1984) is used whereas an interior point boundary projection method (Mulligan and Ahlfeld 1999b) is used for the second stage search. The numerical effects of the unconfined aquifer assumption on the response surface for the optimization problem is described below, along with solution procedures to accommodate both the numerical effects and the head bound constraints.

3.2 Response Surface of an Unconfined Aquifer Problem

The numerical effects of simulating unconfined conditions are shown for the two-dimensional example problem depicted in Figure 3. A turning-bands code (Thompson et al 1989) generated a heterogeneous conductivity field based on a log-normally distributed population with a geometric mean of 3.0 m/d, a variance of 0.8, and a correlation length of 100 m. The entire domain is 1000 m by 1000 m with constant head conditions on the right and left boundaries and no-flow boundary conditions on the top and bottom. In the unconfined aquifer, the average initial saturated thickness throughout the aquifer is 4.5 m. A line source at x = 385 m was simulated using particle tracking to define the plume boundary as shown.

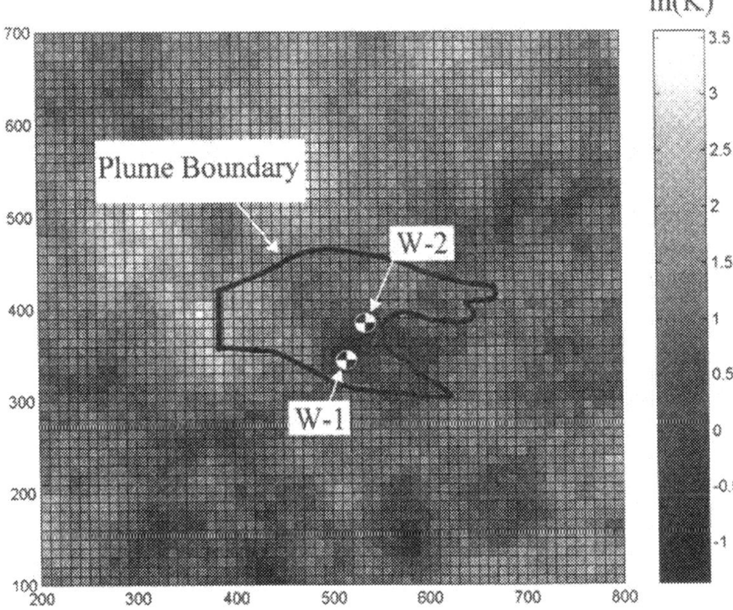

Figure 3. Plan view of two-dimensional example problem. Natural flow is from left to right across the domain.

Pumping rates at the two wells shown are systematically altered and the constraint function p_k for 108 plume control points is determined. The sum of squared constraint values for each set of pumping rates is shown graphically in Figures 4a and 4b for confined and unconfined assumptions, respectively. The minimum saturated thickness at the grid cells containing the two extraction wells is also shown on Figure 4b. The optimal solution identified on the figure for the unconfined problem satisfies a minimum saturated thickness constraint of 2 m.

(a)

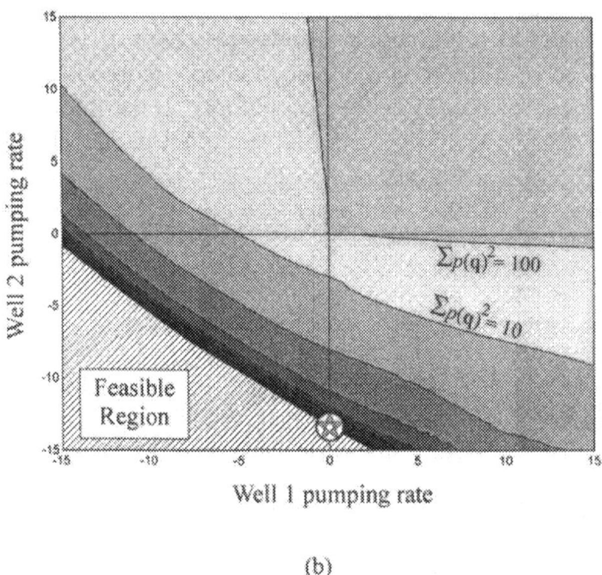

(b)

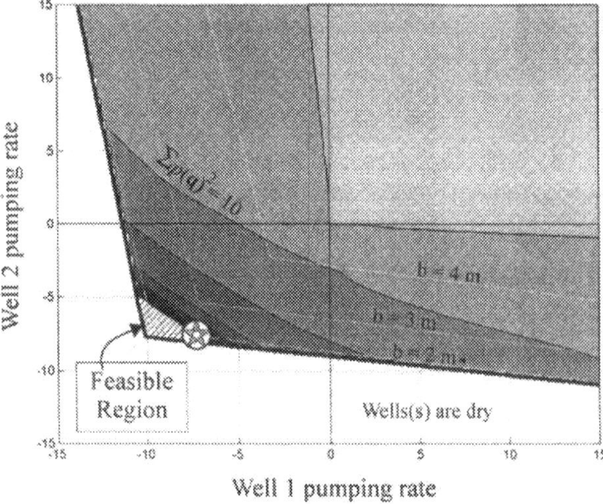

Figure 4. Response surface for (a) confined and (b) unconfined assumptions for the example depicted in Figure 3. The optimal solution is indicated by the star. The minimum saturated thickness b at the wells is contoured in (b). Negative pumping rates imply extraction.

The most striking differences between the two response surfaces are the size of the feasible region and the location of the optimal solution. The optimal solution under the confined assumption is not feasible under unconfined conditions. The infeasible space of the two problems is very similar and without modifications to the optimization solution algorithm, the search trajectory for the unconfined case is similar to that for the confined case. This results because there is no feedback in the infeasible space of the unconfined problem to prevent the search from identifying well W-2 as the more promising of the two wells. Without additional search procedures, the optimization algorithm will fail when extraction at W-2 leads to dewatering.

Although this example results in fairly extreme conditions, it does point the need to alter the gradient based search to avoid well dewatering and algorithmic failure. Because the search trajectory of the optimization algorithm and the size and shape of the feasible region cannot be predicted, additional search procedures must be included for all unconfined aquifers.

3.3 Solution Procedure for Unconfined Aquifers

Traditional nonlinear optimization methods use transition rules to generate a sequence of points that successively improve the objective function. The transition from one point to another can be represented by

$$\mathbf{q}^{w+1} = \mathbf{q}^{w} + \alpha^{w}\mathbf{d}^{w} \tag{7}$$

where w is the iteration number, $\mathbf{d}^w$ is a search direction, and α^w is a positive step size. To maintain feasibility of the head constraints and prevent dewatering, controls must be placed either on the search direction or the step size in (7).

When the head constraints are not binding, cell dewatering can be controlled by placing an upper limit on the step size. An upper bound on the step size can be determined by using the head response coefficients, which represent the change in head at location i with respect to a unit change in pumping at well j. The head at location i under pumping $\mathbf{q}^{w+1}$ can be approximated with a truncated Taylor series expansion about the head under pumping $\mathbf{q}^w$ (Riefler and Ahlfeld 1996):

$$\left\{ h_i(\mathbf{q}^{w+1}) = h_i(\mathbf{q}^{w}) + \sum_{j=1}^{n_w} \left.\frac{\partial h_i}{\partial q_j}\right|_{\mathbf{q}^w} (q_j^{w+1} - q_j^{w}) \right\} \geq h_i^l \tag{8}$$

Substituting (7) into (8), and using matrix notation, the head constraints become

$$\mathbf{h}^{w} + \alpha^{w}\mathbf{A}^{w}\mathbf{d}^{w} \geq \mathbf{h}^{l} \tag{9}$$

where $\mathbf{A}^{w}$ is the head response matrix, bold lower case variables are vectors, and the superscript w implies the values determined at iteration w under the current solution $\mathbf{q}^{w}$. To prevent cell dewatering, the step size α^{w} is bounded by rearranging equation (9).

The truncated Taylor series used in equation (8) is a linear approximation that is equivalent to assuming confined conditions, which will underpredict drawdown in unconfined problems. Therefore, a safety factor must be included to reduce the upper bound on α^{w}. This is done by multiplying the upper bound on α^{w} determined by equation (9) by the ratio of minimum allowable saturated thickness to the saturated thickness at the current solution.

The maximum step-size determined using equation (9) is used in both stages of the solution process to prevent cell dewatering when simulating unconfined aquifers. If one of the head constraints becomes binding, then the second procedure for accommodating cell dewatering constraints is invoked. In this case, the search direction is projected onto the binding constraint (Bertsekas 1995) so that the solution remains feasible with respect to the head bound constraint and dewatering is avoided. In reference to Figure (4b), projecting the search direction onto a binding constraint results in searching along a saturated thickness contour. Projecting the gradient in the squared distance measures onto the 2 m contour leads directly to the feasible region.

4. EXAMPLE APPLICATION TO UNCONFINED AQUIFERS

4.1 Three-dimensional simulation

In this section, a three-dimensional example problem is developed to demonstrate the capabilities of the advective control model. The simulation model domain is 2000 m by 2000 m in the horizontal dimensions and 20 m in the vertical. The domain is discretized into 80 rows, 80 columns, and 5 layers. The rows and columns are 40 m wide along the domain boundaries and 20 m wide near the area of interest. The northern and southern boundaries are no-flow conditions, the eastern boundary has constant head conditions that vary linearly from 18 m in the north to 17 m in the south, and the western boundary has constant flux conditions. The hydraulic conductivity is isotropic in the horizontal and an order of magnitude smaller in the vertical. A low conductivity layer is modeled explicitly in the middle layer, where conductivity values are an order of magnitude smaller than in the other four layers. A river runs from north to south through the domain and extends through the top model layer only. Recharge is applied to the surface and creates substantial vertical gradients.

A line source of contaminants in the top layer was simulated using particle tracking to define the contaminant plume to be contained. Figures 5a and 5b show a portion of the domain in plan view and cross section, respectively. The contaminant plume has not undergone dispersion, but vertical gradients have transported the contaminants down into the second model layer. If remedial action is not undertaken, the plume will continue migrating downwards until upward gradients force the plume to leave the domain through the river. Three potential well locations are identified on Figure 5a and are included in the optimization formulation as candidate wells. Two sets of candidate wells are evaluated for capturing the plume. In the first set, all three wells pump from the second layer to coincide with the vertical location of the plume. In the second test, all wells pump from the top model layer. To ensure that well cell dewatering does not occur, the minimum saturated thickness in a well cell is constrained to be 1.5 m.

The solutions to the two problems are listed in Table 1 along with the head values and grid layer elevations at each pumping well. The optimal solution when wells pump from the second layer calls for pumping from well W-2 only and drawdown within the well is not sufficient to result in a dewatering problem.

To demonstrate the effects of dewatering considerations, the wells are changed to pump from the top layer and the optimization problem is resolved. The optimal pumping rates and final head values for this problem are listed in Table 1. In this case, extraction from well W-2 is sufficient to cause dewatering concerns. The solution now incorporates pumping from wells W-1 and W-2, with approximately 40% of total extraction pumping occurring at well W-1. If the aquifer is simulated as confined and the wells pump only from the top layer, the optimal solution is to extract 192 m^3/day from well W-2 only. In this case, the pumping solution from the confined assumption is quite different from the solution based on unconfined simulation and results in dewatering under unconfined conditions.

In both unconfined cases, well W-2, the most downgradient well, is chosen to be active in the optimal solution. When the drawdown constraint at that well becomes binding, then the next downgradient well is activated.

These results reflect the fact that the downgradient portion of the plume is the most difficult to capture. The solution pumps as much as possible from the most downgradient well until drawdown constraints prevent further pumping.

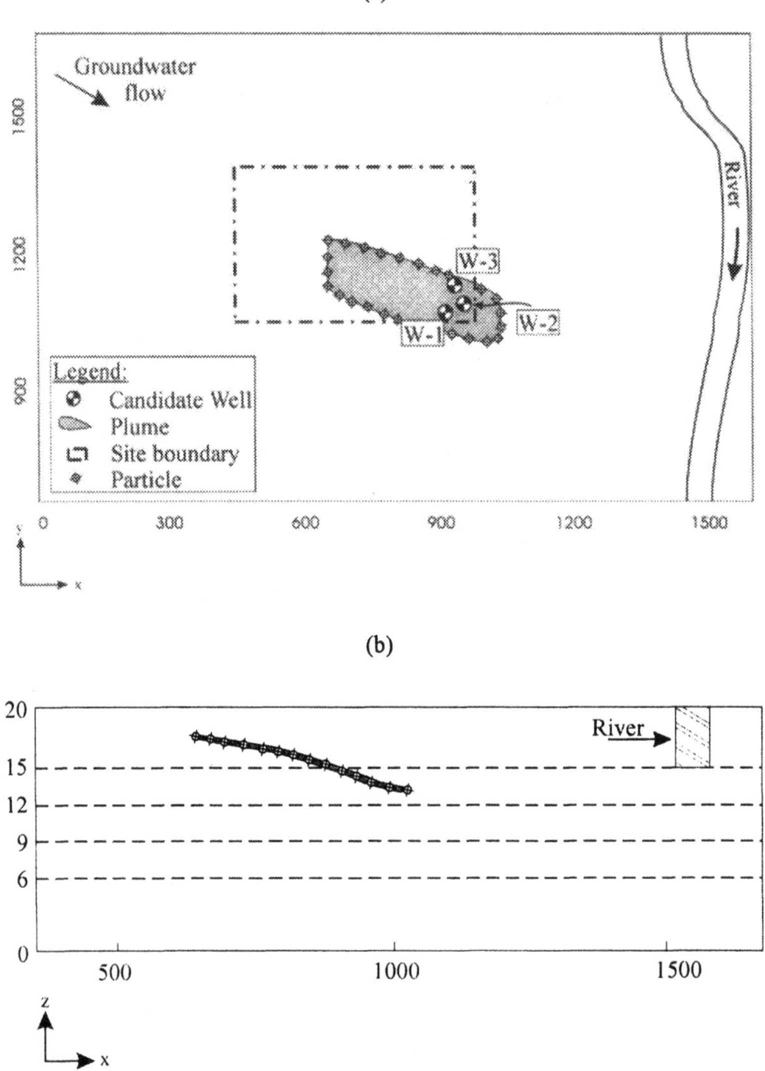

Figure 5. Three-dimensional plume control problem shown in (a) plan view and (b) projection of plume at y = 1100 m. Dimensions on the left side of (b) indicate the vertical elevations of each layer.

Table 1. Solutions for the plume capture problem depicted in Figure 5. The head at pumping wells is constrained to be 1.5 m above the bottom of the cell. Negative pumping rates imply extraction.

Grid layer of well screen placement (Top layer = 1)	Elevation of well cell bottom (m)	Well	Optimal solution, $\mathbf{q}^*$ (m^3/day)	Head in well at $\mathbf{q}^*$ (m)
		W-1	0	17.4
2	12	W-2	-186	16.4
		W-3	0	17.4
		W-1	-80	16.9
1	15	W-2	-122	16.5
		W-3	0	17.4

4.2 Two-dimensional Simulation

A two-dimensional example problem is also developed to demonstrate the advective control model. The example problem is solved for both confined and unconfined conditions and the solutions are compared. In this example problem, the aquifer is homogeneous and isotropic, with no flow conditions imposed at the top and bottom boundaries and constant head conditions along the left and right boundaries. The head on the constant head boundaries slopes downward toward the bottom of the domain. The domain is 3100 m by 3100 m and is discretized into 49 rows and 58 columns. A river runs through the domain as shown in Figure 6.

Both confined and unconfined conditions are simulated, where the hydraulic conductivity is 5 m/day and the initial saturated thickness is approximately 15.4 m for the unconfined aquifer. The transmissivity for confined conditions is 77 m^2/day. The design problem is to contain the two plumes depicted on the figure, with the condition that remedial wells be located on site but not located in the building. The six candidate wells shown are used as potential pumping wells and the minimum total pumping from these wells must be determined such that the two plumes are captured.

The two contaminant plumes are represented in the first stage of the optimization formulation with a set of 110 control points along the plume boundaries. These same control point locations are used as starting points for particles when forward tracking is used in the second stage of the solution process. For the unconfined simulation, additional constraints are included to require a minimum saturated thickness of 1.5 m at each well cell. Both confined and unconfined assumptions are simulated under two sets of penalty parameters. Recall that the solution algorithm uses the penalty method for the plume capture constraints, in which each constraint violation is multiplied by a penalty parameter and added to the objective function.

When the penalty multiplier is large, constraint violations are discouraged and the search algorithm is biased toward satisfying constraints at the expense of minimizing pumping.

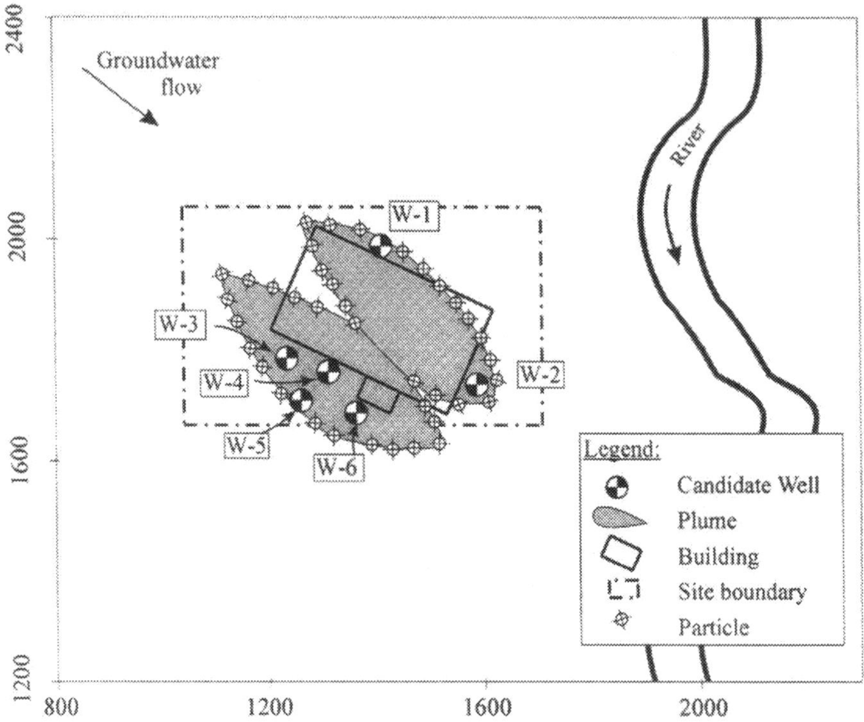

Figure 6. The plume control model must determine the pumping rates from the six candidate wells so that the two plumes are captured. A total of 110 particles are used to represent the plume, a few of which are shown. The horizontal scales are in meters.

The optimal solutions for both confined and unconfined conditions are listed in Table 2. The optimal solutions both require extraction from the two downgradient wells, wells W-2 and W-6, and the solutions are nearly identical. The drawdown at well W-2 in the unconfined case is approximately 1% of the initial saturated thickness. When the drawdown is this small, the head response in the aquifer is nearly linear and the confined approximation does not alter the solution significantly. Nonetheless, the additional search controls for the unconfined aquifer prevented dewatering by controlling the step sizes and slowing down algorithmic progress. When the penalty parameters were small, the algorithm required 12% more function evaluations in the Stage 1 search under unconfined conditions and 68% more evaluations in the Stage 2 search. For larger penalty parameters, which biases the search to reduce constraint violations at the expense of minimizing pumping, the algorithm required 23% more evaluations in Stage 1 and 73% more evaluations in Stage 2.

Table 2. Solution for confined and unconfined conditions for the example depicted in Figure 6.

	Confined Aquifer		Unconfined Aquifer	
	q‡	Cumulative number of function evaluations	q‡	Cumulative number of function evaluations
Starting Solution	(0,0,0,0,0,0)	--	(0,0,0,0,0,0)	--
Small Initial Penalty†				
Stage 1 Solution	(-1.4, -22.0, 0, 0, 0, -5.5)	435	(0, -21.8, 0, 0, -1.0, -5.2)	486
Optimal Solution	(0, -21.2, 0, 0, 0, -8.1)	745	(0, -21.0, 0, 0, 0, -8.0)	1008
Large Initial Penalty†				
Stage 1 Solution	(0, -49.2, 0, 0, 0, 0)	145	(0, -47.8, 0, 0, 0, 0)	178
Optimal Solution	(0, -21.2, 0, 0, 0, -8.1)	281	(0, -21.0, 0, 0, 0, -8.0)	663

‡ Pumping rates $-1 \le q_i \le 1$ are listed as 0 in the stage 1 solutions. The solution vector $\mathbf{q} = (q_1, q_2, q_3, q_4, q_5, q_6)$ lists the pumping rates in m^3/day at the individual wells identified in Figure 6. Negative values imply extraction.

† The penalty method is used for plume capture constraints during the Stage 1 search; squared constraint violations are multiplied by a penalty and added to the objective function. The sets of parameter values are: initial multiplier = 10, which is incremented by a factor of 10 at each iteration; and initial multiplier = 100, which is incremented by a factor of 100 at each iteration.

The algorithmic modifications for unconfined aquifers slow the search process considerably in order to avoid dewatering well cells. Unlike the three-dimensional example presented earlier, the optimization search path for the two-dimensional example did not include solutions with sufficient extraction to approach dewatering and the additional algorithmic steps primarily added to the computational burden. However, when the search becomes biased toward large pumping rates, as can occur when penalty parameters are large, the additional steps for unconfined simulations are important for slowing down the algorithmic progress to reduce some of the bias in the search process.

5. SUMMARY AND CONCLUSIONS

The particle tracking-based advective control model described in this chapter is capable of solving two- and three-dimensional plume capture problems with multiple candidate wells and multiple particles in confined and unconfined aquifers. The formulation provides a direct approach to solving plume control problems and uses both forward and reverse particle tracking to exploit numerical characteristics of the two reference frames.

Under certain conditions, such as an unconfined aquifer with large extraction rates or a small saturated thickness, dewatering of the aquifer can result in a significant reduction in the feasible region of the optimization problem. The effects of aquifer dewatering and the nonlinear head response to pumping are accommodated by controlling the search algorithm. The additional algorithmic steps described and presented in this chapter resulted in a substantial increase in the computational burden. Therefore, if drawdown is expected to be small relative to saturated thickness, then a confined aquifer assumption might be used because solution will be quicker. A possible approach when designing pump-and-treat systems for unconfined aquifers is to first determine the optimal solution under confined conditions and then check this solution under unconfined conditions to check for cell dewatering. If dewatering does not occur, then the confined solution could be used as the starting point for an optimization search using the unconfined assumption. If drawdown is significant however, this approach is likely to result in a solution that is infeasible under unconfined conditions (Mansfield and Shoemaker 1999).

In the example problems presented, the optimization algorithm chose to activate downgradient wells. This indicates that the pumping solutions allow the natural flow patterns to transport contaminants toward extraction wells. The active wells are then used primarily to alter gradients along the lateral and downgradient regions of the plume.

A consequence of the advective control formulation presented here is that many of the particles stay in the aquifer as long as possible while under the control of the pumping scheme. Pumping rates larger than the optimal rates will generally result in increased velocities and hence decreased travel-times. This outcome allows natural processes such as biological decay to act on the contaminants before bringing them to the surface for engineered treatment. While this approach may be acceptable for diffuse low concentration portions of a plume, it may not be a desirable procedure for high concentration areas (hot-spots). Instead, hot-spots may need to be controlled by wells placed within or near the high concentration area. Procedures for constraining hot-spot areas to be captured by a subset of the candidate wells should be investigated.

BIBLIOGRAPHY

Ahlfeld, D. P., and Heidari, M. (1994). "Applications of optimal hydraulic control to ground-water systems." *Journal of Water Resources Planning and Management*, 120(3), 350-365.

Ahlfeld, D. P., and Mulligan, A. E. (2000). *Optimal Management of Flow in Groundwater Systems*, Academic Press, San Diego, CA.

Ahlfeld, D. P., Page, R. H. and Pinder, G. F. (1995). "Optimal ground-water remediation methods applied to a Superfund site: from formulation to implementation." *Ground Water*, 33(1), 58-70.

Ahlfeld, D. P., Riefler, R.G., and Mulligan, A. E. (1998). "A new code for MODFLOW-coupled groundwater management of unconfined aquifers." *MODFLOW'98 Proceedings*, October 4-8, 1998, Golden, Colorado, 431-438.

Ahlfeld, D. P., and Sprong, M. P. (1998). "Presence of nonconvexity in groundwater concentration response function." *Journal of Water Resources Planning and Management*, 124(1), 8-14.

Barlow, P. M., Wagner, B. J. and Belitz, K. (1996). "Pumping strategies for management of a shallow water table: The value of the simulation--optimization approach." *Ground Water*, 34(2), 305-317.

Bertsekas, D. P. (1995). *Nonlinear Programming*. Athena Scientific, Belmont, MA.

Chan Hilton, A., Aksoy, A., and Culver, T. B., (2000). "Dynamic optimal design of groundwater remediation using genetic algorithms." in J. Smith and S. Burns (eds.), *Remediation of Contaminated Subsurface Environments*, Kluwer Academic, New York, NY.

Colarullo, S. J., Heidari, M., and Maddock III, T. (1984). "Identification of an optimal groundwater management strategy in a contaminated aquifer." *Water Resources Bulletin*, 20(5), 747-790.

Gorelick, S. M. (1983). "A review of distributed parameter groundwater management modeling methods." *Water Resources Research*, 19(2), 305-319.

Gorelick, S. M., Voss, C. I., Gill, P. E., Murray, W., Saunders, M. A., and Wright, M. H. (1984). "Aquifer reclamation design: The use of contaminant transport simulation combined with nonlinear programming." *Water Resources Research*, 20(4), 415-427.

Gorelick, S. M. (1990). "Large scale nonlinear deterministic and stochastic optimization: Formulations involving simulation of subsurface contamination." *Mathematical Programming*, 48, 19-39.

Greenwald, R. M., and Gorelick, S. M. (1989). "Particle travel times of contaminants incorporated into a planning model for groundwater plume capture." *Journal of Hydrology*, 107, 73-98.

Jones, L., Willis, R., and Yeh, W. W.-G. (1987). "Optimal control of nonlinear groundwater hydraulics using differential dynamic programming." *Water Resources Research*, 23(11), 2097-2106.

Lefkoff, L. J., and Gorelick, S. M. (1986). "Design and cost analysis of rapid aquifer restoration systems using flow simulation and quadratic programming." *Ground Water*, 24(6), 777-790.

Luenberger, D. G. (1984). *Linear and Nonlinear Programming*. Second Edition, Addison-Wesley, Reading, MA.

Mansfield, C. M., and Shoemaker, C. A., (1999). "Optimal remediation of unconfined aquifers: Numerical applications and derivative calculations." *Water Resources Research*, 35(5), 1455-1469.

Marryott, R. A., Dougherty, D. E., and Stollar, R. L. (1993). "Optimal groundwater management 2. Application of simulated annealing to a field-scale contamination site." *Water Resources Research*, 29(4), 847-860.

McDonald, M.G., and Harbaugh, A. W. (1984). *A Modular Three-Dimensional Finite-Difference Ground-Water Flow Model.* U.S. Geological Survey, Reston, VA.

Mulligan, A. E. (1999). Advective Control of Groundwater Contaminant Plumes Using Simulation and Optimization. PhD dissertation, University of Connecticut, Storrs.

Mulligan, A. E., and Ahlfeld, D. P. (1999a). "Advective control of groundwater contaminant plumes: Model development and comparison to hydraulic control." *Water Resources Research*, 35(8), 2285-2294.

Mulligan, A. E., and Ahlfeld, D. P. (1999b). "An interior point boundary projection method for nonlinear groundwater optimization with zero-derivative constraints." *RCGRD Publication 98-2*, University of Vermont, Burlington, VT.

Pollock, D. W. (1994). User's Guide for MODPATH/MODPATH-PLOT, Version 3: A particle-tracking post-processing package for MODFLOW, the U.S. Geological Survey finite-difference groundwater flow model. U.S. Geological Survey, Reston, VA.

Rastogi, A. K. (1989). "Optimal pumping policy and groundwater balance for the Blue Lake aquifer, California, involving nonlinear groundwater hydraulics." *Journal of Hydrology*, 111(3), 177-194.

Riefler, R. G., and Ahlfeld, D. P. (1996). "The impact of numerical precision on the solution of confined and unconfined optimal hydraulic control problems." *Hazardous Waste and Hazardous Materials*, 13(2), 167-176.

Shafer, J. M. (1987). "Reverse pathline calculation of time-related capture zones in nonuniform flow." *Ground Water*, 25(3), 283-289.

Shafer, J. M., and Vail, L. W. (1987). "Screening method for contaminant plume control." *Journal of Water Resources Planning and Management*, 113(3), 336-352.

Thompson, A. F. B., Ababou, R., and Gelhar, L. W. (1989). "Implementation of the three-dimensional turning bands random field generator." *Water Resources Research*, 25(10), 2227-2243.

Tiedeman, C., and Gorelick, S. M. (1993). "Analysis of uncertainty in optimal contaminant capture design." *Water Resources Research*, 29(7), 2139-2153.

Varljen, M. D. and Shafer, J. M. (1993). "Coupled simulation-optimization modeling for municipal ground-water supply protection." *Ground Water*, 31(3), 401-409.

Wagner, B. J. (1995). "Recent advances in simulation-optimization groundwater management modeling." *Reviews of Geophysics*, Supplement, 1021-1028.

Willis, R., and Finney, B. A. (1985). "Optimal control of nonlinear groundwater hydraulics: Theoretical development and numerical experiments", *Water Resources Research*, 21(10), 1476-1482.

Chapter Three

Palladium Catalysis for the Treatment of Contaminated Waters: A Review

NAOKO MUNAKATA AND MARTIN REINHARD
Stanford University, Stanford, CA 94305-4020

Key words: Palladium, catalysis, halogen, remediation

Abstract: The widespread contamination of waters by halogenated and unsaturated hydrocarbons (e.g. solvents and pesticides) has spurred the investigation of a number of remediation technologies. Treatment via reductive dehalogenation and hydrogenation with Pd catalysts shows promise, offering a number of advantages over other processes. First, a broad range of compounds are reducible in aqueous solution at ambient temperatures by Pd catalysts: halogenated alkanes, ethylenes, and aromatics; PAHs; formate/carbonate; and oxidized nitrogen species. In addition, the halogenated compounds generally react to form the corresponding simple alkanes (or completely dehalogenated aromatics in the case of the halogenated aromatics), with few halogenated intermediates, if any. Finally, reactions are generally extremely rapid, which allows small, in-well reactors. This paper discusses each of these advantages in more detail, as well as some of the challenges encountered in the field. Topics include (i) intermediates and products of the Pd-catalyzed reactions; (ii) the reaction pathways and kinetics; (iii) the effects of the surrounding water matrix and the support on the reaction; and (iv) findings from laboratory columns studies and field tests.

1. INTRODUCTION

Catalysts have long been used for a wide variety of organic reactions, including hydrogenation, dehydrogenation, exchange, oxidation, reduction and hydrogenolysis. (Biswas et al. 1988) This paper focuses on hydrogenation and reductive dehalogenation reactions catalyzed by palladium (Pd) catalysts in the aqueous phase. (Figure 1) These reactions are of interest to environmental engineers seeking to remediate waters contaminated with halogenated and/or unsaturated hydrocarbons (e.g. solvents and pesticides).

(a) $R\text{-}X + H_2 \xrightarrow{\text{catalyst}} R\text{-}H + H\text{-}X$ (b) $R_2\text{-}C{=}C\text{-}R_2 + H_2 \xrightarrow{\text{catalyst}} H_2R\text{-}C\text{-}C\text{-}R_2H$

Figure 1. (a) Reductive dehalogenation (hydrodehalogenation) and (b) hydrogenation.

Catalyst particles generally consist of a metal deposited onto the surface of a support and are denoted by metal/support, e.g. Pd/C indicates palladium metal on a carbon support. Among the metals used for catalysis, Pd is often found to be the most active metal. (Augustine 1965) For example, in the aqueous hydrodechlorination of 1,1,2-trichloroethane, Pd catalysts achieved significantly more conversion than Pt or Rh catalysts. (Kovenklioglu et al. 1992) Catalyst supports can vary in shape, size, porosity and surface area; typical materials include carbon, alumina, silica and zeolites.

In the past, most catalytic research focused on organic or gas phases. However, in recent years, literature has appeared on the treatment of organic compounds in the aqueous phase at ambient temperatures using supported palladium catalysts and a hydrogen source. This research has been spurred by the increasing realization of widespread contamination of groundwaters by halogenated hydrocarbons. These compounds tend to be harmful to human health and recalcitrant to remediation. The traditional treatment method uses granular activated carbon in fixed bed reactors to sorb contaminants from water streams; however, this results in contaminated solids which must be further treated. More recent research has investigated bioremediation and the use of reactive iron barriers; these technologies show potential, but they sometimes result in the formation of partially halogenated intermediates or by-products which are equally or more harmful. Furthermore, bioremediation often requires the addition of hydrocarbon metabolites to the groundwater, while iron barriers require construction of a trench, which restricts economical application to shallow contaminated aquifers. Hydrodehalogenation with Pd catalysts eliminates a number of these issues.

This paper outlines some of the advantages of Pd catalysis. As discussed in the first two sections ("Range of Activity and Reaction Products" and "Reaction Pathways"), the process is amenable to the treatment of a wide variety of compounds, including mixtures such as those found in contaminated groundwater sites. The resulting products are generally simple alkanes; few halogenated intermediates, if any, are detected. The third section discusses reaction kinetics: the reactions are quite rapid, allowing short residence times and small reactors (e.g. that can be placed in treatment wells). The next two sections ("Water Quality Effects" and "The Role of

Catalyst Support") consider the effects of the aqueous matrix on the reaction and the interactions with the catalyst support; optimization of supports for components in the aqueous matrix are touched upon briefly. The final two sections investigate the application of batch study results to laboratory column studies and field studies.

2. RANGE OF ACTIVITY AND REACTION PRODUCTS

A variety of compounds have been tested for catalytic reduction by Pd:

- halogenated alkanes (methanes, ethanes, propane and hexane)
- halogenated alkenes (ethylenes)
- halogenated aromatics (benzenes, phenols and polychlorinated biphenyls (PCBs))
- nonhalogenated aromatics (polycyclic aromatic hydrocarbons (PAHs))
- oxidized carbon species (formate, carbonate)
- oxidized nitrogen species (nitrate, nitrite, nitric oxide).

Almost all reported compounds were amenable to reductive dehalogenation and/or hydrogenation. These compounds are summarized in Tables 1-4, along with the catalyst used in each study, the time allowed for reaction, the reaction intermediates and the final products.

Five catalyst types were used in these product distribution studies: Pd/C, Pd/alumina, pure Pd metal powder, Pd/Fe, and Pd/Cu/support. The last two catalysts are not traditional materials. The Pd/Cu, which was used in the reduction of nitrate, was a bimetallic catalyst on an alumina, silica, or polymer support. Fe is an atypical support in that it has the intrinsic ability to reductively dehalogenate chlorinated compounds by oxidizing to Fe^{2+}; however, the rapid reaction rates associated with Pd/Fe are indicative of Pd-catalyzed reactions, which are much faster than Fe reactions.

The time period over which the reaction was observed varied from 5 minutes to 7 days. This variation in reaction times can affect the reaction products observed. For example, after 5-10 minutes of reaction time, Grittini reported biphenyl as the product in the reduction of PCBs. (Grittini et al. 1995) Schüth investigated the reaction of 4-chlorobiphenyl and found similar results; after 10 minutes, the 4-chlorobiphenyl was reduced primarily to biphenyl, with small amounts of cyclohexylbenzene. However, Schüth continued the reaction and detected dicyclohexyl after 7 hours of reaction; complete conversion to dicyclohexyl was achieved after approximately 7 days. (Schüth and Reinhard 1998) This implies that biphenyl is, in fact, an intermediate; had the Grittini experiment been continued, biphenyl might have been observed as an intermediate, rather than the final product. A number of the tested compounds formed multiple reaction intermediates and products; where available, the maximum intermediate concentration and the final observed concentration (as a percentage of the initial concentration of reactant) are provided in parentheses.

Table 1. Halogenated alkanes transformed by Pd catalysts in aqueous solution

Alkanes	Catalyst (Reaction Time)	Intermediates	Products	Source[1]
CT[2]	Pd/C (300 min)	CF (25%)	methane (35%) CF (~10%) ethane (3%) ethylene (<1%)	1
	Pd/Al_2O_3 (343 min)	CF (21%) ethylene (<1%) propylene (trace)	methane (60%) CF (18%) ethane (14%) propane (1.0%)	2
	Pd powder (113 min)	CF (24%) ethylene (trace) propylene (trace)	methane (51%) CF (23%) ethane (12%) propane (2%)	2
	Pd/Fe ("minutes")	Not reported	methane	3
CF	Pd/Al_2O_3 (291 min)	None detected	methane (80%) ethane (<1%) MeCl (trace)	2
	Pd/Fe (< 1 hr)	Not reported	methane	3
MeCl	Pd/Fe (~400 min)	Not reported	methane (~75% MeCl removal)	3
1,1,2,2-TCA	Pd/C (NR)	Not reported	ethane	1
1,1,2-TCA	Pd/C (3 hr)	CCND[3]	ethane	4
Freon-113	Pd/Al_2O_3 (NR)	Not reported	Not reported	2
DBCP	Pd/Al_2O_3 (130 min)	propylene	propane	5
	Pd/Al_2O_3 (NR)	CCND	CCND	2
Lindane	Pd/Al_2O_3 (18 min)	None detected	cyclohexane (99%)	6

[1]Sources: (1) Schreier 1996; (2) Lowry and Reinhard 1999; (3) Muftikian et al. 1995; (4) Kovenklioglu et al. 1992; (5) Siantar et al. 1996; (6) Schüth and Reinhard 1998.
[2]Chemical abbreviations: CT (carbon tetrachloride); CF (chloroform); MeCl (methylene chloride); TCA (tetrachloroethane or trichloroethane); Freon-113 (1,1,2-trichlorotrifluoroethane); DBCP (1,2-dibromo-3-chloropropane); Lindane (γ-hexachlorocyclohexane)
[3]CCND: Chlorinated compounds not detected.

Table 2. Halogenated alkenes transformed by Pd catalysts in aqueous solution

Alkenes	Catalyst (Reaction Time)	Intermediates	Products	Source[1]
PCE[2]	Pd/C (30 min)	Not reported	ethane (55%)	1
	Pd/Al_2O_3 (30 min)	ethylene (2%)	ethane (80-85%)	1
	Pd/Al_2O_3 (Not reported)	CCND[3]	CCND	2
	Pd/Fe (Not reported)	Not reported	ethane	3
TCE	Pd/C (3 hr)	CCND	CCND	4
	Pd/Al_2O_3 (163 min)	ethylene (<1%)	ethane (97%) ethylene (<1%)	2
	Pd powder (193 min)	ethylene (<1%) 1,1-DCE (<1%) cis-DCE (2%) trans-DCE (<1%) VC (<1%)	ethane (83%) 2 unknowns	2
	Pd/Fe (30 hrs)	ethylene	ethane/ethylene VC (<3%)	5
TCE (20 mg/L)	Pd/Fe (Not reported)	Not reported	ethane	3
TCE (sat. solution)	Pd/Fe (Not reported)	Not reported	ethane butane hexane	3
cis-, trans- & 1,1-DCE	Pd/Al_2O_3 (Not reported)	CCND	CCND	2
	Pd/Fe (Not reported)	CCND	ethane	3
trans-DCE	Pd/Al_2O_3 (Not reported)	ethylene (< ~5%)	ethane (55%)	1
VC, cis-DCE, TCE	Pd/Al_2O_3 (Not reported)	ethylene (< ~5%)	ethane (70-75%)	1

[1]Sources: (1) Schreier and Reinhard 1995; (2) Lowry and Reinhard 1999; (3) Muftikian et al. 1995; (4) Kovenklioglu et al. 1992; (5) Liang et al. 1997.

[2]Chemical abbreviations: PCE (perchloroethylene); TCE (trichloroethylene); DCE (dichloroethylene); VC (vinyl chloride).

[3]CCND: Chlorinated compounds not detected.

Table 3. Halogenated aromatics transformed by Pd catalysts in aqueous solution

Aromatics	Catalyst (Reaction Time)	Intermediates	Products	Source[1]
chlorobenzene	Pd/Al_2O_3 (NR)	Not reported	benzene	1
1,2-DCB[2]	Pd/Al_2O_3 (13 min)	chlorobenzene (<1%)	benzene (>90%)	1
1,2,4-TCB	Pd/C (3 hr)	Not reported	Not reported (18-93% 1,2,4-TCB removed)	2
[3]2-CP, 3-CP, 4-CP	Pd/C (30-90 min)	None detected	Phenol (≥99%)	3
[4]2-CP, 4-CP	Pd/Fe (2-3 min)	Not reported	Phenol	4
[3]2,4-DCP; 2,5-DCP; 2,6-DCP; 3,5-DCP	Pd/C (60-360 min)	None detected	Phenol (≥99%)	3
[3]2,3-DCP	Pd/C (120 min)	None detected	Phenol (≥98%)	3
[3]3,4-DCP	Pd/C (135 min)	None detected	Phenol (91%)	3
[4]2,3-DCP; 2,4-DCP; 2,6-DCP	Pd/Fe (2-3 min)	Not reported	Phenol	4
[3]2,4,5-TCP; 2,4,6-TCP	Pd/C (3-22 hr)	None detected	Phenol (≥99%,)	3
[3]2,3,5-TCP	Pd/C (1.5 hr)	None detected	Phenol (87%)	3
[3]2,3,6-TCP	Pd/C (3 hr)	None detected	Phenol (≥98%)	3
[4]2,3,4-TCP; 2,3,5-TCP; 2,4,6-TCP	Pd/Fe (3-4 min)	Not reported	Phenol	4
[3]PCP	Pd/C (83 hr)	None detected	Phenol (≥98%)	3
[5]PCP	Pd/Fe (2 hr)	Not reported	Not reported (no PCP detected)	4
4-chlorobiphenyl	Pd/Al_2O_3 (7 days)	biphenyl (>90%) cyclohexylbenzene (85%)	dicyclohexyl	1
[6]Araclor 1254, Araclor 1260	Pd/Fe (5-10 min)	CCND	biphenyl (106%)	5

[1]Sources: (1) Schüth and Reinhard 1998; (2) Kovenklioglu et al. 1992; (3) Hoke et al. 1992; (4) Korte et al. 1997; (5) Grittini et al. 1995.

[2]Abbreviations: DCB (dichlorobenzene); TCB (trichlorobenzene); CP (monochlorophenol); DCP (dichlorophenol); TCP (trichlorophenol); PCP (pentachlorophenol).

[3]Experiments conducted in an ethanol solution.

[4]Solvent not specified.

[5]Experiments conducted with a water-diluted 100% ethanol extract of PCP-contaminated soil.

[6]Experiments conducted with a 1:3:1 v/v solution of methanol/water/acetone. Araclor 1254 and 1260 are PCB mixtures containing 54 and 60% Cl by weight, respectively.

Table 4. Other compounds transformed by Pd catalysts in aqueous solution

Species	Catalyst (Reaction Time)	Intermediates	Products	Source[1]
Naphthalene	Pd/Al_2O_3 (40 min)	None reported	1,2,3,4-THP[2]	1
Phenanthrene	Pd/Al_2O_3 (25 min)	1,2,3,4,4a,9,10,10a-OHP; 1,2,3,4,5,6,7,8-OHP; 1,2,3,4-THP; 9,10- DHP; dicyclohexyl	Same as intermediates, equilibrium concentrations obtained	1
Carbonate	Pd/C Pd/Al_2O_3 $Pd/BaSO_4$ Pd metal	None detected	Formate	2-5
Formate	Pd/C	None detected	Carbonate	3
Nitrate	Pd/Cu (NR)	nitrite	nitrogen[3] ammonia	6
	Pd/Cu (NR)	nitrite (20%)	nitrogen[3] ammonia (5%)	7
	Pd/Cu (0.8 g) Pd (0.6 g) (NR)	nitrite (~3%)	nitrogen[3] ammonia (5%)	7
Nitrite	Pd metal (NR)	Not reported	nitrogen[3] ammonia (0.1%)	7
Nitric oxide	Pd/C (500 min)	Not reported	nitrous oxide nitrite	8

[1]Sources: (1) Schüth and Reinhard 1998; (2) Engel et al. 1995; (3) Stalder et al. 1983; (4) Wiener et al. 1988; (5) Wiener et al. 1986; (6) Lüdtke et al. 1998; (7) Hörold et al. 1993; (8) MacNeil et al. 1998.

[2]Chemical abbreviations: OHP (octahydrophenanthrene); THP (tetrahydrophenanthrene); DHP (dihydrophenanthrene).

[3]Formation assumed, not measured.

Several trends are apparent in Tables 1-4. The halogenated alkanes are largely dehalogenated to simple alkanes, although the chlorinated methanes appear to form some chlorinated by-products and larger carbon chains. The halogenated ethylenes are reduced primarily to ethane, with ethylene formed as an intermediate. Vinyl chloride was also formed as a product (only in the case of the Pd/Fe catalyst), and chlorinated intermediates were also observed with the pure Pd powder. It is worth noting that chlorinated intermediates or products were only detected with these two catalysts, which are not traditionally used in catalysis. The chlorinated aromatics are also reduced to the corresponding chlorine-free aromatic compounds. The PAHs formed equilibria with partially and/or fully hydrogenated compounds. Similarly, carbonate and formate reached an equilibrium state with a carbonate: formate ratio of 1-1.5:1. Nitrate and nitrite were reduced to nitrogen gas and ammonia, while nitric oxide was both reduced to nitrous oxide and oxidized to nitrite.

A few compounds have been found to be recalcitrant to the catalytic reaction and are not reported in Tables 1-4: 1,1-dichloroethane (1,1-DCA) and 1,2-dichloroethane (1,2-DCA). (Lowry and Reinhard 1999; McNab and Ruiz 1998) Experiments with 1,1-DCA and 1,2-DCA yielded traces of ethane, but no detectable decreases in DCA concentration. (Lowry and Reinhard 1999) Lowry reports similar results with methylene chloride (MeCl); 78 minutes of reaction yielded traces of methane, but no detectable decrease in the MeCl concentration in the presence of Pd/alumina. However, Muftikian reports the successful transformation of MeCl in the presence of Pd/Fe. (Muftikian et al. 1995) Finally, Hörold reports that nitrate is only reduced by bimetals; Pd alone showed no activity for nitrate reduction. (Hörold et al. 1993)

3. REACTION PATHWAYS

3.1 Halogenated alkanes

The hypothesized transformation pathways of CT and CF to methane are shown in Figure 2. The transformation of CT and CF to methane in the Pd/alumina system, despite the low reactivity of MeCl, indicates that the reactions do not involve sequential dehalogenation of CF to methane (i.e. MeCl is not an intermediate). The formation of C2 and C3 compounds during the transformation of both CT and CF indicates the existence of a radical pathway. However, the production of ethane (12-14%) and CF (18-23%) from CT was much lower than that of methane (51-60%). This implies that the main transformation pathway is a direct reaction of CT to methane, with a secondary pathway involving a trichloromethyl radical which then reacts to form CF and C2 and C3 species. Similarly, the relatively low production of ethane (<1%) from CF indicates that the major pathway for the reaction of CF to ethane occurs through direct transformation to methane, rather than through a dichloromethyl radical species. (Lowry and Reinhard 1999)

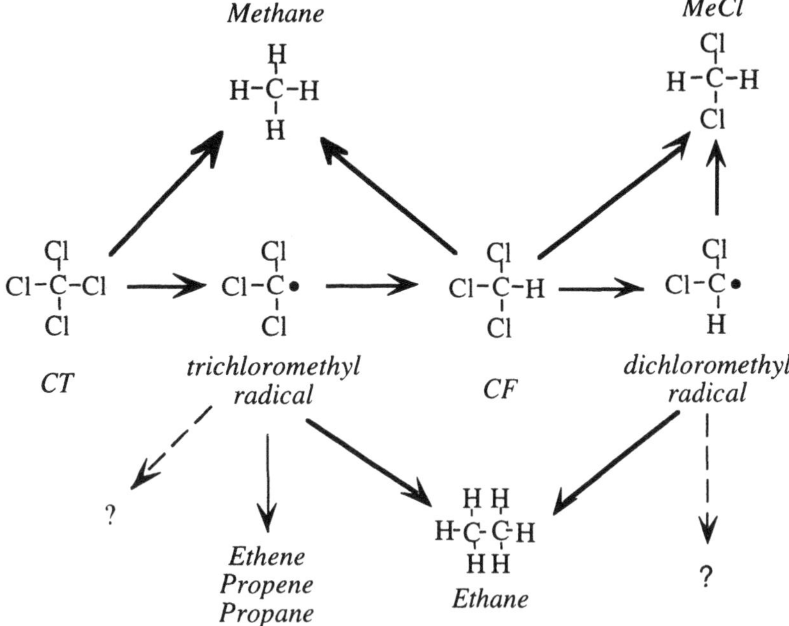

Figure 2. Hypothesized transformation pathways of chlorinated methanes. (based on Lowry and Reinhard, 1999)

Dechlorination of 1,1,2,2-TCA can follow one of the three pathways in Figure 3: elimination and formation of a partially chlorinated ethylene; sequential dechlorination; or direct transformation to ethane. The first pathway is unlikely, given the lack of chlorinated ethylene intermediates and the fact that the transformation rate of 1,1,2,2-TCA is an order of magnitude lower than that of PCE, which has a similar rate to TCE, the DCEs and VC. (Schreier 1996) The recalcitrance of the DCA isomers to transformation (Lowry and Reinhard 1999; McNab and Ruiz 1998) implies that sequential dechlorination through DCA does not occur, but this pathway cannot be ruled out because the DCA tests were conducted using Pd/alumina catalysts, rather than the Pd/C used for the 1,1,2,2-TCA and 1,1,2-TCA tests. However, the available data (lack of chlorinated intermediates and the low reactivity of the DCA isomers) suggest that direct transformation is the most probable pathway.

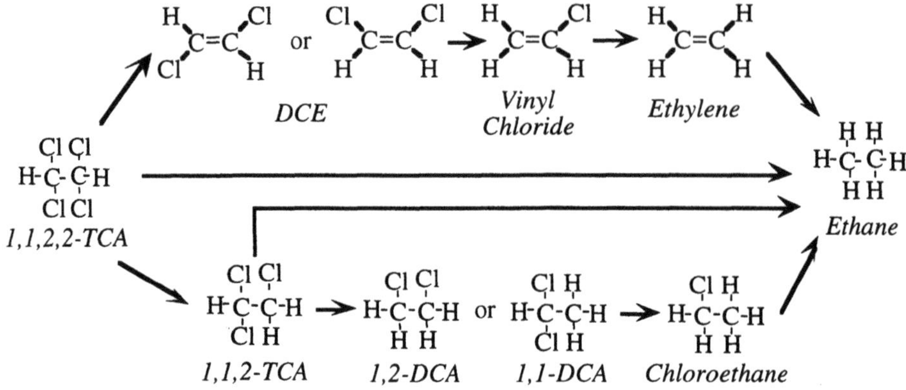

Figure 3. Possible transformation pathways of 1,1,2,2-TCA.

Similarly, the transformation of DBCP to propane can occur (i) through sequential dehalogenation or (ii) through elimination of the halogens to form propylene, followed by hydrogenation to propane. In this case, the observation of propylene in Siantar's system indicates that the latter pathway does occur. (Figure 4, Siantar et al. 1996) This pathway is further supported by data showing the reaction rate of DBCP to be similar to that of the alkene PCE (Lowry and Reinhard 1999; Siantar et al. 1996) and the other chlorinated ethylenes (Lowry and Reinhard 1999), which react much faster than the alkane 1,1,2,2-TCA (Schreier 1996).

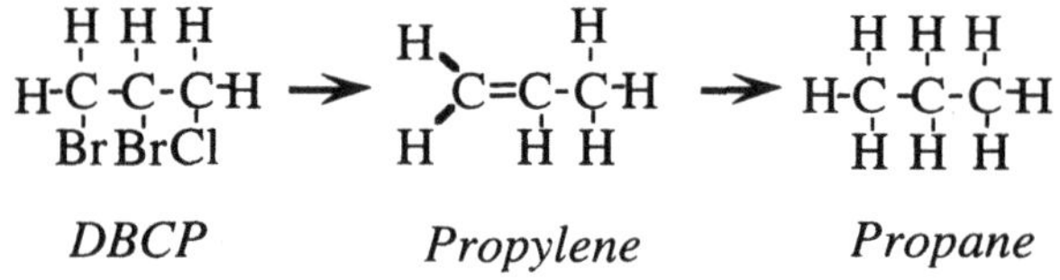

DBCP → *Propylene* → *Propane*

Figure 4. Possible transformation pathway of DBCP. (based on Siantar et al., 1996)

In the case of lindane, the lack of chlorinated intermediates indicate direct transformation to benzene. (Figure 5, Schüth and Reinhard 1998)

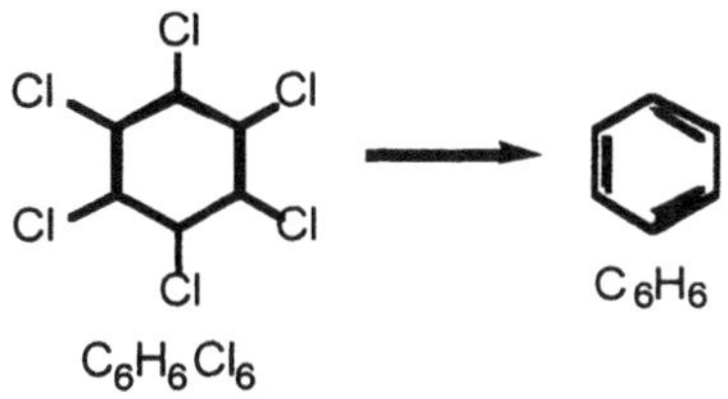

$C_6H_6Cl_6$ *lindane* → C_6H_6 *benzene*

Figure 5. Transformation pathway of lindane to benzene. Reprinted from Applied Catalysis B: Environmental, Vol. 18, Schüth and Reinhard, "Hydrodechlorination and Hydrogenation of Aromatic Compounds over Palladium on Alumina in Hydrogen-Saturated Water," pp. 219, Copyright 1998, with permission from Elsevier Science.

3.2 Halogenated alkenes

The chlorinated alkenes are likely to follow one of three basic reaction pathways.

Pathway 1: formation and reaction of radical ethyl species

Pathway 2: dehalogenation, followed by hydrogenation of unsaturated bonds

Pathway 3: hydrogenation of the double bond, followed by dehalogenation.

Pathway 1 is implied by Muftikian's observation of butane and hexane from a saturated aqueous TCE solution with Pd/Fe. (Muftikian et al. 1995) However, these products were only observed in this singular case, indicating that the radical reaction may occur only under special conditions (e.g. saturated solution).

Pathway 2 (shown below in Figure 6) is probable for most cases, given that chlorinated ethanes (i.e. from Pathway 3) have not been observed. In addition, Schüth reports that hydrodechlorination generally occurs faster than hydrogenation for halogenated aromatics (Schüth and Reinhard 1998); if the same holds true for halogenated alkenes, Pathway 2 would be kinetically favorable over Pathway 3. This is corroborated by Schreier, who found that the reaction rate of 1,1,2,2-TCA was an order of magnitude slower than that of PCE (Schreier 1996); this indicates that saturation of the double bond is not the first step in the transformation of PCE to ethane. The common observation of ethylene as a reactive intermediate is further evidence for Pathway 2. From the two cases where DCE and/or VC were observed (with Pd/Fe and pure Pd powder), it is clear that sequential dehalogenation associated with the second pathway does occur under some conditions. However, Lowry noted that, in the case of metallic Pd powder, DCE and VC were produced and disappeared simultaneously; this implies multiple dechlorination of TCE to both DCE and VC, rather than sequential reaction. (Lowry and Reinhard 1999) Multiple dechlorination is also supported by the fact that even when chlorinated intermediates occurred, their concentrations were lower than expected for sequential dehalogenation. Kinetic data fitted for the removal of TCE and the formation of intermediates indicated that the low levels of intermediates implied a direct reaction pathway from TCE to ethane, for both Pd powder and the supported Pd/alumina catalyst. (Lowry and Reinhard 1999)

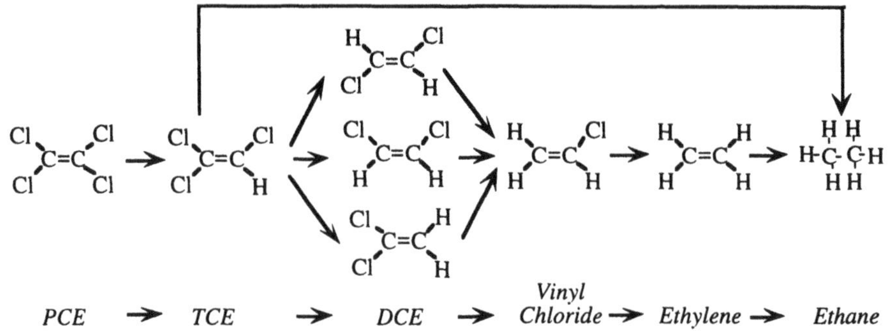

Figure 6. Sequential dehalogenation of chlorinated ethylenes (based on Lowry and Reinhard, 1999).

3.3 Halogenated aromatics

In order to determine the reaction pathway for 1,2-dichlorobenzene, Schüth fitted kinetic data and found that the primary pathway was direct reaction to benzene with a parallel reaction of sequential dechlorination through chlorobenzene to benzene. (Figure 7) These pathways were further supported by independent determination and verification of the reaction rate constant for the second hydrodechlorination step. (Schüth and Reinhard 1998)

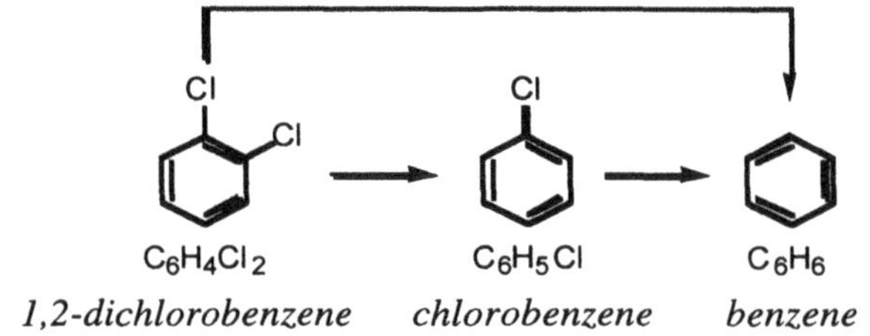

Figure 7. Transformation pathway of 1,2-dichlorobenzene. Reprinted from Applied Catalysis B: Environmental, Vol. 18, Schüth and Reinhard, "Hydrodechlorination and Hydrogenation of Aromatic Compounds over Palladium on Alumina in Hydrogen-Saturated Water," pp. 218, Copyright 1998, with permission from Elsevier Science.

The lack of partially chlorinated by-products in the reactions of the Araclor PCB congeners (Grittini et al. 1995) and of all the chlorophenols (Hoke et al. 1992) may indicate direct transformation of these compounds to biphenyl and phenol, respectively; these were the final observed products of the reactions. Schüth also observed the rapid dechlorination of 4-chlorobiphenyl to biphenyl. The first benzene ring of biphenyl was then gradually hydrogenated to form cyclohexylbenzene, then the second benzene ring was hydrogenated to form the final product of dicyclohexyl. The best fit of the kinetic data was obtained by including a direct reaction of 4-chlorobiphenyl to cyclohexylbenzene. (Figure 8, Schüth and Reinhard 1998)

$C_{12}H_9Cl$	$C_{12}H_{10}$	$C_{12}H_{16}$	$C_{12}H_{22}$
4-chlorobiphenyl	*biphenyl*	*cyclohexylbenzene*	*dicyclohexyl*

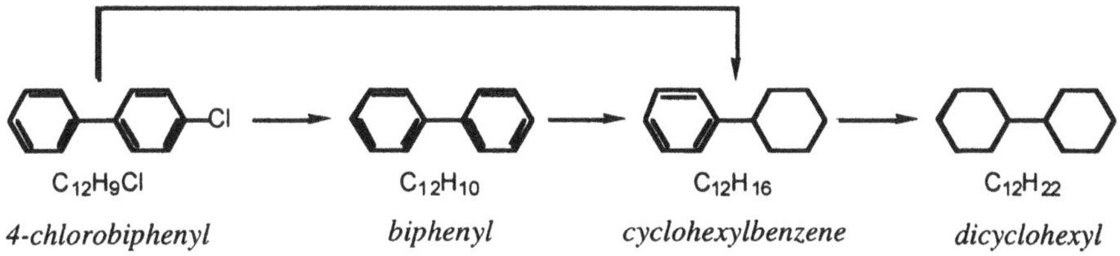

Figure 8. Transformation pathway of 4-chlorobiphenyl. Reprinted from Applied Catalysis B: Environmental, Vol. 18, Schüth and Reinhard, "Hydrodechlorination and Hydrogenation of Aromatic Compounds over Palladium on Alumina in Hydrogen-Saturated Water," pp. 219, Copyright 1998, with permission from Elsevier Science.

3.4 Nonhalogenated PAHs

Phenanthrene undergoes a complex series of hydrogenation reactions (pathways not shown), leading to a variety of partially and fully hydrogenated products. Mass balance was incomplete, and the pathways determined were insufficient to model the kinetic data, indicating the possibility of unidentified pathways. However, the identified products did appear to reach equilibrium concentrations. After the batch system had come to apparent equilibrium, it was respiked to the original concentration of phenanthrene. This resulted in similar conversions at similar rates, and the final phenanthrene concentration was double that before the respike, as would be expected for a system at equilibrium. (Schüth and Reinhard 1998)

Naphthalene underwent hydrogenation of one of its two aromatic rings to form 1,2,3,4-tetrahydronaphthalene. Fits of the kinetic data indicated that the reverse reaction is possible; this was verified with separate experiments with 1,2,3,4-tetrahydronaphthalene as the initial reactant. (Figure 9, Schüth and Reinhard 1998)

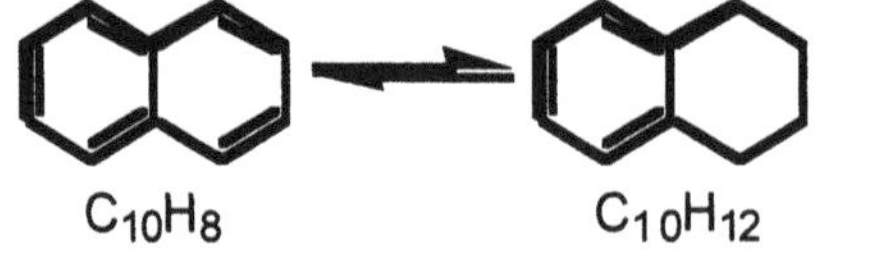

naphthalene *1,2,3,4-tetrahydronaphthalene*

Figure 9. Transformation pathway of naphthalene. Reprinted from Applied Catalysis B: Environmental, Vol. 18, Schüth and Reinhard, "Hydrodechlorination and Hydrogenation of Aromatic Compounds over Palladium on Alumina in Hydrogen-Saturated Water," pp. 219, Copyright 1998, with permission from Elsevier Science.

4. REACTION KINETICS

The kinetics associated with catalytic transformations in reactor systems are a complex function of many variables, such as reactor configuration and operation (e.g. stirring mechanisms within the reactor, relative volumes of liquid and gas phases, etc.) and catalyst parameters (support material, quantity and distribution of metal, etc.). For example, Grittini reported that the rate of PCB hydrodechlorination on Pd/Fe depended on amount of Pd/Fe added to the system, Pd loading on the Fe, and the aqueous solution composition. (Grittini et al. 1995) As a result of the complicated nature of the kinetics, it is extremely difficult to compare results across multiple studies; however, information can be obtained from comparisons made within a single system. This is evident in Table 5, where first order reaction rate constants for chlorinated ethylenes are reported by Lowry and Schreier, who used different reactor systems and different catalysts; the values vary by orders of magnitude, but the trends are similar within each set of experiments. Similarly, it is reasonable to expect that these trends will hold true as the process is scaled up from laboratory systems to field units, although the absolute rates of reaction will be strongly system-dependent. General trends which have been observed include

- hydrodechlorination generally occurs faster than hydrogenation (Schreier 1996; Schüth and Reinhard 1998)
- dehalogenation rates are often extremely rapid with half-lives on the order of seconds to (<10) minutes (Grittini et al. 1995; Lowry and Reinhard 1999; McNab et al. 2000; Munakata et al. 1998; Schreier 1996; Schreier and Reinhard 1995; Schüth and Reinhard 1998). It should be noted that some hydrogenation reactions have been observed with slower rates, e.g. the hydrogenation of cyclohexylbenzene to dicyclohexyl had a half-life of approximately one day.
- reactions are often observed to be first order with respect to the reactant hydrocarbon (Hoke et al. 1992; Liang et al. 1997; Lowry and Reinhard

1999; Schreier and Reinhard 1995; Schüth and Reinhard 1998; Siantar et al. 1996; Yu 1991)

- polyaromatic compounds react faster than monoaromatic compounds (Schüth and Reinhard 1998), and chlorinated alkenes react faster than their corresponding alkane (Lowry and Reinhard 1999; Schreier 1996)
- reaction rate often increases with increasing degree of chlorination, as observed with chlorinated methanes (Lowry and Reinhard 1999; Muftikian et al. 1995) and phenols. (Hoke et al. 1992)

It is interesting to note that the chlorinated ethylenes do not appear to follow this trend of increasing rates with increasing chlorination. (Lowry and Reinhard 1999; Schreier 1996) This may be due in part to the extremely fast rates of these reactions, which increase the relative importance of mass transfer limitations. For very fast reactions, mass transfer of compounds to the catalyst surface, rather than the intrinsic catalytic reaction rate, may determine the rate of disappearance of hydrocarbons and the resulting apparent rate constants.

Table 5. First order reaction rate constants (L water/g Pd-min) for chlorinated ethylenes

	Lowry and Reinhard (1999)	Schreier (1996)
PCE	53	800
TCE	64	1900
cis-DCE	83	2500
trans-DCE	78	2400
1,1-DCE	70	2040

5. THE ROLE OF CATALYST SUPPORT

The catalyst support can have a significant impact on the rate of a catalyzed reaction, the reaction pathway (quantities and species of intermediates and products) and the resistance of the catalyst to deactivation. This section summarizes some observed effects of the catalyst support on these parameters.

5.1 Effects on Reaction Rate

Siantar compared the rates of reaction of DBCP on several 1% w/w Pd catalysts: Pd/alumina powder, Pd/PEI/silica beads and powder, and Pd/PAC. The PEI/silica beads had the lowest first order reaction rate constant, followed by the PAC, the PEI/silica powder, and the alumina powder, which showed the highest activity. (Siantar et al. 1996)

Kovenklioglu found that supports which sorbed substrates more strongly had higher reaction rates. The hydrodehalogenation of 1,1,2-TCA on three catalysts with 5% Pd loading was compared in two batch reactor systems: Pd/C-ESC110 and Pd/C-ESC12 in Reactor 1, and Pd/C-ESC110 and Pd/alumina in Reactor 2. (Table 6) TCA and chloride concentrations were monitored throughout the experiment. Conversion of TCA was calculated based on Cl production, and the remaining TCA loss was attributed to sorption. The ESC110 support sorbed TCA most rapidly, despite a 30% smaller surface area than ESC12; the alumina support did not adsorb TCA. The initial reaction of TCA was highest for ESC110, followed by ESC12, and was lowest for alumina. (Kovenklioglu et al. 1992)

Table 6. Effect of Support on Conversion of 1,1,2-TCA.

Reactor	Catalyst	Reaction time	Conversion	% Removal from Solution
1	Pd/ESC-110	10 min	40%	98%
1	Pd/ESC-12	10 min	20%	54%
2	Pd/ESC-110	9 min	35%	92%
2	Pd/alumina	9 min	4.4%	4.4%

Similarly, in electrochemical studies of TCE, the reduction on pure Pd electrodes was slower than that on Pd/Fe; this was attributed to strong sorption of TCE to Fe, which then allowed reduction by atomic H in the Pd matrix. (Li and Farrell 2000)

5.2 Effects on Reaction Pathway

In the transformation of CT, the Pd powder and Pd/alumina catalysts formed similar intermediates and products in similar quantities. (Lowry and Reinhard 1999) In contrast, Pd/C gave rise to ethylene (seen only as an intermediate in the other two cases) and produced no propane. In addition, overall product yields were lower with the C support, even though the intermediate CF concentration was similar to that of the other supports. (Schreier 1996)

With PCE transformation, the product (ethane) was the same for all supports but yields (measured by both ethane and Cl production) varied: 50-55% ethane yield was obtained on C, 68% yield on Pd/PEI/silica, and 80-85% yield on alumina. The possibility that lower yields resulted from PCE sorbed to the support was considered; the C catalyst was therefore heated to 180°C in an attempt to desorb any species. However, only a few nanomoles of PCE (tenths of a percent of the original mass) and traces of lesser chlorinated ethylenes were detected. This suggests that the low ethane yields may be due to side reactions occurring on the C and PEI/silica supports. (Schreier 1996)

All studies of intermediates and products in the reaction of TCE report ethylene as an intermediate and ethane as the major product. In the case of Pd/alumina, these were the only intermediates and products. (Lowry and Reinhard 1999) However, additional intermediates and minor products were observed with other supports. With Pd powder, all DCE isomers and VC were detected as intermediates, and two unidentified compounds were detected as products. (Lowry and Reinhard 1999) With Pd/Fe, VC was detected as a final product in addition to ethane/ethylene, which were not differentiated from each other. (Liang et al. 1997)

In the reduction of nitrite, Hörold found that supports with lower surface areas showed improved selectivity for nitrogen over ammonia. The same study found that silica supports had much higher selectivity (i.e. much less ammonia formation) than alumina supports. (Hörold et al. 1993)

5.3 Effects on Resistance to Deactivation

After finding that sulfite reduced the catalytic activity of Pd/alumina, Schüth investigated the effects of support on deactivation. Applying sulfite as the model poison and 1,2 dichlorobenzene as a model substrate, batch tests were conducted with a variety of zeolite supports loaded with 1% Pd by weight. The results indicate that pore size and hydrophobicity must both be considered for optimal catalyst performance. Larger pore sizes allowed faster diffusion of the substrate to the active Pd sites and therefore faster reaction; however, the larger pore sizes also allowed more access to the poison sulfite, which resulted in faster catalyst deactivation. The optimal pore size was one which was large enough to allow the substrate molecule to enter the zeolite pores, but small enough to limit sulfite access to the Pd. When this optimal pore size was used in conjunction with a hydrophobic support, the overall catalyst activity was maximized. (Schüth et al. 1998; Schüth and Reinhard 1997)

6. WATER QUALITY EFFECTS

Water quality and the components present in the solution matrix can affect the reaction rates and product distribution, and can cause the catalyst to deactivate by slowing or eliminating the desired reaction. For example, Muftikian found ethane as the sole product from a 20 mg/L solution of TCE, but observed butane and ethane from a saturated TCE solution. (Muftikian et al. 1995) The next two subsections provide other examples and discuss the effects of pH and water solutes on the catalytic reaction.

6.1 Effects of pH

The pH of the solution can also play an important role. For nitrate reduction, reaction was faster in lower pH solutions (Hörold et al. 1993; Lüdtke et al. 1998); however, acidic solutions can dissolve the catalyst metal and negate the effects of the faster rates. The Cu in the bimetallic nitrate-reducing catalyst dissolved at pH values < 5 and deactivation was observed (Hörold et al. 1993); Pd dissolution in a monometallic catalyst became significant below a pH of 4. (Munakata et al. 1998)

In contrast, some reactions are promoted by basic solutions. Kovenklioglu found that MeCl reacted faster in basic solutions than in neutral solutions, presumably due to neutralization of acid produced by the reaction or due to alteration of the catalyst surface in basic conditions. (Kovenklioglu et al. 1992) Similarly, Hoke found that the addition of base was required for the reaction of chlorinated phenols. (Hoke et al. 1992) The type of base affected the reaction rate, with NH_4OH being the best, followed by NaOH (which had 1/3 the reaction rate with NH_4OH), followed by sodium acetate (which had a rate initially equal to that with NH_4OH, but which decreased rapidly), and finally triethylamine, for which there was no observable reaction. Hoke also found that reactions were faster for 4-CP, 2,4-DCP and 2,4,5-TCP (but not for PCP) in 50/50 mixture of ethanol/water than in ethanol alone; the anomaly with PCP was probably due to its low solubility in water. Hoke observed that the catalyst maintained higher activity levels in ethanol/water mixture as well; this was attributed to the observed lesser degree of Pd oxidation and lower Pd losses from the catalyst support.

6.2 Effects of Solutes

For the application of catalytic hydrodehalogenation to groundwater treatment, it is important to know the effects of common groundwater components. The results of three studies which investigated these effects are given in Table 7.

As stated earlier, the reaction kinetics and decreases in reaction rates depend on many factors, including solute concentrations, the nature of the reactant compound, the catalyst used, and the reactor system. As a result, it is difficult to reach definitive conclusions across studies (e.g. about the relative resistance of different supports to the solutes). However, the trends are similar for the results of both Siantar and Schreier. For the reaction of

DBCP on a 1% w/w catalyst, on a mass basis, it appears that sulfite has a stronger effect than nitrate or sulfate, which in turn has a stronger effect than chloride. Similarly, for the reaction of PCE on a 1% w/w Pd/PAC catalyst, the effect of solutes in decreasing order of strength are bisulfide, nitrite, nitrate, sulfate, and chloride. From these results, it is clear that solutes present in the water can affect the reaction kinetics; the Schüth results indicate that catalysts can be tailored to minimize the negative effects of solutes on the reaction process.

Table 7. Effects of Water Solutes on Reaction Rates

Source	Catalyst	Reactant	Solute	Conc. (mg/L)	Reaction Rate Decrease
(Siantar et al. 1996)	Pd/Al_2O_3	DBCP	Sulfite	28	90%
			Sulfate	28	49%
			Nitrate	28	47%
			Chloride	100	53%
			Oxygen	saturated	60%
			Groundwater	NA	28%
(Schreier and Reinhard 1995)	Pd/PAC	PCE	Bisulfide	66	100%
			Nitrite	8.4	50%
			Nitrate	11	25%
			Sulfate	18	15%
			Chloride	350	0%
(Schüth and Reinhard 1998)	Pd/Al_2O_3	1,2-DCB	Sulfite	20	>99%
	Pd/zeolite[1]		Sulfite	20	>99%
	Pd/zeolite[2]		Sulfite	20	60%

[1]This category includes five zeolite supports, with varying pore sizes and hydrophobicities.
[2]This zeolite support optimized pore size and hydrophobicity for the reaction of DCB in the presence of sulfite, as described in the section on "The Role of Catalyst Support."

The results described in the above sections were obtained in batch studies, which generally run for less than one day. Long-term effects of water quality on catalyst activity are better observed through continuous flow columns which can operate for months or years. These column experiments generally use packed bed reactors (as in the field) and provide better simulations of field conditions. However, because the Pd technology is relatively new, few column studies have been conducted thus far; results of published studies are discussed in the following section in more detail. Note that for both the column and field studies, the most relevant parameters are residence time, conversion data and pore volumes treated. The residence time is a measure of the reaction time in the column and allows calculation of kinetic rates when used in conjunction with conversion data. The number of pore volumes treated allows comparison between systems; two units which are identical in every aspect but size should exhibit similar behavior (e.g. with respect to deactivation) after treating the same number of pore volumes, though the total volume of water in each system will be different.

7. COLUMN STUDIES WITH SUPPORTED PALLADIUM

7.1 Yu Study

Work on aqueous phase hydrodechlorination in columns with supported Pd was first published by Yu. (Yu 1991) Yu employed a cocurrent downward flow trickling bed reactor, containing 31.35 mg of 1% w/w Pd/C particles which were ground and sieved to a diameter of approximately 0.45 mm. The gas feed consisted of pure hydrogen, and the liquid stream consisted of water containing 100-190 mg/L of 1,1,2-TCA or TCE. Residence times in the reactor varied between approximately 0.02 and 0.45 hours. Using a plug flow model, an apparent first order rate expression with respect to the halogenated hydrocarbons was calculated. The "steady-state" reaction rate constants were 2.4/hr for 1,1,2-TCA and 6.0/hr for TCE. These results are consistent with the trend seen by Schreier in batch reactions; the reaction rate of the chlorinated alkene, TCE, is higher than that of the alkane, TCA. (Schreier 1996)

Catalyst deactivation was also observed. Initially in the TCA reaction, the reaction rate constant was 7.14/hour; however it dropped to a constant ("steady-state") value of 2.4/hr after 10 hours of operation, and then decreased rapidly to 0.85/hour after 70 hours of operation. The deactivation was attributed to chloride poisoning, which has been observed in the gas phase. (Simone et al. 1991) An attempt to regenerate the catalyst through water washing resulted in detection of a small amount of chloride in the effluent wash water, and a small increase in catalyst activity. Regeneration was also attempted by drying the catalyst at room temperature and then exposing it to vacuum for three days. The initial TCA conversion was quite high, but decreased rapidly (within four hours) to the level of the deactivated catalyst. The final method of regeneration was through catalyst washing with a NH_4OH solution, upon which the reaction rate constant increased to 2.0/hour. The decrease relative to the "steady-state" value was attributed to loss of the Pd metal during the NH_4OH washing. (Yu 1991)

7.2 Stanford Study

The Munakata study utilized an upflow packed bed column filled with 0.1-2.0 g of 1% w/w Pd/alumina catalyst crushed to a 0.3-0.8 mm size fraction. (Munakata et al. 1998) The influent water stream consisted of deionized water containing 1.5-25 mg/L of TCE. Hydrogen was supplied by purging the influent water supply with hydrogen gas and pressurizing it to 0.5 atm. As in the Yu study, a plug flow model was used and an apparent first order rate constant with respect to TCE was calculated.

With an average residence time of approximately one minute, TCE was >99% removed with a average first order rate constant of 436/hour and no noticeable decrease in activity over 5 months, or 200,000 pore volumes (though it should be noted that at high removals, losses in catalyst activity are difficult to detect). However, the addition of 44 mg/L of nitrate to the influent stream decreased TCE removal to 52%; this is consistent with batch study results in which nitrate slowed the dehalogenation reaction. The removal efficiency of TCE recovered when the nitrate was removed from the influent stream, which indicates competition by nitrate for hydrogen and/or catalyst sites, but no permanent effects (i.e. deactivation). The addition of sodium carbonate (280 mg/L) to the DI influent water decreased conversion by 80% after 130 hours (3400 pore volumes), carbon dioxide (purged into the influent water supply with a 90% H_2/10% CO_2 gas mixture) completely deactivated the catalyst after 13 days (27,000 pore volumes); and phosphate (100 mg/L) caused minimal deactivation of the catalyst. Regeneration of the sodium carbonate-deactivated catalyst through exposure at 300°C to 0.07 mm Hg vacuum followed by oxygen flow appeared to return the catalyst to original activity levels. (Munakata et al. 1998)

7.3 Livermore Study

The Livermore study used a 1.8 L upflow reactor, packed with a 1% w/w Pd/alumina catalyst, nominally 0.32 cm in diameter. (McNab and Ruiz 1998) The water supply consisted of groundwater from a well at the Livermore site and contained PCE (5-7 μg/L), TCE (500-600 μg/L), 1,1-DCE (15-25 μg/L), CT (8-15 μg/L), CF (4-10 μg/L), and 1,2-DCA (4-10 μg/L). The water was first passed through an electrolyzer, which served as a source of hydrogen. With an average residence time of 2.3 minutes, the column showed >95% removal of chlorinated ethylenes (PCE, TCE and 1,1,-DCE) and CT. The apparent first order rate constants for all four compounds were approximately 72/hour. Chloroform was removed at a much slower rate (≤32/hour) and 1,2-DCA removal was not significant. These results are also consistent with the batch studies of Lowry which showed rapid reaction of

PCE, TCE, DCE and CT, with slower reaction of CF and virtually no reaction of 1,2-DCA. (Lowry and Reinhard 1999)

Oxygen appeared to compete with the hydrocarbons; removing the anodic (oxygen-rich) stream of the electrolyzer and feeding only the cathodic (hydrogen-rich) stream to the reactor roughly doubled the rate of reaction. Deactivation was observed over a period of hours (high levels of carbonate caused greater deactivation), and effluent concentrations increased. Unlike in the Yu study, the catalyst activity appeared to recover when the column was soaked in DI water for hours to days. (McNab and Ruiz 1998)

8. FIELD STUDIES WITH SUPPORTED PALLADIUM

Because the application of Pd catalysts to the treatment of contaminated water is relatively new, only one major field study (at Lawrence Livermore National Laboratories) has been conducted and published thus far. (McNab et al. 2000) Other studies, such as that in Bitterfeld, Germany, are currently underway. The Bitterfeld site operates at a residence time of 15 minutes, with a flow of approximately 100 pore volumes/day and uses a zeolite-supported Pd catalyst, which was optimized in laboratory experiments. In the initial tests in the field, the catalyst was deactivated, apparently by sulfide-producing bacteria. Treating the column with 10 g/L of hydrogen peroxide for 2 hours each week (approximately 8 pore volumes of peroxide solution per 700 pore volumes of water treated) resulted in column operation for 15 weeks with 90-99% removal of chlorobenzene and without any apparent catalyst deactivation. (Weiss et al. 1999) As the Pd technology develops further, more field tests are expected.

8.1 Livermore Study

The Pd technology is particularly appropriate for the Livermore site because the water table is fairly far below the surface (20-30 m deep), which makes trenching (e.g. for iron wall barriers) difficult. In addition, the rapid reaction rates and small reactors afforded by Pd catalysis permit the system to be contained within a well bore, which keeps the tritiated water at the site largely below the ground surface. The facility at the Livermore field site consists of two packed bed reactors, place in series in a well-bore. The flow rate is 4 L/min, which yields a residence time of five minutes in the first column and six minutes in the second column. A membrane is used to diffuse hydrogen gas into the influent stream prior to the first reactor. The influent water stream contains the following contaminants: PCE (366-370 μg/L), TCE (3612-3777 μg/L), 1,1-DCE (130-180 μg/L), cis-DCE (0.6-0.7 μg/L), CT (18-21 μg/L), CF (167-235 μg/L), and 1,2-DCA (26-28 μg/L). (McNab et al. 2000)

Initially, the facility operated for four hours per day, followed by purges with three pore volumes of nonhydrogenated water and then drainage; this is equivalent to approximately 50 pore volumes/day. The system successfully dechlorinated PCE (>99%), TCE (>99%), 1,1-DCE (>99%), carbon tetrachloride (>98%), and chloroform (>91%); however, 1,2-DCA concentrations were not significantly changed. Chlorinated intermediates were not observed in the effluent, except for low levels of cis-DCE, which may have been present in the influent. Analysis of effluent from the first reactor indicated that almost all removal occurred in the first reactor; only CF showed further reductions in concentration across the second column. (McNab et al. 2000)

Increased operation to eight hours/day (approximately 90 pore volumes/day) resulted in reduced removal efficiencies: PCE (>82%), TCE (>93%), 1,1-DCE (>96%), carbon tetrachloride (>98%), and chloroform (>57%). In addition, daughter products VC and cis-DCE appeared in the effluent stream. Decreasing operating time back to four hours/day resulted in recovery of catalyst activity, and a subsequent increase to 5-6 hours/day (approximately 65 pore volumes/day) yielded stable performance. An effort to increase operating time by running with upward flow for 4-5 hours/day followed by 4-5 hours of downward flow was successful, with removal efficiencies roughly equivalent to running five hours/day in the mode of upward flow only. The column has now been successfully operating for over one year. (McNab et al. 2000)

9. SUMMARY

As detailed in this chapter, a wide variety of substances are reducible in aqueous solution at ambient temperatures by Pd catalysts: halogenated alkanes, ethylenes, and aromatics; PAHs; formate/carbonate; and oxidized nitrogen species. The halogenated compounds generally react to form the corresponding simple alkanes (or completely dehalogenated aromatics in the case of the halogenated aromatics); chlorinated intermediates and by-products are only observed occasionally under certain conditions. Products indicative of radical reactions are also seen under some circumstances, particularly with the chlorinated methanes. The PAHs formed partially hydrogenated compounds, and carbonate and formate reached an equilibrium state.

Transformation of many compounds appears to be direct, without detectable chlorinated intermediates. In some cases, direct transformation occurs in parallel with other reactions, such as the formation of radicals or sequential dehalogenation. Some alkanes (1,1,2,2-TCA) appears to be dehalogenated sequentially, while others (DBCP) first undergo elimination to form an alkene, followed by complete dehalogenation and hydrogenation. Hydrogenation of PAHs is reversible; products form equilibrium ratios.

The kinetics associated with catalytic reactions are complex; however, some general trends can be determined. Reactions are often first order with respect to the reactant, and the rates of hydrodechlorination are faster than hydrogenation. Polyaromatic compounds react faster than monoaromatic compounds, and chlorinated alkenes react faster than their corresponding alkanes. Finally, the reaction rate often increases with increasing degree of chlorination, though this does not hold true for the chlorinated ethylenes.

Water quality and components present in the solution matrix affect the catalytic reaction. Acids dissolve the Pd metal, while bases promote some reactions. Batch tests showed that oxygen, sulfate, nitrate, and nitrite slowed reaction somewhat; sulfite slowed it significantly, and bisulfide deactivated the catalyst completely. Column tests of TCE reaction on a Pd/alumina catalyst with DI water caused no observable deactivation, but deactivation was seen with the reaction of TCA on Pd/C. In the Pd/alumina column, the addition of nitrate to DI water did slow the reaction; phosphate, carbonate, and carbon dioxide caused some deactivation. Regeneration through evacuation and oxidation of the catalyst improved the activity.

The catalyst support impacts the rate of a catalyzed reaction, the reaction pathway (quantities and species of intermediates and products) and the resistance of the catalyst to deactivation. In DBCP reactions, powders had higher rate constants than beads, presumably due to reduced mass transfer limitations; alumina yielded a faster rate than C, which had a faster rate than PEI/silica. Sorptive capabilities of the supports may also play an important role: Kovenklioglu found that supports which sorbed 1,1,2-TCA more strongly had higher reaction rates, and Farrell concluded that TCE sorption to Fe cause higher reaction rates on Pd/Fe electrodes than on pure Pd electrodes. It is also clear that supports influence reaction products, but the correlation between a given support and pathways/products it promotes is not yet understood. The choice of support can also affect its resistance to deactivation; this implies that catalyst supports may be tailored to maximize activity over the long term.

Column studies have confirmed the catalytic capabilities of Pd. Although laboratory tests have indicated some potential issues with deactivation, field studies offer encouraging results for the Pd technology. With periodic regeneration or optimization of the catalyst material, field units have been successfully operated (in one case, for more than one year) without signs of long-term deactivation.

In conclusion, reductive dehalogenation with Pd catalysts offers a number of advantages: it can treat a wide variety of compounds, including mixtures; it generally results in simple alkanes, with few halogenated intermediates, if any; and it is extremely rapid, which allows small, in-well reactors. As more studies are conducted, the applicability to a broad range of conditions will be tested and will provide opportunities to better understand the process. This will facilitate optimization of catalyst parameters and column operation for the most effective remediation under a variety of field conditions.

ACKNOWLEDGEMENTS

This work was funded in part by the Western Region Hazardous Substance Research Center as sponsored by the United States Environmental Protection Agency under Assistance ID number R825689-01, and by a National Science Foundation Student Fellowship.

BIBLIOGRAPHY

Augustine, R. L. (1965). Catalytic Hydrogenation; Techniques and Applications in Organic Synthesis, M. Dekker, New York.

Biswas, J., Bickle, G. M., Gray, P. G., Do, D. D., and Barbier, J. (1988). "The Role of Deposited Poisons and Crystallite Surface Structure in the Activity and Selectivity of Reforming Catalysts." Catalysis Reviews - Science and Engineering, 30(2), 161-247.

Engel, D. C., Versteeg, G. F., and Van Swaaij, W. P. M. (1995). "Reaction Kinetics of Hydrogen and Aqueous Sodium and Potassium Bicarbonate Catalysed by Palladium on Activated Carbon." Chemical Engineering Research and Design, 73(A6), 701-706.

Grittini, C., Malcomson, M., Fernando, Q., and Korte, N. (1995). "Rapid Dechlorination of Polychlorinated Biphenyls on the Surface of a Pd/Fe Bimetallic System." Environmental Science and Technology, 29(11), 2898-2900.

Hoke, J. B., Gramiccioni, G. A., and Balko, E. N. (1992). "Catalytic Hydrodechlorination of Chlorophenols." Applied Catalysis B: Environmental, 1, 285-296.

Hörold, S., Vorlop, K.-D., Tacke, T., and Sell, M. (1993). "Development of Catalysts for a Selective Nitrate and Nitrite Removal from Drinking Water." Catalysis Today, 17, 21-30.

Korte, N., Liang, L., Muftikian, R., Grittini, C., and Fernando, Q. (1997). "The Dechlorination of Hydrocarbons: Palladised Iron Utilised for Ground Water Purification." Platinum Metals Review, 41(1), 2-7.

Kovenklioglu, S., Cao, Z., Shah, D., Farrauto, R. J., and Balko, E. N. (1992). "Direct Catalytic Hydrodechlorination of Toxic Organics in Wastewater." AIChE Journal, 38(7), 1003-1012.

Li, T., and Farrell, J. (2000). "Reductive Dechlorination of Trichloroethene and Carbon Tetrachloride Using Iron and Palladized Iron Cathodes." Environmental Science and Technology, 34(1), 173-179.

Liang, L., Korte, N., Goodlaxson, J. D., Clausen, J., Fernando, Q., and Muftikian, R. (1997). "Byproduct formation During the Reduction of TCE by Zero-Valence Iron and Palladized Iron." Ground Water Monitoring and Remediation, 17(1), 122-127.

Lowry, G. V., and Reinhard, M. (1999). "Hydrodehalogenation of 1- to 3-Carbon Halogenated Organic Compounds in Water Using a Palladium Catalyst and Hydrogen Gas." Environmental Science and Technology, 33(11), 1905-1910.

Lüdtke, K., Peinemann, K.-V., Kasche, V., and Behling, R.-D. (1998). "Nitrate Removal of Drinking Water by Means of Catalytically Active Membranes." Journal of Membrane Science, 151, 3-11.

MacNeil, J. H., Berseth, P. A., Westwood, G., and Trogler, W. C. (1998). "Aqueous Catalytic Disproportionation and Oxidation of Nitric Oxide." Environmental Science and Technology, 32, 876-881.

McNab, W. W. J., and Ruiz, R. (1998). "Palladium-Catalyzed Reductive Dehalogenation of Dissolved Chlorinated Aliphatics Using Electrolytically-Generated Hydrogen." Chemosphere, 37(5), 925-936.

McNab, W. W. J., Ruiz, R., and Reinhard, M. (2000). "In Situ Destruction of Chlorinated Hydrocarbons in Groundwater Using Catalytic Reductive Dehalogenation in a Reactive Well: Testing and Operational Experiences." Environmental Science and Technology, 34(1), 149-153.

Muftikian, R., Fernando, Q., and Korte, N. (1995). "A Method for the Rapid Dechlorination of Low Molecular Weight Chlorinated Hydrocarbons in Water." Water Research, 29, 2434-2439.

Munakata, N., Roberts, P. V., Reinhard, M., and McNab, W. W. J. (1998). "Catalytic Dechlorination of Halogenated Hydrocarbon Compounds Using Supported Palladium: A Preliminary Assessment of Matrix Effects." IAHS Publication(250), 491-496.

Schreier, C. G. (1996). "The Destructive Removal of Chlorinated Organic Compounds from Water Using Zero-Valent Metals or Hydrogen and Supported Palladium," Ph.D. Thesis, Dept. of Civil Engineering, Stanford University, Stanford, CA.

Schreier, C. G., and Reinhard, M. (1995). "Catalytic Hydrodehalogenation of Chlorinated Ethylenes Using Palladium and Hydrogen for the Treatment of Contaminated Water." Chemosphere, 31(6), 3475-3487.

Schüth, C., and Reinhard, M. (1997). "Catalytic Hydrodehalogenation of Some Aromatic Compounds Using Palladium on Different Support Materials." *Proc.,* 213th ACS Meeting, San Francisco, CA, 173-175.

Schüth, C., and Reinhard, M. (1998). "Hydrodechlorination and Hydrogenation of Aromatic Compounds over Palladium on Alumina in Hydrogen-Saturated Water." Applied Catalysis B: Environmental, 18, 215-221.

Siantar, D. P., Schreier, C. G., Chou, C., and Reinhard, M. (1996). "Treatment of 1,2-Dibromo-3-chloropropane and Nitrate-Contaminated Water with Zero-Valent Iron or Hydrogen/Palladium Catalysts." Water Research, 30(10), 2315-2322.

Simone, D. O., Kennelly, T., Brungard, N. L., and Farrauto, R. J. (1991). "Reversible Poisoning of Palladium Catalysts for Methane Oxidation." Applied Catalysis, 70, 87-100.

Stalder, C. J., Chao, S., Summers, D. P., and Wrighton, M. S. (1983). "Supported Palladium Catalysts for the Reduction of Sodium Bicarbonate to Sodium Formate in Aqueous Solution at Room Temperature and One Atmosphere of Hydrogen." Journal of the American Chemical Society, 105, 6318-6320.

Weiss, H. , Dauss, B., Teutsch, G. (1999). "SAFIRA: 2.Statusbericht." *UFZ-Report No. 17/1999.* Environmental Research Center Leipzig/Halle (UFZ), Leipzig, Germany .

Wiener, H., Blum, J., Feilchenfeld, H., Sasson, Y., and Zalmanov, N. (1988). "The Heterogeneous Catalyst Hydrogenation of Bicarbonate to Formate in Aqueous Solutions." Journal of Catalysis, 110, 184-190.

Wiener, H., Sasson, Y., and Blum, J. (1986). "Palladium-Catalyzed Decomposition of Aqueous Alkali Metal Formate Solutions." Journal of Molecular Catalysis, 35, 277-284.

Yu, G. (1991). "Hydrodechlorination of Trichloroethane and Trichloroethylene in a Trickle-Bed Reactor," Master of Engineering Thesis, Stevens Institute of Technology, Castle Point, Hoboken, NJ.

Chapter Four

Electrochemical and Biogeochemical Interactions under dc Electric Fields

AKRAM N. ALSHAWABKEH[1] and KRISHNANAND MAILLACHERUVU[2]

[1] *Department of Civil and Environmental Engineering, Northeastern University, Boston, MA 02115*
[2] *Department of Civil Engineering and Construction, Bradley University, Peoria, IL 61625*

Key words: electrostimulation, electrokinetic, electrochemical, bioremediation

Abstract: Direct electric currents generate electrochemical and biochemical changes that impact the subsurface environment and can be engineered for transporting and mixing contaminants, biostimulants, and possibly bioaugmentation inoculants to enhance *in situ* bioremediation. A review of transport processes under dc fields, electrolysis reactions, and microbial adhesion and transport is provided in this chapter, followed by an evaluation the impact of dc fields on microbial activity. In general, dc electric field intensities tend to produce complex effects on the activity of mixed microbial cultures. Anaerobic cultures seem to experience an "environmental shock" when exposed to electric field intensities greater than 1.5 V/cm. However, these cultures are able to recover their activities once the electric currents are switched off. Aerobic cultures did not seem to be adversely affected by exposure to field intensities less than 0.28 V/cm. Higher dc field intensities, up to 1.14 V/cm, seem to stimulate aerobic cultures during the first 24 hours of exposure. However, continuous exposure to dc fields (in the range of 1.14 V/cm) following the first 24 hours seems to retard the growth of aerobic cultures.

1. INTRODUCTION

Organic contaminants have been known to be present in many hazardous waste sites (USEPA, 1991; DOE, 1995; Kelsh and Parsons, 1997). Economical restoration of these contaminated sites to environmentally acceptable conditions is an important challenge facing the scientific and technical community. Current *in situ* soil remediation technologies depend on hydraulic and air flow for effective remediation of soils and are not as effective in the clean-up of lower hydraulic conductivity soils (less than 10^{-5} cm/sec) such as fine sands, silts and clays. *In situ* bioremediation is an attractive and often cost-effective option to remediate soil and groundwater contaminated with organics. Successful implementation of *in situ* bioremediation is dependent upon presence, or effective injection, of electron acceptors and nutrients into the porous medium. Microbial processes require an electron donor, macronutrients (e.g, nitrogen and phosphates), micronutrients, trace nutrients and an electron acceptor. Microbially-mediated cometabolite transformations require an additional external electron donor as well. In some cases, inducers may be needed to trigger the microbial transformation reaction. Effective introduction and transport of these additives is hindered by low soil permeability, preferential flow paths (channeling), biological utilization, and chemical reactions in the soil. Complications of site geohydrology, additive transport and associated reactions, coupled with the observed inefficiency in the field, have been mostly approached by gross overinjection of the additives. Excessive dosing coupled with the shortcomings of the hydraulically-driven transport processes can result in nutrient rich areas with excessive biological growth (biofouling). Biofouling adversely impacts system implementation due to reduced conductivity by microbial growth plugging the flow paths. Several surveys have concluded that ineffective transport of remediation additives is the primary cause of system failure for some *in situ* bioremediation efforts (Zappi *et al.*, 1993; NRC 1993).

A technology for uniform introduction of nutrients and electron acceptors/donors has been the principal bottleneck in the successful field implementation of *in situ* bioremediation (Suflita and Sewell, 1991; Zappi *et al.*, 1993). An emerging technology for treating such hazardous waste sites using *in situ* bioremediation methods is through the application of electric fields to transport the nutrients as well as bacteria. Electric fields could be used to overcome problems associated with additives injection into heterogeneous and/or low permeability soils. Uniform and accelerated delivery of nutrients, and electron acceptors may be achieved using electric injection instead of hydraulic injection. Uniform transport of ions under electrical fields is controlled by the charge and ionic mobility of available species in the pore fluid. Accelerated and uniform transport rates of nutrients, electron acceptors and microorganisms could be achieved in heterogeneous or low hydraulic conductivity soils by electric fields. The technique could also employ electroosmotic flow into fine-grained soils to enhance additive injection. This method can be used to stimulate bioremediation under aerobic or anaerobic conditions.

The applied electric field may affect the electrochemical and geochemical processes and phenomena in the soil. Figure 1 shows some of the key processes and interactions that occur in soils when an electric field is applied. The conditions generated under dc electric fields and their impact on the porous media are discussed. The chapter describes the effect of dc fields on ions transport in soils, electrolysis and geochemical reactions, microbial adhesion and transport, and microbial activity. The interest in these in these processes is derived from the potential to develop strategies to enhance *in situ* bioremediation. A review of ion transport mechanism in porous media under dc fields is provided to address the potential of injecting and transporting nutrients and biostimulants by ion injection and electroosmosis and also the possibility of injecting bioaugmentation innoculants by electrophoresis. A discussion is provided for electrolysis reactions and their effects on pH and dissolved oxygen (DO) values, which in turn affect microbial survival. Electrolysis reactions are also important as they cause the production of oxygen and hydrogen and also may cause abiotic degradation of contaminants. A discussion is provided for the factors that may impact microbial adhesion and transport under dc fields.

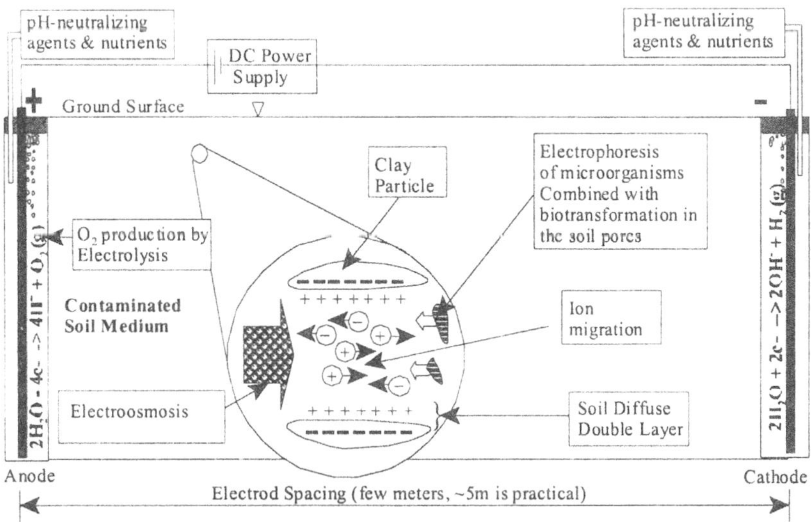

Figure 1. A schematic of the bio-geochemical pocesses under dc fields.

A review of studies that investigate electrokinetic extraction of organic contaminants is also provided. Finally, the results of studies that evaluate the general impacts of electric fields on anaerobic and aerobic cultures are summarized.

2. TRANSPORT UNDER ELECTRIC FIELDS

Injection and transport of bioremediation additives by electric fields are controlled by electrokinetic transport mechanisms in an ion exchange medium. These transport mechanisms include electroosmotic advection (or electroosmosis), electromigration (or ionic migration), electrophoresis and to a lesser extent diffusion. Electroosmosis is mobilization of the pore fluid in the soil, usually from the anode (positive electrode) toward the cathode (negative electrode), while electromigration is the separation of anions (negative ions) and cations (positive ions) by their migration to the anode and cathode, respectively. Electromigration does not require any fluid flow as the ions can migrate in a stationary pore fluid. Electrophoresis describes the transport of charged particles, such as clay particles or bacteria, under the applied dc electric field to the electrode opposite in polarity. The total mass flux of a charged species/particle (that can serve as a nutrient, electron acceptor or donor) is evaluated by the addition of the diffusion, ionic migration and electroosmosis components. The total mass flux can be given by (Alshawabkeh and Acar 1992, 1996).

$$J_i = -D_i^* \frac{\partial c_i}{\partial x} - c_i(u_i^* + k_e)\frac{\partial \phi}{\partial x} \tag{1}$$

where J_i ($MT^{-1}L^{-2}$) is the total mass flux of a charged species/particle i, D_i^* (L^2T^{-1}) is the effective diffusion coefficient of i, u_i^* ($L^2T^{-1}V^{-1}$) is the effective (ionic or electrophoretic) mobility of i, Φ (V) is the electric potential in volt, k_e ($L^2T^{-1}V^{-1}$) is the coefficient of electroosmotic permeability of the soil, and c_i (ML^{-3}) is the concentration of i. The rate of electroosmotic flow is controlled by the value of k_e, which is a measure of the rate of fluid flow per unit area under a unit electric gradient. The value of k_e is given as a function of zeta potential of soil particle surface, viscosity of the pore fluid, porosity and electrical permittivity of the soil medium. Hunter (1981) provides a description of zeta potential in colloid science and a discussion of the factors that impact its value.

The effective ionic mobility is a measure of rate of electromigration or electrophoresis of a charged particle under a unit electric gradient. Higher the charge and smaller the size of the particle or ion will result in higher mobilities. The values of u_i^* have been estimated for ions by multiplying the value of ionic mobility in solution at infinite dilution, u_i, by the soil tortuousity and porosity. The value of u_i is evaluated by assuming that Nernst-Townsend-Einstein relation between D_i, the molecular diffusion coefficient, and u_i, holds for ions in the pore fluid of soils (Holmes, 1962). The value of u_i^* is then evaluated by,

$$u_i^* = \frac{D_i^* z_i F}{R T} = u_i \tau n \qquad (2)$$

where u_i ($L^2T^{-1}V^{-1}$) is the ionic mobility of i at infinite dilution, z_i is the charge of i, n is the soil porosity, τ is a factor accounting for the soil tortuousity, R is the universal gas constant (8.3144 J/K.mole), and T is the absolute temperature. Acar and Alshawabkeh (1993) reported values of effective ionic mobilities in clays in the order of few centimeters per day under a unit electric field (1 V/cm).

Acar et al. (1996; 1997) showed that ionic migration could be used for injection and transport of anionic and cationic additives. In a bench-scale experimental setup, ammonium hydroxide (NH_4OH) was introduced at the anode compartment and sulfuric acid (H_2SO_4) at the cathode compartment. The electric field caused migration of nitrate ion from anode towards the cathode and sulfate ion from cathode towards the anode. The study reported transport rates of 5 to 20 cm/day in fine sand and kaolinite soil specimens and consequent soil saturation of ammonium and sulfate ions. The study concluded that ion migration under dc fields can be used to inject nutrients, electron acceptors/donors to enhance *in situ* bioremediation.

3. ELECTROLYSIS

Application of direct electric currents in saturated soils results in electrolysis reactions at the electrodes. If inert electrodes (such as graphite) are used, water oxidation generates an acid (H^+) and oxygen gas at the anode while water reduction produces a base (OH-) and hydrogen gas at the cathode (Acar and Alshawabkeh, 1993),

$$\tfrac{1}{2} H_2O \Longrightarrow e^- + H^+ + \tfrac{1}{4} O_2 \qquad \text{Anode} \qquad (3)$$

$$H_2O + e^- \Longrightarrow OH^- + \tfrac{1}{2} H_2 \qquad \text{Cathode} \qquad (4)$$

Based on Faraday's law for equivalence of mass and charge, the rate of electrolysis reactions depends on the total current applied. For a specific reaction, e.g., water oxidation at the anode, the rate of electrolysis is given by,

$$J = \frac{I}{z_i F} \tag{5}$$

where J is the rate of oxidation or reduction by electrolysis (MT^{-1}), I is the current (A), z is the charge of the ion (for hydrogen z = 1) and F is Faraday's constant (96,485 C / mole). Assuming that water electrolysis occurs at the electrodes with 100% efficiency, one amp current will oxidize 1/2F mole H_2O into 1/F mole of H^+ and 1/4F mole O_2 at the anode per second. The same current will reduce 1/F mole H_2O into 1/F mole OH^- and 1/2F mole H_2 per second at the cathode. It should be indicated that other electrolysis reactions may occur and limit the water electrolysis reaction. The type of electrolysis reaction depends upon the chemistry of the electrolyte, pH and the standard electrode potential for ions in the electrolyte.

Oxidation and reduction generate complex boundary conditions at the electrodes that can either enhance or retard microbial activity and contaminant degradation. Oxidation at the anode generates oxygen that stimulates aerobic degradation. An acid (H^+) is also produced, which could drop the pH at the anode to below 2. The acid migrates under the applied electric field toward the cathode and may acidify the soil. The soil resistance to pH changes depends upon the soil buffering capacity. However, a pH drop is not favored for microbial growth. Optimum pH for bacterial growth is near neutrality with minimum and maximum pH values for growth near 5 and 9, respectively (Gaudy and Gaudy, 1988). A relatively neutral pH is needed to optimize biological activity in the soil. Soil acidification may also cause dissolution of the soil minerals. Electrolytes conditioning that neutralize acid generation and enhance delivery of proper additives for bioremediation should be considered. As an example, the use of ammonium hydroxide at the anode and sulfuric acid at the cathode was successful in (a) neutralizing the acid and base at the anode and cathode and (b) injection of ammonium and sulfate ions into the soil (Acar et al. 1996).

Oxygen production is another critical boundary condition generated at the anode. The oxygenated water at the anode may be introduced into the soil at efficient rates by electroosmotic flow to enhance aerobic conditions. Preliminary results indicated that the efficiency of gas production by electrolysis is 75%. Other mechanisms, such as air sparging, in-well aeration, hydrogen peroxide, oxygen releasing solid compounds, or even cryogenic oxygen generators, can be used to supplement oxygen requirement. However, the advantage of dc fields is that they serve for both oxygen generation and transport in low hydraulic conductivity soils. It should be noted that while this oxygen will enhance conditions for aerobic microorganisms, it is likely to adversely impact the anaerobic microorganisms. If anaerobic bioremediation is the target treatment method, measures to remove oxygen at the anode using chemical scavengers or gas-stripping (eg. nitrogen gas) are needed to prevent adverse impacts on anaerobic microorganisms. Alternatively, an aerobic-anaerobic treatment could be designed to take advantage of the availability of oxygen in a limited subsurface zone

4. ORGANICS EXTRACTION BY DC FIELDS

Electric fields have been used for extraction of the contaminants from the subsurface. Bench-scale tests and limited field studies demonstrated the use of electric fields for extraction of heavy metal and radionuclides from soils (Lageman *et al.*, 1989; Hamed *et al.*, 1991; Pamukcu and Wittle, 1992; Acar and Alshawabkeh, 1993; Probstein and Hicks, 1993; Runnels and Wahli, 1993; Eykholt and Daniel, 1994; Alshawabkeh and Acar, 1996; Pamukcu *et al.*, 1997). Electric fields have also been used for electroosmotic extraction of organic contaminants from soils (Shapiro *et al.*, 1989; Acar *et al.*, 1992; Bruell *et al.*, 1992). Recently, the Department of Energy, Environmental Protection Agency (EPA), Monsanto, General Electric, and Dupont applied electric fields for electroosmotic extraction of TCE from a site in Paducah, Kentucky using layered horizontal electrodes by the "Lasagna™" process (Ho *et al.*, 1997; 1999a; 1999b). The Lasagna™ process uses mainly electroosmosis for extraction of TCE out of the soil into electrode or treatment wells.

Electric fields use in soil restoration has been focused on contaminant extraction by their transport under electroosmosis and ionic migration. Contaminant extraction by electric fields is a successful technique for removal of ionic or mobile contaminants in the subsurface. However, this technique might not be effective in treatment of soils contaminated with immobile and/or trapped organics, such as dense non aqueous phase liquids (DNAPLs). For such organics, it is possible to use electric fields to stimulate *in situ* biodegradation under either aerobic or anaerobic conditions. It is necessary to evaluate the impact of dc electric fields on the biogeochemical interactions prior to application of the technique. It is not clear yet how dc electric fields will impact microbial adhesion and transport in the subsurface. Further, the effect of dc fields on the activity of microorganisms in a soil matrix is not yet well understood.

5. MICROBIAL ADHESION AND TRANSPORT

The presence of appropriate microorganisms (depending on electron-acceptor and nutrient availability) at the actual site of contamination (sometimes at a micro-scale) has long been recognized as a key factor in determining if biodegradation/biotransformation will occur, as well as in influencing the rate of biodegradation. Microorganisms can be present in the subsurface in suspension (in the pore fluid), as microcolonies or as a biofilm. Bacterial adhesion to porous media often influences the nature and extent of colonization of a subsurface medium. The biofilm development model (Characklis and Wilderer, 1988), summarized below, can provide some understanding into bacterial transport and adhesion in subsurface environments. The "first three steps" in biofilm development are: (1) Surface conditioning, (2) Transport of microoganisms to the conditioned surface and, (3) Sorption of microorganisms to the surface. In biofilm literature, it is recognized that the nature of the bacterial cell surface is a key parameter in determining its adhesion to any media and eventually the biofilm architecture (Reynolds *et al.*, 1989; Characklis and Wilderer, 1988; van Loosdrecht *et al.*, 1987). The sorption of microorganisms is due to reversible sorption (governed by charge/electrostatic interactions) and irreversible sorption (due to production of extracellular exoploymers and formation of matrix). The dependency of reversible sorption of bacteria on charge/electrostatic interactions indicates that an applied electric field may play a significant role in bacterial adhesion and transport in subsurface environments. Further, the role of bacterial surface (eg., lipid content and surface charge) itself could influence the electrophoretic mobility of bacteria in porous media. Data available in literature indicate that the presence of heavy metals alter the electrokinetic properties of bacteria, although the context of these researchers was not hazardous waste decontamination (Collins and Stotzky, 1992). The imposed dc electric field is also expected to affect the electrokinetic properties, adhesion and transport of microorganisms. It is not clear what will be the extent of this effect; however, the authors noted that microorganisms tend to stick and attach themselves to the electrodes in experiments conducted using diluted sludge samples under dc fields. The type or activity of microorganisms attached to the electrodes was not evaluated. Electrode polarity did not seem to have an effect on adhesion as microorganisms were attached not only to the anode but also to the cathode. This might indicate that the adhesion to electrode surface is not necessarily due to the electric attraction between the negatively charge microrganisms and the positively charged anode. Further evaluation is needed to verify if this adhesion is due to reversible sorption, irreversible sorption, or due to electrode charge.

Electric fields will not only impact microorganisms adhesion but will also affect their transport in porous media. As microbes are generally negatively charged, dc fields will cause their transport towards the anode. DeFlaun and Condee, (1997) demonstrated electrokinetic transport of a pure bacterial culture in bench-scale soil samples. Generally, the rate of transport is related to the effective electrophoretic mobility in soils and is affected by soil physical parameters such as porosity, pore size distribution and tortousity. The results of DeFlaun and Condee (1997) demonstrate the potential of using microorganisms electrophoresis for the purpose of bioaugmentation to enhance *in situ* bioremeidation. It is also necessary to note that microorganisms transport in porous media under electric field might not be strictly governed by electrophoresis alone, since microbes, as living entities, may be subject to other influences or "attractors" and also tend to form colonies and attach to the soil particle surface.

6. MICROBIAL ACTIVITY UNDER DC FIELDS

Recently, there has been an increasing interest in the bioelectrochemical processes in medicine and biosensor fields. Studies showed that low level ac (alternating currents) and dc electric fields (in the range of volts/cm and up to few hundred Hz) stimulate the metabolic processes (Berg and Zhang, 1993; McLeod *et al.*, 1992, Blank *et al.*, 1992) in a nonlinear way so that only a specific range (or so-called window) of field strength and frequencies can cause a significant impact (Tsong, 1992; Fologea *et al.*, 1998). This "electrostimulation" process has been explored in areas that include enzyme activation, biopolymer synthesis, membrane transport, and proliferation (Berg, 1993). Furthermore, bioelectrochemical devices (or biosensors) are being developed for manipulation of bacteria, viruses and genetic material using sophisticated microelectrodes (Buerk, 1993; Ramsey, 1998). Electrostimulation and biosensors research fields are at the micro-scale (might reach the nano-meter) level, focus on ac fields, and have not yet been employed in the soil bioremediation area. It is necessary to evaluate the dc field intensities that microorganisms can sustain and also the "window" of dc fields that may stimulate microbial activity.

Electric fields also introduce environmental changes that affect microbial growth. As discussed earlier, electrolysis reaction impact pH, Dissolved Oxygen (DO) and other geochemical conditions. Furthermore, electric fields may produce an increase in temperature. Most microorganisms grow rapidly at temperatures between 20 and 45 °C and are capable of growing over a range of 30 to 40°C (Gaudy and Gaudy, 1988). Temperature increase to above 45 °C will significantly limit the growth of most microbes (some could survive high temperatures). Temperature increase due to current application will depend upon field strength and resistivity of the medium.

Acar and Alshawabkeh (1996) reported 10 °C increase in temperature in an unenhanced (no additives were used to control electrolyte pH) large-scale test on extraction of lead from kaolinite.

6.1 Anaerobic Microbial Activity

The impact of electric current on the environmental conditions and the anaerobic microbial activity in completely mixed fed-batch reactors was studied at various electric field strengths (Maillacheruvu and Alshawabkeh, 1999). Experiments were conducted in bio-electrokinetic (BioEK) reactors, which consist of plexiglass boxes with titanium-coated mesh electrodes mounted at both ends. Electric fields of 1.5 V/cm through 6 V/cm were applied. Unacclimated anaerobic cultures obtained from a mesophilic anaerobic digester were used in these experiments.

Limited pH changes may occur if water electrolysis reactions (Equations 3 and 4) occur at the same rate and efficiency. In a completely mixed reactor, the proton produced at the anode should neutralize the hydroxyl ion produced at the cathode. However, the results indicated that the pH decreased to less than 5.5 even under completely mixed conditions in fed-batch reactors. The pH drop indicate less hydroxyl production at the cathode, either because different electrolysis reactions occurred (other than Equation 4) or because of biochemical reactions in the reactor. The type and concentrations of ions in the solution will impact the pH changes and require further investigation. Sodium bicarbonate was used and was effective in buffering the system for the range of electric field strengths studied.

Dissolved oxygen was produced in the experimental reactors, proportional to electric field strength used, where no oxygen scavengers were used. Dissolved oxygen was, however, controlled by addition of sodium sulfite to the system. In an actual soil, there may exist niches, which are devoid of oxygen where anaerobic bacteria may survive even if some dissolved oxygen was produced and not eliminated completely using oxygen scavengers. Further studies in soils are needed to evaluate this hypothesis. Some anaerobic bacteria (particularly sulfate-reducing bacteria or SRB) may indeed have survived even in the presence of relatively high concentrations of oxygen.

Electric current in the reactor generally increased slightly with time (5% at 1.5 V/cm to about 15%-18% at 4.5 V/cm for an exposure period of about 140 hours) at different electric field strengths. While this suggests that the ionic composition of the electrolyte medium changed over time, results from this preliminary study indicate that dissolved organic carbon (DOC) removal efficiency and microbial activity do not appear to be significant, especially at the lower end of electric field strengths.

Microbial activity was estimated as the capacity of the culture to recover from exposure to electric currents. Microbial activity was measured as a function of the ability of the anaerobic microorganisms to consume readily degradable acetate (measured as DOC). A sample of culture was withdrawn from a fed-batch experimental reactor at regular intervals. One part of the sample was immediately analyzed (designated the "t_0 sample") for DOC concentration while the other part allowed to "recover" for a period of 24 hours in an evacuated glass vial to preserve anaerobic conditions. After 24 hours had elapsed the DOC was measured again. The second part of the sample was designated the "t_{24} sample". If the t_{24} sample shows a decrease in DOC as compared to the t_0 sample, it is indicative of an active culture. Figure 2 shows the microbial activity data for experiments at electric field strengths of 1.5 V/cm and 4.5 V/cm. These data indicate that, before the electric current was applied, the t_{24} sample showed more removal (about 20%) of DOC than the t_0 sample -- indicative of an active culture as noted earlier. From Figure 2, it is apparent that the initial "shock" exposure to the electric current results in a decrease in activity for the t_{24} sample as compared to the t_0 sample. This trend continued for several hours during which period the total DOC percent removal in the experimental reactor also dropped by about 10 to 20% in all the experiments tested. However, after a period of several hours of exposure, the t_{24} sample gradually showed an increase in microbial activity by exhibiting higher DOC removal than the t_0 sample -- even during the application of the electric field. This is an interesting phenomenon since it indicates a certain degree of acclimation of the culture to electric current, and a tendency for the anaerobic culture to recover from the initial shock load in terms of changes in environmental conditions due to application of the electric current. Once the electric current application was stopped, the rate of DOC percent removal in the experimental reactors gradually improved. Eventually, after the removal of the electric current, the 24 hour sample showed about 20% higher removal of DOC as compared to the t_0 sample. These data also suggested that there was essentially no difference in recover of microbial activity 1.5 V/cm and 4.5 V/cm experiments. Results from 3.0 V/cm and 6.0 V/cm showed the same trends.

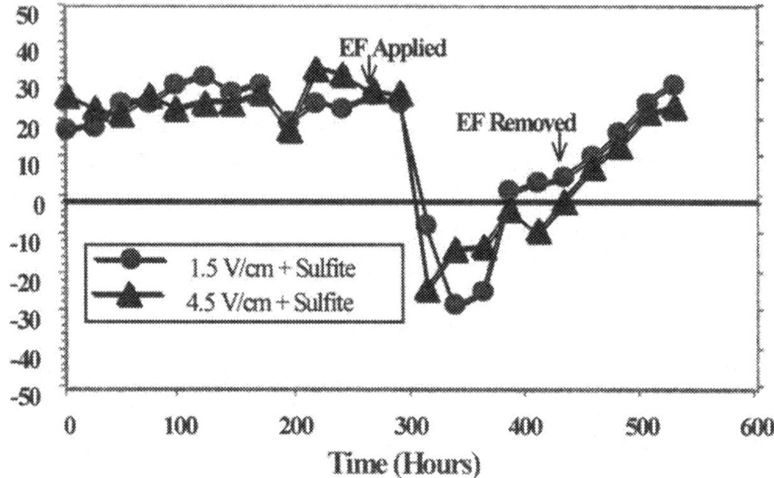

Figure 2. Variation in anaerobic microbial activity under electric fields (Maillacheruvu and Alshawabkeh, 1999).

6.2 Aerobic Microbial Activity

Another preliminary study was conducted to evaluate the impact of dc fields on aerobic microbial activity in a completely mixed and aerated reactor. The study used a sludge sample from the aeration tank of Deer Island Wastewater Treatment Plant, MA. The sludge was placed in two identical polyethylene carboys and aerated by gas diffusing stones connected to an air source. A batch-fed system was used to maintain the sludge. Percent increment of volatile suspended solids (VSS) was used to measure growth rate and the sludge was maintained until steady state growth rates were achieved after about 20 days. Samples were then taken from the sludge and placed in Bioelectrokinetic (BioEK) reactors (Figure 3). The reactors consist of acrylic boxes (14-cm length × 14-cm width × 10-cm height), which have two holes on the top: one was used for sampling and for measurement probes (DO, pH, conductivity and temperature probes), while the other hole was used for the aeration tubing. Two titanium-coated mesh electrodes were fixed on the inner side of the box as anode and cathode (Fig. 3).

Samples were then taken from the sludge and placed in the BioEK reactors (Figure 3). The reactors were modified to allow aeration of sludge. Electric dc fields of 4, 8, and 16 volts, which reflect 0.28, 0.57, and 1.14 V/cm dc electric fields, respectively, were applied. These electric field strengths were less than those used in the anaerobic tests, which makes it easier to control the environmental changes, such as pH, produced by electrolysis. Initial testing conditions are summarized in Table 1.

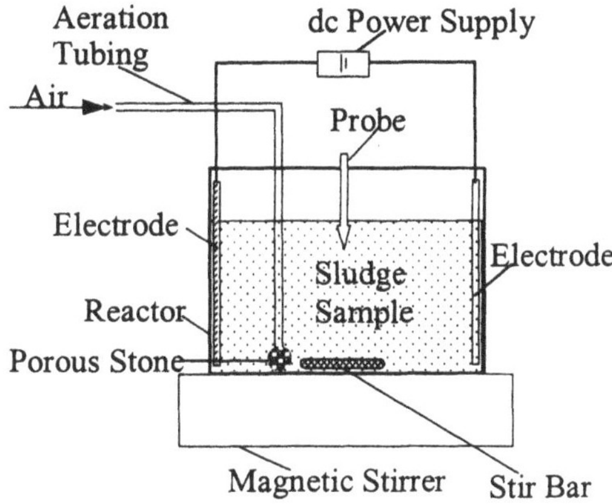

Figure 3. A schematic of BioEK reactor

Table 1. Initial testing conditions for aerobic tests.

Sludge Volume	~ 1.5L
Volatile Suspended Solids (VSS)	~ 280 mg/l
Chemical Oxygen Demand (COD)	390~414 mg/l
Dissolved Oxygen (DO)	~ 8 mg/l
pH	~ 8
Conductivity	0.5~1.0 mS/cm
Voltage (No. of Tests)	0 (3); 4V (3); 8V (2); 16V (2)

Complete aeration and mixing maintained dissolved oxygen around saturation during exposure to electricity. Temperature measurements did not show any significant change during testing. Although there was a slight decrease in pH, sludge mixing allowed electrolysis reactions at the electrodes to neutralize each other, thus minimizing significant pH changes. Anaerobic tests showed a drop in the pH to about 5.5. The slight drop in pH in the aerobic tests compared to the anaerobic tests may be related to the difference in processing time (which was shorter in the aerobic tests) and the constituents of the anaerobic and aerobic cultures. It is possible that another electrolysis reaction, other than water reduction, has occurred at the cathode in the anaerobic tests thus limiting OH- production. Another possible reason is that the currents used in the aerobic tests are smaller than those used in the anaerobic tests, which may limit the changes due to electrolysis in the aerobic tests when compared to the anaerobic tests. In any case, this is an issue that needs further evaluation.

With most variables (pH, DO and temperature) were controlled, any changes in microbial activity can only be attributed to electric currents. Percent change in chemical oxygen demand (COD) of the sludge was used as an indication of microbial activity. Other measures of substrate utilization may be used, such as TOC, DOC, BOD, and/or VSS. However, COD was used because it is easier and faster than some of these measures and because of our interest in the general response. Some concerns may rise because microorganisms are organic in nature and COD might not accurately reflect microbial activity. However, microorganisms consume more organic matter than they synthesize and the ratio of consumed to synthesized organic matter is about 10:1 under normal aerobic condition (Gaudy and Gaudy, 1988). Therefore, COD values provide a good indication of microbial activity. Furthermore, COD values were used for comparison with control test (no electric field) results and not as absolute values.

Average COD changes are summarised in Figure 4. The behavior can be divided into 2 groups: one describing the low voltage (LV) of 0 and 4 V tests and the other describing high voltage (HV) of 8 and 16 V tests. After 24 hours, the LV group showed less drop (about 17% difference) in COD, when compared to the HV group. This is an indication that tests with higher voltage gradients (8 and 16 V) resulted in more degradation of the organic matter. After 48 hours of exposure to electricity, both LV tests showed the same behavior, where COD dropped to around 42-45% of the initial value. This is an indication that application of 4 volts (0.28 V/cm) may not be high enough to produce any changes, when compared to tests with no electricity. On the other hand, the test with 8 volts seems to cause the most significant drop in the COD value (61% after 48 hours exposure). The impact of the highest voltage (16 V) seems to diminish after the first 24 hours. While the 16 V tests showed less COD drop when compared to the 8 V test, it still showed more COD drop when compared to the LV tests. The change in COD drop between the tests may be attributed to an increased microbial activity.

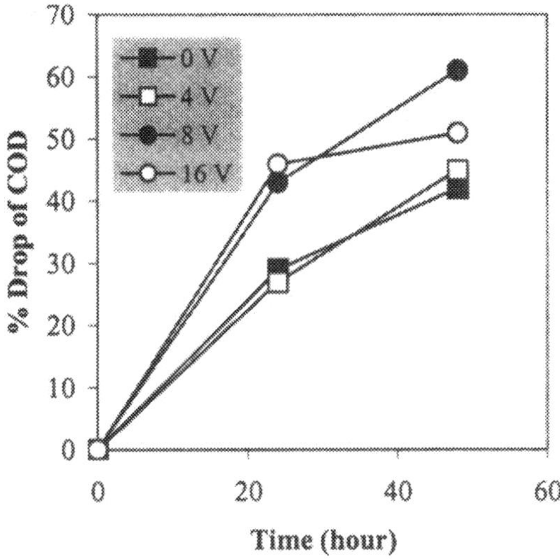

Figure 4. Average COD drop in aerobic tests
(electrode spacing =14 cm)

The results indicate that dc electric fields (up to 1.14 V/cm) do not have an adverse effect on mixed aerobic cultures. In fact, an increase in the degradation rate may occur due to increasing the voltage gradient up to a certain value (0.58 V/cm in this study), beyond which this increase may diminish. This behavior is similar to electrostimulation by ac electric fields reported in the bioelectrochemistry research area (Fologea *et al.*, 1998). For the conditions of the preliminary study, the window of significant response of bacteria stimulation is somewhere between 0.27 and 1.14 V/cm. Further, the results show that this window of dc range is also affected by time of exposure. This can be concluded as the impact of the 1.14 V/cm field diminished after 24 hours. However, to confirm this conclusion, it is necessary to separate the effects of any abiotic processes.

The tests were conducted in an open, mixed and aerated reactor to maintain constant values of pH, DO, and temperature. Thus the difference in COD drop may not be related to pH, temperature. Aeration and mixing maintained DO around saturation in all tests, thus the effect of oxygen production at the anode is minimized. The only other process (other than microbial activity) that may relate to COD drop is abiotic transformation by electrolysis reactions at the electrodes. If abiotic redox of the organic content occurs in this study, then increasing the current density should increase the

COD drop. The results do not show this pattern, but show a peak followed by a decrease in COD drop with increasing current density (Figure 4). Accordingly, microbial activity may be the major factor in the difference in COD drop obtained in tests with different current densities.

7. SUMMARY AND CONCLUSIONS

This chapter describes the effects of dc fields on ions transport in soils, electrolysis and geochemical reactions, microbial adhesion and transport, and microbial activity. The interest in these in these processes is derived from the potential of using electric fields for transporting and mixing contaminants, biostimulants, and bioaugmentation inoculants to enhance *in situ* bioremediation.

The focus of this chapter is the response of microbial cultures to dc fields. The results showed that dc electric field intensity tends to have a complex effect on the activity of mixed microbial cultures. Anaerobic cultures seem to experience an "environmental shock" when exposed to electric field intensities greater than 1.5 V/cm. However, these cultures are able to recover their activities once the electric currents are switched off. Aerobic cultures did not seem to be affected by exposure to field intensities less than 0.28 V/cm. Higher dc field intensities, up to 1.14 V/cm, seem to stimulate aerobic cultures during the first 24 hours of exposure. However, continuous exposure to dc fields (in the range of 1.14 V/cm) following the first 24 hours seems to retard the growth of aerobic cultures. The authors identify the need for further evaluation and assessment of the impact of dc fields on the physicochemical and biochemical interactions in porous media.

BIBLIOGRAPHY

Acar, Y. B. and Alshawabkeh, A. (1993) "Principles of Electrokinetic Remediation," *Environmental Science and &Technology*, 27(13), pp. 2638-2647.

Acar Y. B. and Alshawabkeh A. (1996) "Electrokinetic Remediation: I. Pilot-scale Tests with Lead Spiked Kaolinite," *Journal of Geotechnical Engineering*, ASCE, 122(3), pp.173-185.

Acar, Y. B., Li, H., and Gale, R. J. (1992). "Phenol Removal from Kaolinite by Electrokinetics," *Journal of Geotechnical Engineering*, ASCE, 118(11), pp. 1837-1852.

Acar, Y. B., Ozsu, E., Alshawabkeh, A. N., Rabbi, F. M., and Gale, R. (1996) "Enhanced Soil Bioremediation with Electric Fields," *CHEMTECH*, ACS, 26(4), pp. 40-44.

Acar, Y. B., Rabbi, M. F. and Ozsu, E. (1997) "Electrokinetic Injection of Ammonium and Sulfate Ions into Sand and Kaolinite Beds," *Journal of Geotechnical and Geoenvironmental Engineering*, ASCE, 123(3), pp. 239-249.

Alshawabkeh, A. N. and Acar, Y. B. (1992) "Removal of Contaminants from Soils by Electrokinetics: A Theoretical Treatise," *Journal of Environmental Science and Health*, A 27 (7), pp. 1835-1861.

Alshawabkeh, A., Acar, Y. B., (1996) "Electrokinetic Remediation: II. Theoretical Model," *Journal of Geotechnical Engineering*, ASCE, 122(3), pp. 186-196.

Berg, H. (1993) "Electrostimulation of Cell Metabolism by Low Frequency Electric and Electromagnetic Fields," *Bioelectrochemistry and Bioenergetics*, 31, pp. 1-25.

Berg, H. and Zhang, L. (1993) "Electrostimulation in Cell Biology by Low Frequency Electromagnetic Fields," *Electro Magnetbiology*, 12, 147-163.

Blank, M., Soo, L., Henderson, A. S., Goodmann, R. (1992) "Changes in Transcription in HL-60 Cells Following Exposure to Alternating Currents from Electric Fields," *Bioelectrochemistry and Bioenergetics*, 28, pp. 301-309.

Bruell, C. J., Segall, B. A., and Walsh, M. T. (1992), "Electroosmotic Removal of Gasoline Hydrocarbons and TCE from Clay," *Journal of Environmental Engineering*, ASCE, 118(1), 68-83.

Buerk, D. G. (1993), *Biosensors: Theory and Applications*, Technomic Publishing Co., 221p.

Characklis W. and Wilderer, P., (1988), *Biofilms*, John Wiley and Sons, New York.

Collins, Y.E., and Stotzky, G. (1992). Heavy metals alter the electrokinetic properties of bacteria, yeasts, and clay minerals. Applied and Environmental Microbiology, vol. 58, p 1592-1600.

DeFlaun, M. F. and Condee, C. W. (1997) "Electrokinetic Transport of Bacteria," *Journal of Hazardous Materials*, Special Edition on Electrochemical Decontamination of Soil and Water, Edited by Yalcin B. Acar and Akram N. Alshawabkeh, pp. 263-278.

DOE/EM-0232 (1995) "Estimating the Cold War Mortgage, Volume I" The 1995 Baseline Environmental Management Report, US Department of Energy, March 1995.

Eykholt, G. R. and Daniel, D.E. (1994), "Impact of System Chemistry on Electroosmosis in Contaminated Soil," *Journal of Geotechnical Engineering*, ASCE, 120(5), pp.797-815.

Fologea, D., Vassu-Dimov, T., Stoica, I., Csutak, O., and Radu, M. (1998) "Increase of *Saccharomyces Cerevisiae* Plating Efficiency After Treatment with Bipolar Electric Pulses," Short Communication, *Bioelectrochemistry and Bioenergetics*, 46, pp. 285-287.

Gaudy, A. F. and Gaudy, E. T. (1988), *Elements of Bioenvironmental Engineering*, Engineering Press, Inc., San Jose, California, 592p.

Hamed, J., Acar, Y. B., and Gale, R. J. (1991) "Pb(II) Removal from Kaolinite Using Electrokinetics," *Journal of Geotechnical Engineering*, ASCE, 117(2), pp. 241-271.

Ho, S. V., Athmer, P.W., Sheridan, P. W. and Shapiro, A. (1997) "Scale-up Aspects of the Lasagna Process for in-situ Soil Remediation," *Journal of Hazardous Materials*, Special Edition on Electrochemical Decontamination of Soil and Water, Edited by Yalcin B. Acar and Akram N. Alshawabkeh, pp. 39-60.

Ho, S. V., Athmer, C., Sheridan, P. W., Hughes, B. M., Orth, R., McKenzie, D., Brodskey, P. H., Shapiro, A. M., Sivavec, T. M., Salvo, J., Schultz, D., Landis, R., Griffith, R., and Shoemaker, S. (1999a) "The Lasagna Technology for In Situ Soil Remediation. 1. Small Field Test," *Environmental Science and Technology*, 33(7), pp. 1092-1099.

Ho, S. V., Athmer, C., Sheridan, P. W., Hughes, B. M., Orth, R., McKenzie, D., Brodskey, P. H., Shapiro, A. M., Thornton, R. Salvo, J., Schultz, D., Landis, R., Griffith, R., and Shoemaker, S. (1999b) "The Lasagna Technology for In Situ Soil Remediation. 2. Large Field Test," *Environmental Science and Technology*, 33(7), pp. 1092-1099.

Holmes, P. J. (1962), *The Electrochemistry of Semiconductors*, Academic Press, London, 396 p.

Hunter, R. J. (1981), *Zeta Potential in Colloid Science: Principles and Applications*, Academic Press, London, 386p.

Kelsh, D. J. and Parsons, M. W. (1997) "Department of energy sites suitable for electrokinetic remediation" *Journal of Hazardous Materials*, Special Edition on Electrochemical Decontamination of Soil and Water, Vol 55, No. 1—3.

Lageman, R., Wieberen, P. and Seffinga, G (1989). "Electro-reclamation in Theory and Practice," *Chem. Ind.* London, 9, pp. 585 – 590.

Maillacheruvu K. and Alshawabkeh, A. N. (1999) "Anaerobic Microbial Activity under Electric Fields," Book Chapter in "*Emerging Technologies in Hazardous Waste Management VIII*, Kluwer Academic/Plenum Publishers, in press.

McLeod, B. R., Liboff, A. R., and Smith, S. D. (1992) "Biological Systems in Transition: Sensitivity to Extremely Low Frequency Field, *Electro Magnetbiology*, 11, 29-42.

Mitchell, J. K. (1993), *Fundamentals of Soil Behavior*, John Wiley and Sons, New York, 437 p.

NRC (1993), *In Situ Bioremediation: When Does it Work?*, National Research Council, National Academy Press, Washington D.C., 207 p.

Pamukcu, S., and Wittle, J. K. (1992) "Electrokinetic Removal of Selected Heavy Metals From Soil," *Environmental Progress*, AIChE, 11(4), 241-250.

Pamukcu, S., Weeks, A. and Wittle, J. K. (1997) "Electrochemical Extraction and Stabilization of Selected Inorganic Species in Porous Media," *Journal of Hazardous Materials*, Special Issue on Electrochemical Decontamination of Soil and Water, Edited by Y.B. Acar and A.N. Alshawabkeh, V. 55, No's. 1—3, pp. 1-22.

Probstein, R. F. and Hicks, R. E. (1993), "Removal of Contaminants from Soils by Electric Fields," *Science*, 260, pp. 498-504.

Ramsey, G. (1998), *Commercial Biosensors: Applications to Clinical, Bioprocess, and Environmental Samples*, John Wiley and Sons, Inc., 304p.

Reynolds, P. J., Sharma, P., Jenneman, G. E., Mclnerney, M.J. (1989). Mechanisms of microbial movement in subsurface materials. Applied and Environmental Microbiology, vol. 55, p 2280-2286.

Runnells, D. D., and Wahli, C. (1993) "In Situ Electromigration as a Method for Removing Sulfate, Metals, and Other Contaminants from Groundwater," *Ground Water Monitoring & Remediation*, 13(1), 121-129.

Shapiro, A. P., Renauld, P., and Probstein, R. (1989), "Preliminary Studies on the Removal of Chemical Species from Saturated Porous Media by Electro-osmosis," *Physicochemical Hydrodynamics*, 11(5/6), pp. 785-802

Suflita, J. M. and Sewell, G. W. (1991) "Anaerobic Biotransformation of Contaminants in the Subsurface," EPA, Robert S. Kerr Environmental Laboratory, Ada, OK, EPA/600/M-90/024.

Tsong, T. Y. (1992) "Molecular Recognition and Processing of Periodic Signals in Cells: Study of Activation of Membrane ATPases by Alternating Electric Fields," *Biochem. Biophys. Acta*, 1113, pp. 53-70.

USEPA, 1991, Groundwater Issue. Report No. EPA/540/4-91/002, 21 p.

Van Loosdrecht, M.C.M, Lyklema, J., Norde, W., Schraa, W., and Zehnder, A.J.B. (1987). Electrophoretic mobility and hydrophobicity as a measure to predict the initial steps of bacterial adhesion. Applied and Environmental Microbiology, vol. 53, p 1898-1901.

Zappi, M, Gunnison, D., Pennington, J., Cullinane, J., Teeter, C. L., Brannon, J. M., and Myers, T. (1993) "Technical Approaches for In Situ Biological Treatment Research: Bench-Scale Experiments," US Army Corps of Engineers, Waterways Experiment Station, Vicksburg, MS, August 1993, Technical Report No. IRP-93-3.

Chapter Five

Transport of Trichloroethylene (TCE) in Natural Soil by Electroosmosis

SOUHAIL R. AL-ABED[1] and JIANN-LONG CHEN[1]

[1]National Risk Management Research Laboratory, USEPA, 26 W. Martin Luther King Dr., Cincinnati, OH 45268

Key words: electroosmosis; trichloroethylene; dechlorination

Abstract: Contamination in low permeability soils poses a significant technical challenge to in situ remediation, primarily due to low mobilization of the contaminants and difficulty in uniform delivery of treatment reagents. An alternative approach using electroosmosis (EO) is used to mobilize Trichloroethylene (TCE) in soil. However, the EO approach causes significant chemical changes in the soil which may affect transport and/or chemical transformation of TCE. Laboratory experiments and mathematical modeling were used to characterize the transport and chemical transformation of TCE in undisturbed soil cores during EO. A contamination zone (CZ) were located 1 cm below the anode. Electroosmotic fluid flow was vertically downwards from anode to cathode. A voltage gradient of 1.4 V/cm was applied to the soil for 4 weeks. More than 95% of the TCE was mobilized toward the cathode in the soil over a period of 672 hrs. The advective velocity of TCE was approximately 1.3×10^{-5} cm/sec and the dispersion coefficient is two times the diffusion coefficient of 6.9×10^{-6} cm^2/sec. We observed dichloroethylene (cis-1,2-DCE) indicating dechlorination of TCE. Dechlorination occurred in parts of the soil column where reducing conditions (Eh-pH conditions) are dominant. The most significant reductive dechlorination of TCE occurred near the cathode, a source of electrons during electroosmosis. Results show the need to include a decay term in the transport equations. The results show that potential chemical transformation of chlorinated organic compounds could enhance the remediation efficiency during EO.

1. INTRODUCTION

The contamination of soil and groundwater with organic contaminants is widespread throughout the US and the world. Chlorinated hydrocarbons are persistent in the environment and rapidly transported in groundwater (Fusillo et al., 1985). Trichloroethylene (TCE) is a chlorinated organic solvent that has been widely used in metal processing, electronics, dry cleaning, paint and many other industries. Because of leaks, spills, and dumping, TCE has found its way into many subsurface environments (Olsen and Kavanaugh, 1993). It is the most frequently reported contaminant at hazardous waste sites on the National Priority List of the U.S. Environmental Protection Agency (USEPA, 1985). TCE, and similar chlorinated alkenes present a serious groundwater contamination problem because they are suspected carcinogens and generally resist biodegradation in the environment.

TCE partition in a contaminated aquifer depends on its solubility in water, partitioning into air (Henry's Law), and sorption onto soil particles. Most attempts at the remediation of aquifers containing TCE have involved pump and treat (P&T), i. e. pumping the water containing the contaminant or vacuuming the air containing the contaminant to some above-ground treatment facilities. Even under the best conditions, P&T is limited by hydraulic conductivity, preferential flow, sorption onto soil particles, and residual saturation or pockets of contamination. Generally, P&T techniques have limited removal capability and, after pumping is stopped, the contaminant concentrations might increase again (Mercer et al., 1990).

Many in situ soil remediation processes require mobilizing and transporting contaminants to a treatment/collection zone or delivering nutrients, microorganisms, and chemicals to degrade the contaminant. For soils with high hydraulic permeability, treatment solutions can be delivered hydraulically to the contaminated zones. Mobilized contaminants and degradation products can be removed in the same manner. In sandy soils such as aquifers that yield significant quantities of water, P&T strategies have been shown successful (USEPA, 1992). However, pressure driven hydraulic delivery/removal of contaminants in low permeability soils, such as clays, is impractical because the removal efficiency is limited by hydraulic conductivity, soil matrix effects, desorption, and residual saturation of TCE. Thus, alternative means of moving contaminant mass are required for successful remediation.

One method of transporting solutions and compounds in low permeability soils is the application of an electric current to the soil using a process called Electroosmosis (EO). EO fluid flow is a result of ions movement in the double layer of clay surfaces. For this reason, EO is ideally suited to fine-grained, clay-rich soils. The magnitude of electroosmotic fluid flow is proportional to the applied electrical gradient, just as the rate of hydraulic flow is proportional to the hydraulic gradient. EO thus provides a potential transport mechanism for organic contaminants, while dissolved ions and other charged species may be transported by electromigration (Probstein and Hicks, 1993). Unlike hydraulic flow, electroosmotic flow is relatively independent of pore size, so problems associated with preferential paths are reduced (Probstein and Hicks, 1993). This technology has seen widespread use in engineering applications, and recently has shown promise for efficient, cost-effective in-situ remediation of both organic and inorganic contaminants in fine-grained soils (Acar et al., 1992 and 1995; Ho et al., 1995; Powell et al., 1995; Ho et al., 1999). EO has been successfully applied to the mobilization of heavy metals like copper (Runnels and Larson, 1986), zinc (Pamukcu and Wittle, 1992), cadmium (Acar et al., 1992), and lead (Hamed et al., 1991). Organic contaminant removal has been demonstrated for compounds like benzene, toluene, xylene, phenol, and chlorinated solvents (Bruell et al., 1992; Acar et al., 1992; Shapiro and Probstein, 1993).

Most laboratory experiments demonstrating the utility of EO transport of organic compounds were conducted with kaolinite as the model clay-rich soil medium. Shapiro et al. (1989) used EO to transport phenol in kaolinite. Bruell et al. (1992) have shown that TCE can be transported down a slurry column by electroosmotic fluid flow, and more recently, Ho et al. (1995) demonstrated electroosmotic movement of p-nitrophenol in kaolinite. Kaolinite is a pure clay mineral, which has a very low cation exchange capacity and is generally a minor component of the silicate clay mineral fraction present in most natural soils. It is not, therefore, representative of most natural soil types, particularly those which are common in the midwestern United States. The clay content can impact the optimization and effectiveness of electroosmosis in field-scale applications, as has recently been discussed by Chen et al. (1999).

The objective of this study was to demonstrate the physical transport of TCE by EO through cores of undisturbed soil. While research approaches have been performed on packed columns of pure clay (e.g. kaolinite), few have used native soils, and only in the form of slurries. At this time, no information is available for transport of TCE by EO through intact cores of natural soil. Therefore, the results of EO experiments using undisturbed soil are more applicable to actual site conditions than using single mineral soil. Parameters governing TCE transport in the soil are used in a one dimension advective model to describe TCE transport during the experiment.

2. MATERIALS AND METHODS

2.1 Soil and Field Site

Intact soil cores (6.7 cm i.d.) were taken with spilt spoon at depths of 1 to 2 meters from a field test site located approximately 50 km east of Cincinnati, Ohio. The soil in this interval consists mainly of quartz (60%) and clay minerals (35%) with minor amounts of plagioclase and potassium feldspar. The majority of clay is illite and smectite, with minor amount of kaolinite. Soil chemical properties were analyzed prior to, and after, electroosmosis, in order to evaluate the effects of electroosmosis on the distribution of elements within the soil column. Sampled cores were wrapped in aluminum foil and stored at 12°C until the EO cell was assembled.

2.2 Electroosmosis Cell (EC)

Electroosmosis soil column is a 29 cm long acrylic tubes (5.8 cm i.d.) with tight fitting screw caps at the top. Each column contained 18 cm of undisturbed soil core. All ECs were assembled as shown in Figure 1. Adsorption of TCE to the surface of column was checked at a concentration used in the experiments and was found non-detect (< 0.10 μg/g).

A total of thirteen ECs were prepared. Four ECs were used as controls, i.e., no electrical current was applied. One column was designed so that in-situ Eh-pH conditions could be monitored, while the remaining ECs were part of the experiments designed to model the electrosmotic transport of TCE as a function of time. In these experiments, soil columns were subjected to EO for a period of time ranging from one to four weeks. Three columns (two test and one control) were dissected weekly, sampled and TCE concentrations measured. In this way, profiles of TCE concentration were collected as a function of distance from the anode and as a function of time. The TCE and DCE isomers in the Eh-pH column was analyzed at the end of the experiment (672 hr).

After loading the cores, columns were saturated with DI water with a peristaltic pump for 48 hours. After saturation, clean and TCE contaminated slurries were prepared to complete the assembly. Because it is not possible to uniformly contaminate dense soil cores, it was necessary to start with clean soil cores and introduce TCE contamination in a discrete zone.

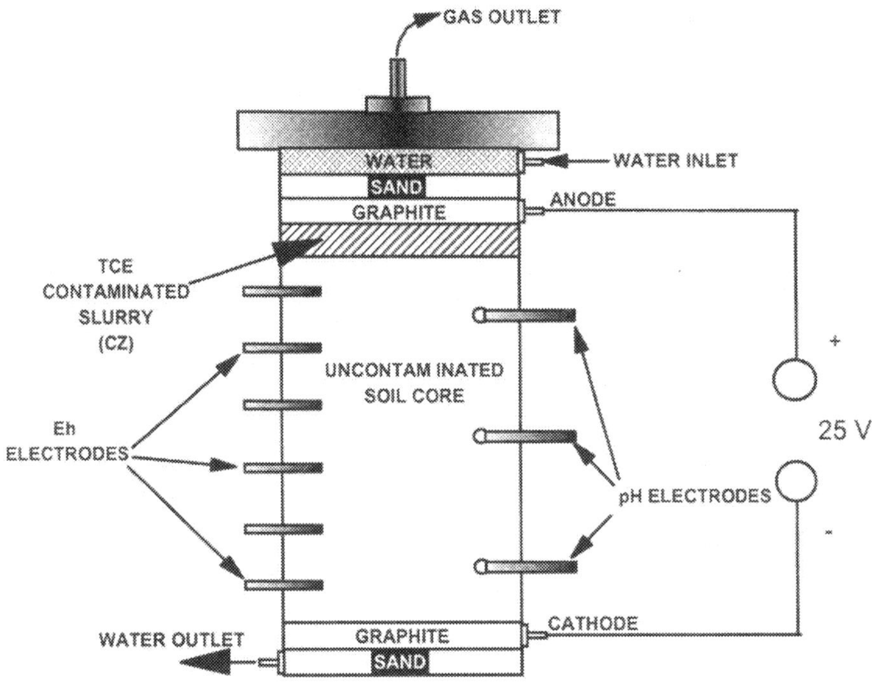

Figure 1. Schematic diagram of the electroosmosis cell (EC) showing the horizontal configuration of graphite electrodes used in this experiment. Note the proximity of the contaminated zone (CZ) to the anode. Length of undisturbed soil core was 15 cm.

Soil slurry was prepared with a representative sample of the site soil. The sample was dried, grinded, and sieved through a ASTM #40 mesh. Pure TCE (Fisher Scientific) and TCE saturated solution were added to 70 g of soil slurry to simulate contaminated groundwater conditions. Sodium azide solution (250 mg/L) was added to the TCE solution to inhibit bioactivity in the column. After homogenizing, 60 g of the slurry was placed over the clean soil core in the EC. This resulted in a 1 cm layer contamination zone which was immediately covered by 1 cm layer of the clean slurry. The clean slurry layer served to isolate the contaminated layer from the graphite in the anode, thereby preventing the graphite from adsorbing TCE prior to initiation of EO. Three samples from each slurry batch was extracted by methanol and analyzed using GC/PID. The final TCE concentrations averaged 111 μg/g for week 1 ECs; 109 μg/g for week 2 ECs; 100 μg/g for week 3 ECs; and 103 μg/g for week 4 ECs.

Granular graphite layers were used as the electrodes. Electrical contact was made by inserting one-eighth inch graphite rods through holes drilled into the column and into the electrode material (Figure 1). Above and below each electrode was a layer consisting of 30 g of coarse sand. This was to ensure that fluid flow was unimpeded into and out of the anode and cathode zones during EO.

A perforated stainless steel plate and spring were placed over the anode and the cap screwed on to apply pressure on top of the column. An inlet at the anode end of the column allowed clean water to be introduced to replace water moved by EO. The water level was maintained constant during electroosmosis by use of a Mariotte Bottle filled with DI water containing 250 mg/L Sodium Azide solution and connected to the column inlet with a teflon tubing. An outlet at the bottom of the cathode was connected back up to the height of the inlet and filled with clean water to the level of the Marriotte Bottle so that the column was not under any hydraulic gradient. In this way, there were no sidewall effects, and water movement in the ECs could be attributed only to EO. EO was initiated with an applied potential of 25 V. Initially, this represented an estimated voltage gradient of 1.4 V/cm across the length of the core. During EO, fluid flow was measured and effluent samples were collected and refrigerated for later analysis. The current in milliamps was also measured daily. Gases emitted from the top of each EC was connected to a carbon trap to measure volatile TCE. All traps were analyzed for TCE at the end of the experiment. No TCE was found to be adsorbed in these traps.

2.3 Redox Potential and pH Measurements

One of the objectives of this experiment was to monitor in-situ redox and pH conditions in the EC. Knowledge of these two parameters is important in any interpretation of degradation reactions involving chloro-organic compounds in which electrons and protons are involved, as in reductive dechlorination, for example.

Commercial gel-filled pH probes were inserted into the soil column at 3 cm intervals, located between the anode and cathode (Figure 1). These were inserted through butyl rubber septa placed into holes drilled into the sides of the column. The pH probes were calibrated using standard buffer solutions before insertion into the soil column. Readings were taken daily for the duration of the experiment, by placing a reference electrode into a well at the top of the column, and connecting the reference and in-situ pH probes to an Orion Model EA 940 Expandable Ion Analyzer. All pH measurements were temperature compensated to 25 °C.

Amorphous black platinum Eh probes were inserted into the soil column at 1.5 cm intervals, located between the anode and the cathode (Figure 1). These probes were manufactured by us following the procedures outlined by Jackson (1958). The probes contain a 1 cm length of 1 mm diameter bright platinum wire set with epoxy inside a 6 cm long end-piece of a glass pipette tip. The inside end of the platinum wire is soldered to a copper wire lead. The tips of the bright platinum wire are polished to a flat mirror surface, washed in concentrated HCl and rinsed in distilled water. Two such electrodes are then immersed into a platinizing solution consisting of platannic acid and lead acetate. The electrodes are then connected to two 1.5 V DC batteries in series, which results in the deposition of a dense black coating of amorphous platinum on the exposed ends of the bright platinum wire tips. The polarity of the DC current is reversed every 30 seconds until a satisfactory coating is achieved, usually after 3 to 5 minutes. In both soil and water systems these probes were well poised. Reproducible Eh measurements were routinely obtained using ASTM Standard Method C 1498 to measure the response of these black platinum electrodes.

It is not possible to make in-situ Eh-pH measurements while EO is in progress. The very small currents measured by the probes are severely interfered with by the voltage gradient in the EC. For this reason, the current was turned off during the daily Eh-pH measurements.

2.4 TCE and Chemical Analysis

In order to determine the initial TCE concentrations prior to initiation of EO, samples of the contaminated slurry were collected at the same time they were loaded into individual EC. A 5 g slurry sample was loaded into a previously weighed, teflon-capped borosilicate vial. After reweighing to obtain the actual sample weight, the vial was opened and reagent grade methanol added to the top of the vial and the teflon cap replaced without headspace in the vial. The vial was again re-weighed to obtain the amount of methanol added. A second slurry sample was taken in the same manner.

At the termination of each experiment, 5 to 8 g soil samples were taken at 3 cm intervals along the 18 cm length of the soil column using a stainless steel push tube sampler that fit the access ports of the Eh-pH probes. Using a teflon-tipped rod, the soil samples were extruded from the tube directly into 5 mL teflon-capped vials, and filled to zero headspace with methanol. TCE concentrations were determined using purge and trap gas chromatography (methods 8021B and 5030B), with analyses being performed on an SRI 8610 Gas Chromatographer (GC), equipped with a Photoionization Detector (PID). Method Detection Limits for liquid samples and extracted soils were at 0.05 μg/L and 0.10 μg/g for TCE respectively.

TCE and other chloro-organic compounds encountered in the EC (Eh-pH EC) were extracted as described above. Analyses were performed using gas chromatography on an HP 5890 Gas Chromatography equipped with a 5971 series Mass Selective Detector (GC/MS), using ultra high purity Helium as the carrier gas. Detector temperature was 280°C and inlet temperature was 160°C. Manual injection of 1 μL of methanol sample was made into a VOCOL capillary column (60 m x 0.25 mm x 1.5 μm). The temperature program was as the following: initial temperature of 50°C for 4 minutes, ramped to 210°C at 20°C per minute, and held at 210°C for 6 minutes.

Calibrations were carried out for the GC/PID or the GC/MS daily. Calibration standards were prepared based on standard reference materials obtained from Supelco Chromatography products. A check standard was analyzed every ten samples to assure calibration and accuracy. A reagent blank was included in each analytic batch of samples. Blanks were made from reagent or make-up water and matrix similar to the sample. A spiked sample was analyzed every twenty samples. This was done by splitting an appropriate sample into two subsamples and adding a known quantity of TCE to one of the split samples. The purpose of a spiked sample is to determine the extent of matrix bias or interference on TCE recovery and sample to sample precision. Accuracy was assessed by analysis of external reference standards (separate from calibration standards) and by percent recoveries of spiked samples. Precision was assessed by means of replicate sample analysis. It is expressed as relative percent difference (RPD) in the case of duplicates or relative standard deviation (RSD) for triplicate (or more) analyses. Recovery was 96% or more for all spiked samples, and RPD/RSD are less than 7% for all samples.

In addition to taking soil samples for analyses of organic components, samples were also taken for chloride ion analysis. Samples were oven dried, grounded in an alumina mortar and sieved through ASTM#40 mesh. Aliquots of the sieved soil (1:1) were then weighed into nalgene bottles containing deionized distilled water, and agitated on a shaker overnight. Aliquots were then filtered through a 0.20 micron nylon membrane polypropylene particle filter. Chloride ion concentrations in the filtered solutions were then determined using a Waters Capillary Ion Analyzer.

Cation exchange capacity (CEC) of soil samples taken prior to, and after, EO was analyzed by CLC LABS in Westerville, Ohio according to the method described by Brown and Wamcke (1988). This method uses 1 M ammonium acetate to extract cations from the soil. Cations concentrations were measured with an Atomic Adsorption Spectrometer (AAS).

3. MODEL DEVELOPMENT

The transport of TCE through a porous medium, such as soil, can be described by the following 1D advection-dispersion equation (Fetter, 1993).

$$\frac{\partial C}{\partial t}=\frac{D_L}{R}\frac{\partial^2 C}{\partial x^2}-\frac{\upsilon}{R}\frac{\partial C}{\partial x} \tag{1}$$

where C = concentration of TCE in pore water, D_L = effective longitudinal dispersion coefficient of TCE, v = average pore fluid velocity, and R = retardation factor. Equation (1) assumes that there is no decay of TCE that might be due to oxidation-reduction reactions and microbial cell synthesis.

Assuming that the sorption-desorption of TCE follows a linear Freundlich isotherm (Bruell et al., 1992), the retardation factor can be expressed as

$$R = 1 + K_d \frac{\rho_b}{n} \tag{2}$$

where K_d = distribution coefficient of TCE, ρ_b = bulk density of the soil, and n = effective porosity of the soil.

The boundary and initial conditions for equation (1) are

$$\begin{aligned} &-\frac{D_L}{R}\frac{\partial C}{\partial x}+\frac{v}{R}C\Big|_{x=0^+}=0 \\ &-\frac{D_L}{R}\frac{\partial C}{\partial x}+\frac{v}{R}C\Big|_{x=L^-}=\frac{v}{R}C_e \\ &C(x,0)=\frac{M}{A}\delta(0) \end{aligned} \tag{3}$$

where Ce = effluent concentration, M = mass released at time t = 0, A = the cross-sectional area of the column, and $\delta(x)$ is the Dirac delta function.

Letting $D=D_L/R$ and $u=v/R$, equations (1) and (3) can be written as

$$\frac{\partial C}{\partial t}=D\frac{\partial^2 C}{\partial x^2}-u\frac{\partial C}{\partial x} \tag{4}$$

$$
\begin{aligned}
& -D\frac{\partial C}{\partial x}+uC\Big|_{x=0^{+}}=0 \\
& -D\frac{\partial C}{\partial x}+uC\Big|_{x=L^{-}}=uC_{e} \\
& C(x,0)=\frac{M}{A}\delta(0)
\end{aligned}
\tag{5}
$$

The condition at x = 0 states that the net flux of solute at the boundary is zero. To solve for the unknown *Ce*, we use the approach of van Genuchten and Parker (1984) by assuming that the solute concentration is continuous across the boundary at x=L. Furthermore, if we let y = x-ut, the problem is the same as diffusion in a stagnant fluid when viewed in a coordinate system moving at a speed *u* (Fisher et al., 1979). Consequently, the solution to equations (4) and (5) can be obtained by substituting y = x-ut to the solution of equations (6) and (7).

$$
\frac{\partial C}{\partial t}=D\frac{\partial^{2}C}{\partial y^{2}} \tag{6}
$$

$$
\begin{aligned}
& \frac{\partial C}{\partial y}\Big|_{y=0^{+}}=0 \\
& C(\infty,t)=0 \\
& C(y,0)=\frac{M}{A}\delta(0)
\end{aligned}
\tag{7}
$$

As the coordinate is moving at a velocity of *u*, the second term in the boundary condition at x=0 in equation (5) is zero and the boundary condition at y = 0 still represents zero flux across the boundary. The second boundary condition is obtained by assuming the concentration to be continuous at y = L and extending the concentration profile to infinity.

The resulting solution is

$$
C(x,t)=\frac{M/A}{\sqrt{4\pi Dt}}\exp\left[\frac{-(x-ut)^{2}}{4Dt}\right]+\frac{M/A}{\sqrt{4\pi Dt}}\exp\left[\frac{-(x+ut)^{2}}{4Dt}\right] \tag{8}
$$

It should be pointed out that equation (8) was obtained by applying the method of superposition (Fisher et al., 1979), where two instantaneous plume, M/A, are initially released at x = 0 and travel at the speed of *-u* and *u*, respectively, to satisfy the no flux boundary at x = 0. Equation (8) can be simplified to equation (9) when there is no advective transport of the solute.

$$C(x,t) = \frac{2M / A}{\sqrt{4\pi t D_d / R}} \exp\left[\frac{-x^2}{4tD_d / R}\right] \tag{9}$$

where D_d = effective diffusion coefficient of TCE.

To apply equations (8) and (9) to the TCE concentration profiles, the effective dispersion and diffusion coefficients need to be quantified first. The effective dispersion and diffusion coefficients can be obtained by (Fetter, 1993; Fischer et al., 1979).

$$\sigma^2 = 2Dt \tag{10}$$

where σ^2 = variance of the longitudinal spreading of the plume. Plotting σ^2 versus $2t$ should yield a line with a slope of D.

4. RESULTS AND DISCUSSION

4.1 Soil Chemical Characterization

Redox potential (Eh) and pH measurements were made daily for the four week duration of the experiment. The power was turned off for a few minutes, and each probe was read in sequence, vertically downwards from anode to cathode. Eh values plotted in Figure 2 represent in situ mV readings with respect to the potential of the standard hydrogen electrode. The data show that the Eh conditions at the anode remained constant at 750 mV for the duration of the experiment. Eh at the rest of the column showed a sharp decrease from positive mV to approximately -750 mV after 4 days of electroosmosis (Figure 2).

Figure 3 is a plot of in situ pH profiles as a function of depth in the column and time. It agrees with the results of several previous laboratory studies which showed that an acid front generated by electrolysis of water at the anode progressively moves through the soil column towards the cathode (Acar and Ashawabkeh, 1993; Eykholt and Daniels, 1994; Hicks and Tondorf, 1994).

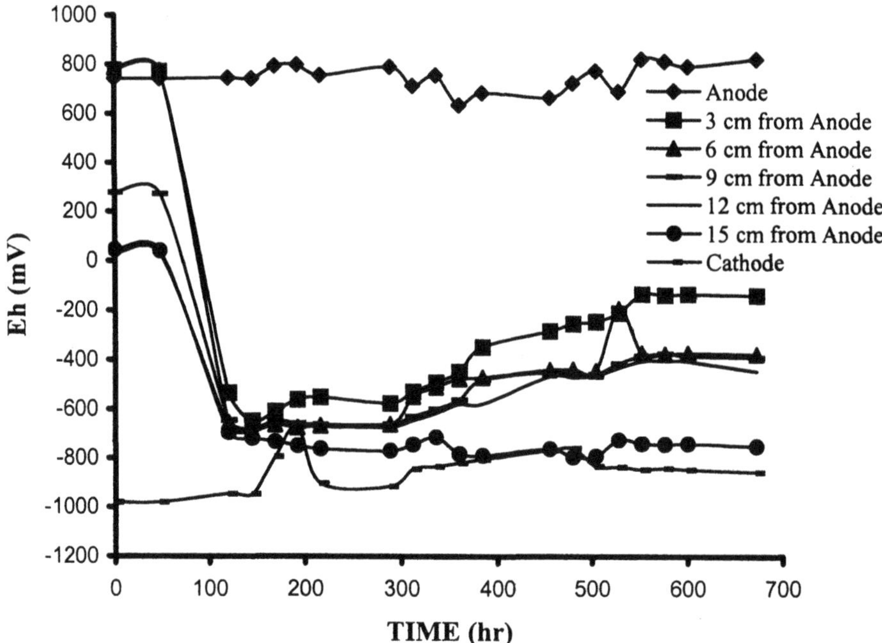

Figure 2. Profile of Eh as a function of time and position along the EC. Note the uniformly stable conditions at the anode and cathode.

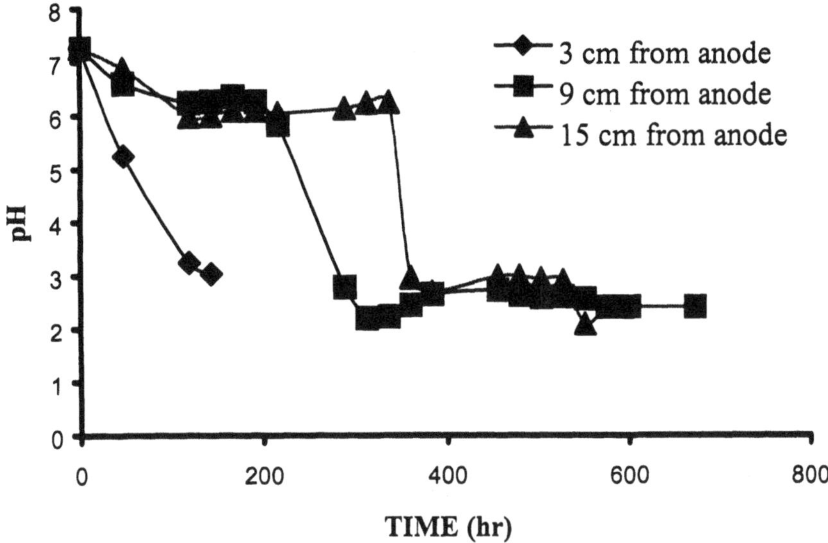

Figure 3. Profile of pH as a function of time and position along the EC. The pH electrode at 3 cm was broken after 180 hrs.

Cation Exchange Capacity (CEC) (Table 1) shows an increase in CEC of the soil as a result of EO. The initial CEC of soil in the EC was 19.3 meq/100 g. Chen et al. (1999) observed that during EO, the pH front generated at the anode progressively displaces cations from clay mineral surfaces. Data in Table 1 show the same effect. The increase in CEC observed is consistent with H^+ ions displacing cations adsorbed onto negatively charged clay mineral surfaces. The CEC data are consistent with data in Figure 3, which show that after 380 hrs of EO, the pH front had travelled 15 cm towards the cathode end of the soil column.

Recent results by Chen et al. (1999) show, however, the transport velocity of H^+ in natural soils during EO is considerably more retarded compared to the migration velocity in pure low CEC clays such as kaolinite. In EO experiments containing only pure kaolinite, very rapid electromigration of H^+ is observed relative to natural soils because kaolinite is essentially a stoichiometric clay mineral and has limited capacity in absorbing cations. The transport of H^+ is therefore rapid.

Table 1. Cation Exchange capacity (CEC), Redox Potential (Eh), and pH in soil after 672 hrs of EO. CEC of soil prior to EO was 19.3 meq/100 g; and initial soil pH was 7.1.

Depth (cm)	CEC meq/100g	Eh mV	pH
Anode			
1.0			3.28
1.5		+829	3.16
2.5	40.4		
3.0		-131	3.37
4.5	38.8		
6.0		-371	3.30
7.5	37.5		
9.0		-382	3.46
10.5	38.0		
12.0		-404	3.54
13.5	36.0		
15.0		-741	4.88
17.5	28.1		
18.0		-846	10.15
Cathode			

In natural soils which commonly contain illite and smectite, there can be a significant charge imbalance between Si^{4+} and Al^{3+} in the structures of these clay minerals. This results in a net negative charge on the clay mineral surfaces, resulting in more adsorption of mobile cations. When an acid front encounters these adsorbed mobile cations, they are very easily displaced by the H^+ ion, which by virtue of its small size is strongly adsorbed to the clay surface. As a result, the measured CEC of such clay-bearing soils is predicted to increase, as we have observed in our experiments.

The electroosmotic conductivity (*Keo*) of all ECs were determined by measuring flow rate through the soil column under a known electrical gradient. The hydraulic head on the boundaries of the soil column was kept constant as the fluid flow freely across the boundaries. The volume of effluent fluid was monitored with time to determine the rate while electrical gradient is applied to the soil column. *Keo* decreased from 5.0 x 10^{-6} to 2.4x 10^{-6} $cm^2V^{-1}s^{-1}$ in four weeks of EO. This reduction in flow is attributed to lowering Zeta potential when the pH of soil decreased during EO.

4.2 TCE Distribution

The TCE distribution in the control column after 672 shows a decline in TCE concentration from the contaminated zone proximal to the anode (0 cm), vertically downwards in the soil column, and most likely represents the diffusion of TCE (Figure 4). The TCE concentration profiles in ECs define a broad distribution and indicate that the bulk of the TCE has moved towards the cathode in response to electroosmotic fluid flow. After one week of processing, the bulk of contaminant has moved more than 5 cm toward the cathode. After two weeks of EO, the center of the plume has moved more than 10 cm toward the cathode. After three and four weeks of EO, there are no definite peaks within the column, however, the concentrations of TCE remained highest near the cathode (Figure 4).

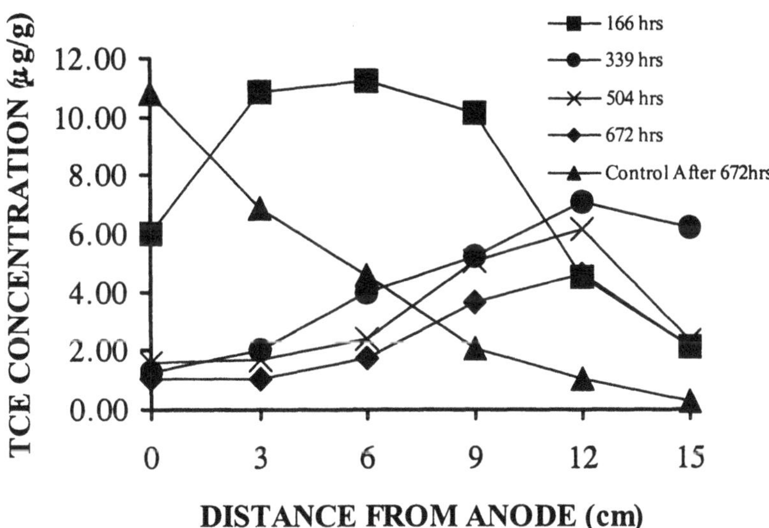

Figure 4. TCE concentration profiles in the control and EC.

It is apparent that the TCE which started out in a small zone, was transported through the column within 339 hrs of EO. This profile shows definitive movement of TCE in a manner unlike that in the control column.

4.3 Comparison of Experimental Results and Model Simulations

In this discussion equation (8) is used to model the transport of TCE during the test. The parameters needed for the model include the effective dispersion/diffusion coefficients, retardation factor, and the pore fluid velocity.

The profiles of TCE concentration of the control and test columns were used to determine the effective dispersion/diffusion coefficients. As the TCE concentrations closed to the cathode were near zero during four weeks without EO (Figure 5), the variance of the four TCE concentration profiles in the control columns were used to determine the values of D_d/R of TCE. The estimated D_d/R for TCE is 6.9 x 10^{-6} cm^2/sec (equation (10)) with r-square of (0.99). Only the TCE profiles of the first week of the test column was used to determine the value of D as the peak of the TCE plume had advanced out of the boundary after two weeks of testing (Figure 6). The estimated D is 1.4 x 10^{-5} cm^2/sec. The results indicate that the effective dispersion coefficient is twice of the effective diffusion coefficient (D_d/R).

Table 2. Summary of Parameters Used in the Model.

Parameter	Value
D (cm^2/sec)	1.4 x 10^{-5}
D_d/R (cm^2/sec)	6.9 x 10^{-6}
R	1.85
n*	0.2
ρ (g/cm^3)*	2.1
v (cm/sec) 1st week	1.3 x 10^{-5}
2nd week	1.0 x 10^{-5}
3rd week	9.6 x 10^{-6}
4th week	9.6 x 10^{-6}

* Measured in laboratory on soil core samples

To obtain the retardation factor, the velocity of the TCE plume is compared to the velocity of the pore water after a week of EO (Fetter, 1993). The velocity of the TCE plume was assumed to be the velocity of the peak, whereas the velocity of the pore fluid was calculated by dividing the flow rate by the cross-sectional area of the column and the porosity, 0.2 (Table 2).

The peak of the TCE plume was at 7.7 cm from the anode after 166 hr of EO, indicating a velocity of 1.3 x 10^{-5} cm/sec. The pore water velocity during this period was approximately 2.4 x 10^{-5} cm/sec. Thus, the retardation factor is 1.85. Accordingly, by substituting 0.2 as the effective porosity and 2.1 g/cm^3 as the bulk density into equation (2), we obtain a K_d

of 0.08 mL/g. In addition to being a parameter of the model simulation, the retardation factor was used to convert the measured TCE concentration (μg per g of solid) to TCE concentration in pore water (μg/cm^3) according to equation (11).

$$C = \frac{\rho_b}{nR} C_m \tag{11}$$

where C_m = measured TCE concentration (μg/g of solid).

The parameters (Table 2) were substituted into equations (8) and (9) to obtain the TCE concentration as a function of distance from the anode at a given time. The simulated TCE concentration profiles for the control columns are similar in magnitude and pattern with the measured values (Figure 5). The simulated concentrations close to the anode, however, deviate more from the measured values than at other locations. This could be due to that the measured values represented the average concentration over 3 cm of the column. In the vicinity of the anode, where the TCE concentration gradient is highest, averaging the concentration could cause more discrepancy between the simulated and measured values.

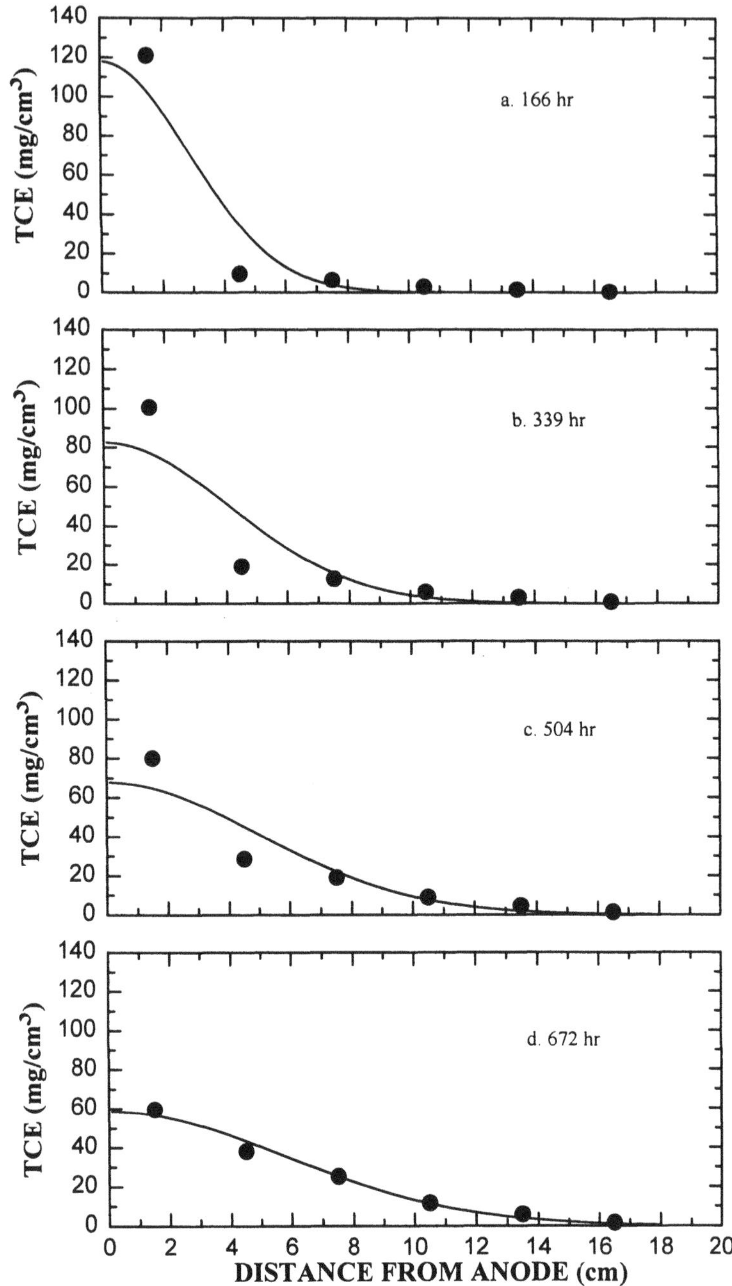

Figure 5. Observed (circle) and Model Output (line) of TCE concentration profiles of the Control.

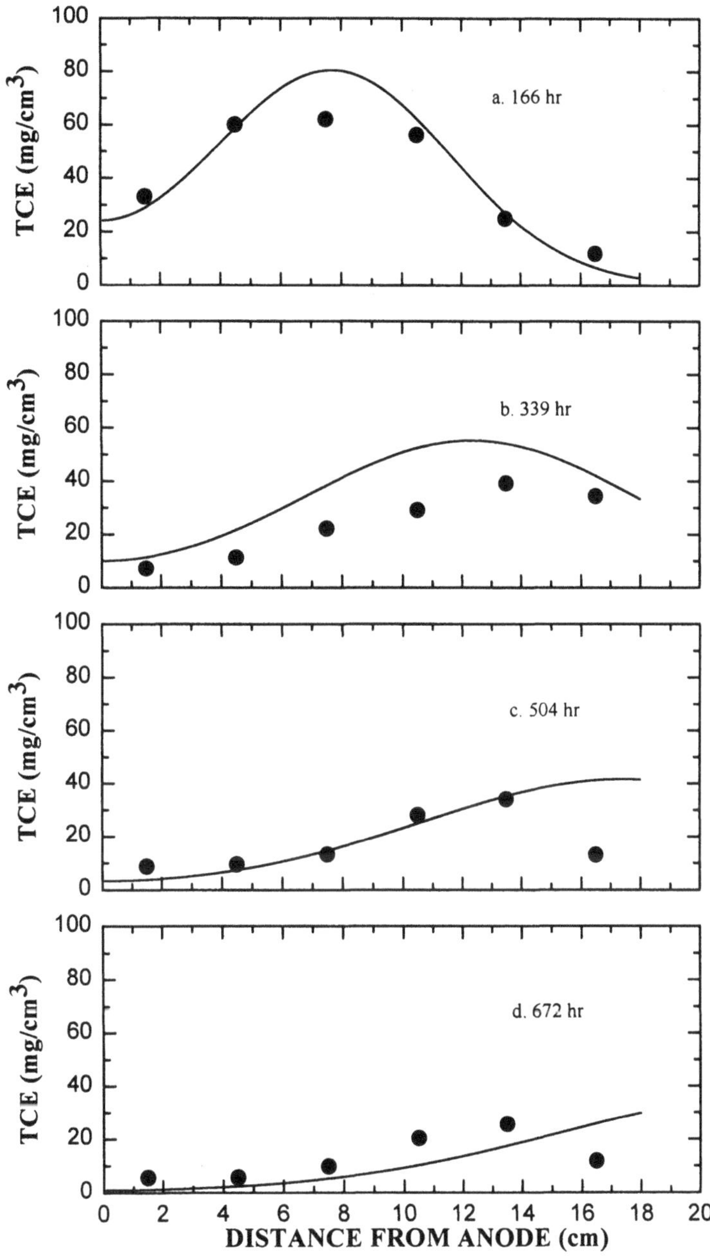

Figure 6. Observed (circle) and Model Output (line) of TCE concentration profiles of the ECs.

The simulated TCE profiles of the test columns agrees reasonable well with the observed values after 339 hrs of EO (Figure 6). In the vicinity of the cathode, however, the observed values deviate from the simulated concentration with time. Mass balance calculation of TCE during EO shows that the amount of TCE loss increased approximately linearly with time (Figure 7). The discrepancy suggests that TCE was consumed either by chemical reactions or microbial synthesis. The latter was unlikely as the solution contained 250 mg/L of sodium azide. Therefore, the loss of TCE close to the cathode can be attributed to chemical destruction. Furthermore, dichloroethylene (cis-1,2-DCE) was detected in the test column after 672 hrs of EO, suggesting that TCE was reduced. The mechanism of TCE reduction is unclear, however, based on the low redox potential and ample electrons available at the cathode, one can hypothesize that TCE was reduced at the interface between the soil and the graphite cathode. The preliminary results of an ongoing research in our laboratory show that the DCE isomers mostly is composed of cis-1,2-DCE with minor amount of trans-1,2-DCE and 1,1-DCE (unpublished results).

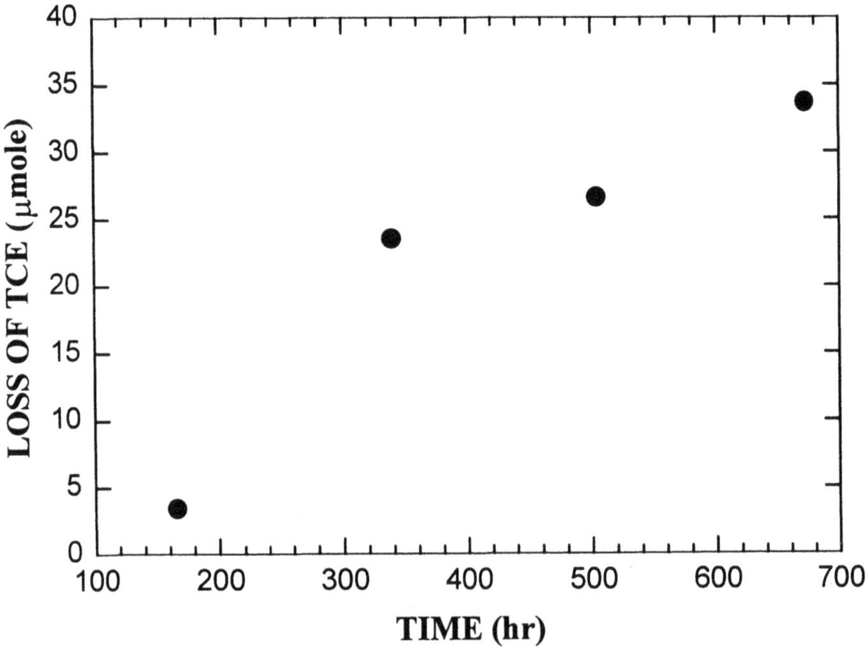

Figure 7. Loss of TCE in the ECs as a function of time.

4.4 TCE Degradation

cis-1,2-DCE was detected at various positions along the length of the EC at the termination of EO. The main mass of TCE had clearly reached the cathode region by the termination of the experiment. At the top of the cathode there is a decrease in TCE concentration from 6.2 to 3.5 μg/g , while the of cis-1,2-DCE concentration shows an increase from 0 to 3.56 μg/g, as does the chloride concentration from 1.1 μg/g originally in the soil to 2.43 μ g/g after 672 hrs of EO. The data indicates that reductive dechlorination of TCE to cis-1,2-DCE has occurred right at the cathode, the most reduced part of the soil column.

pH and Eh show that low redox potential and high H^+ activity were established in the ECs during the four week duration of the experiment (Fig. 2 and 3). The relative stabilities of water and chloro-organic compounds, including TCE and its degradation products are functions of pH and redox potential, Eh. During EO, TCE originating in the CZ was transported down the soil column (Figure 4), and encountered electrons generated at the cathode.

It is quite plausible that some of the cis-1,2-DCE observed by us was produced by the reductive dechlorination of TCE, with the graphite at the cathode being the primary source of electrons, as in the following reaction:

$$C_2HCl_3 + 2\,e^- + H^+ \rightarrow C_2H_2Cl_2 + Cl^-$$

As shown by Figs. 2 and 3, fluid moving towards the cathode experienced low Eh and pH values, which may favour reductive dechlorination of TCE to cis-1,2-DCE.

5. CONCLUSIONS AND PRACTICAL IMPLICATIONS

This study has shown that the application of EO can move TCE from a contaminated zone across an intact column of tight soil. The output from a 1D diffusion-advection equation agrees well with the observed TCE concentration profiles, indicating that the transport equation is an appropriate model.

EO could be used as an alternative technology for remediation of contaminants in low permeability clay-rich soils, and represents an in situ, abiotic physical and chemical treatment process which utilizes the Eh-pH gradients established in the subsurface as a result of electrolysis reactions at the anode and cathode. Results of the present study have several implications for remedial applications of EO in the subsurface, which include a horizontal configuration of graphite electrodes. Foremost the steep Eh-pH changes in the vicinity of the cathode and direct cathodic reduction during EO may enhance the dechorination of chlorinated compounds. The dechlorination process could be optimized by selective placement of electrodes relative to the contaminant hot-spot. It should be possible to optimize TCE degradation in the subsurface by placing the treatment zone (abiotic or biotic) beneath the TCE hot-spot so that TCE moving through the soil is degraded at the treatment zone, and then moved into the region of the cathode where degradation can continue. By configuring electrode dimensions and spacing, it is possible to increase the contact time between contaminants with the cathode so that complete degradation of TCE can be achieved.

ACKNOWLEDGEMENTS

This Chapter has not been subjected to internal review by the US Environmental Protection Agency. Therefore, the research results presented herein do not, necessarily, reflect Agency policy. Mention of trade names of commercial products does not constitute endorsement or recommendation for use.

BIBLIOGRAPHY

Acar, Y. B., Alshawabkeh, A. N., and Gale, R. J., (1992). "A review of fundamentals of removing contaminants from soils by electrokinetic soil processing." *Proceedings of the Mediterranean Conference on Environmental Geotechnology*, May 25 - 27, 1992, Cesme, Turkey, Balkema Publishers, Inc., Amsterdam, Holland.

Acar, Y. B., Gale, R. J., Alshawabkeh, A. N., Marks, R. E., Puppala, S., Bricka, M., and Parker, R., (1995). "Electrokinetic remediation: Basics and technology status." *Journal of Hazardous Materials*, 40, 117-137.

Bruell, C. J., Segall, B. A., and Walsh, M. T. (1992). "Electroosmotic removal of gasoline hydrocarbons and TCE from clay." *Journal of Environmental Engineering*, 118(1), 68-83.

Chen, J.-L., Al-Abed, S. R., Bryndzia, L. T., and Murdoch, L. C. (1999) "Cation Transport and Patitioning During a Field Test of Electroosmosis." *Water Resources Research*, 35(12), 3841-3851.

Eyklot, G. R., and Daniel D. E. (1994). "Impact of System Chemistry on Electroosmosis in Contaminated Soil." *J. Geotech. Eng.*, 120 (5), 797-815.

Fetter, C. W. (1993). *Contaminant Hydrogeology*. Macmillan Pub., New York, NY, 458 pp.

Fischer, H. B., List, E. J., Koh, R. C. Y., Imberger, J., and Brooks, N. H. (1979). *Mixing in Inland and Coastal Waters*. Academic Press, New York, NY, 483pp.

Fusillo, T. V., Hochreiter, J. J., Jr., and Lord, D. G. (1985). "Distribution of Volatile Organic Compounds in a New Jersey Costal Plain Aquifer System." *Groundwater*. 23, 354-360.

Hamed, J., Acar, Y. B., and Gale R. J. (1991) "Pb(II) Removal from Kaolinite by Electrokinetics." *J. Geotech. Eng*. 117, 241-271.

Ho, S. V., Sheridan, W, Athmer, C. J., Heitkamp, M. A., Brackin, J. M., Weber,D., and Brodsky, P. H. (1995). "Integrated In Situ Soil Remediation Technology: The Lasagna Process." *Environmental Science and Technology* . 29, 2528-2534.

Ho, S. V. , Athmer, C. J., Sheridan, W, Sheridan P. W., Orth, R., Macenzie D., Brodsky, P. H., Shapiro A., Thornton R., Salvo J., Schultz D., Landis R., Griffith R., and Shoemaker S. (1999). "The Lasagna Technology for In situ Soil Remediation. 1. Small Field Test." *Environmental Science and Technology* . 29, 2528-2534.

Jackson, M. L. Soil Chemical Analysis, *Prentice Hall, Inc*., Englewood Cliffs, N.J., 1958.

Mercer, J. W., Sipp, D. C., and Giffin, D. (1990). "U.S. Environ. Protect. Agency Report " *EPA/600/8-90/003*.

Pamukcu, S., and Wittle, J. K., (1992). "Electrokinetic removal of selected heavy metals from soil." *Environmental Progress*, 11(3), 241-250.

Probstein, R. F., and Hicks, R. E., (1993). "Removal of contaminants from soil by electric fields." *Science*, 260, 498-503.

Runnells, D., and J. L. Larson. (1986). "A Laboratory Study of Electromigration as Possible Field Techniques for the Removal of Contaminants from Groundwater." *Ground Water Monitoring Review*. 6(3), 85-88.

Segall, B. A., and Bruell, C. J., (1992). "Electoosmotic contaminant removal processes." *Journal of Environmental Engineering*, 118(1), 84-100.

Shapiro, A. P., Renaud, P. C., and Probstein, R. F. (1989). "Preliminary studies on the removal of chemical species from saturated porous media by electroosmosis." *PhysicoChemical Hydrodynamics*, 11(5/6), 785-802.

Shapiro, A. P., and Probstein, R. F. (1993). "Removal of contaminants from saturated clay by electroosmosis." *Environmental Science and Technology*, 27, 283-291.

USEPA, (1992). "A Technology Assessment of Soil Vapor Extraction and Air Sparging." *EPA/600/R-92/173*.

USEPA, (1990). "Ground Water Volume I: Groundwater and Contamination." *EPA/625/6-90/016a.*

Van Genuchten, M. Th., and Parker, T. C. (1984). "Boundary Conditions for Displacement Experiments through Short Laboratory Soil Columns." *Soil Sci. Soc. Am. J.*, 48, 703-708.

Chapter Six

Sorbing Vertical Barriers

ALAN J. RABIDEAU[1], JOHN VAN BENSCHOTEN[1], ASHUTOSH KHANDELWAL[2], and CRAIG R. REPP[3]

[1] *Department of Civil, Structural, and Environmental Engineering, University at Buffalo, Buffalo, NY 14260*
[2] *S.S. Papadopulos & Associates, Bethesda, MD*
[3] *West Valley Nuclear Services Company LLC, West Valley, NY*

Key words: barriers, sorption, slurry wall, zeolite, ion exchange

Abstract: This chapter presents an overview of vertical barrier technologies that use sorbing materials to remove contaminants from groundwater. Two classes of system are considered: 1) low-permeability earthen barriers, in which sorbing additives are used to reduce the diffusive flux of organic contaminants, and 2) high-permeability treatment walls designed to remove contaminants under advection-dominated natural groundwater conditions. The focus of the discussion is on the performance assessment of strongly sorbing barrier materials using laboratory tests. Emphasis is placed on the design and analysis of column studies to characterize the barrier sorption capacity and the appropriate formulation of mathematical models to extrapolate long-term barrier performance. Two case studies are considered: the amendment of soil-bentonite slurry walls with an organic-rich additives, and the use of natural zeolite to remove strontium-90 from groundwater.

1. INTRODUCTION

The severe technological obstacles to the restoration of contaminated aquifers have been well documented (e.g., NRC, 1994) and have led to a renewed interest in low-permeability containment barriers and the development of high-permeability treatment walls. Studies of both classes of barriers have included designs for which sorption is the primary mechanism for reducing contaminant flux, although few actual field installations of sorbing barriers have been reported. Reliance on in situ sorption as the primary remedial measure is complicated by the considerable cost and logistical difficulty of replacing a large barrier when the sorption capacity is exhausted. For this reason, the ability to predict the time-to-replacement is critical in evaluating whether the high capital cost of a barrier system is justified.

This chapter considers two particular types of sorbing barriers: 1) low-permeability slurry walls amended to promote the sorption of hydrophobic compounds (hydraulic conductivity [K] ~ 10^{-7} cm/s), and 2) high-conductivity zeolite treatment walls designed to remove inorganic compounds (K ~ 10^{-1} cm/s). These examples are selected, in part, because they represent the systems that have received the most attention from researchers and practitioners, but also because considering them together highlights the importance of conceptual issues common to both types of systems.

For any type of sorbing barrier, the development of a reliable predictive model necessitates accurate characterization of the sorption capacity of the barrier material. A basic premise of this work is that laboratory column tests are the most appealing tool for this task. However, obtaining meaningful column data for strongly sorbing materials can be quite difficult if the experiments are to be performed under groundwater conditions typical of actual contaminated sites. Over the past 5 years, our research group at the University of Buffalo, NY has conducted several studies related to the development of sorbing and reactive barriers. In performing these studies, we have considered several experimental and modeling issues common to low- and high-permeability barriers, and have noted similarities and differences between the two applications. This chapter presents some of these insights in the context of a general approach to barrier performance assessment.

2. CONCEPTUAL AND ANALYTIC FRAMEWORK

Despite the extensive literature on the measurement of sorption parameters, methods for characterizing the behavior of strongly sorbing barrier materials are not well developed. Although laboratory columns are established tools for environmental and soil scientists, strongly sorbing materials represent a challenge because of the long time required for contaminant effluent breakthrough to occur. Two methods of circumventing this difficulty are: 1) the use of artificially high column flow rates, and 2) the analysis of the *spatial* contaminant distribution at the end of the experiment (prior to achieving full breakthrough). As discussed below, we have adopted the latter approach for both low- and high-permeability systems, primarily to minimize possible artifacts due to nonequilibrium sorption effects caused by elevated flow rates.

Contaminant transport in treatment walls and soil columns is commonly analyzed using the one-dimensional advective-dispersive-reactive equation:

$$\frac{\partial C_i}{\partial t} = -v\frac{\partial C_i}{\partial x} + D_i\frac{\partial^2 C_i}{\partial x^2} - \mu_{ai}C_i - \frac{\rho_b}{n}\frac{\partial q_i}{\partial t} - \frac{\rho_b}{n}\mu_{si}q_i \qquad (1)$$

where C_i is the dissolved phase contaminant concentration for solute i, t is time, x is distance from the entrance of the column or barrier, v is the fluid velocity, D_i is the dispersion coefficient (includes hydrodynamic dispersion and molecular diffusion), μ_{ai} and μ_{si} are first-order decay coefficients for the aqueous and sorbed phases, respectively, ρ_b is the bulk (dry) density, n is the porosity, and q_i is the sorbed phase mass fraction.

The solution of Eq. 1 requires specification of boundary conditions (BCs) and additional equation(s) that describe the sorption reaction. The assumptions reflected in these choices strongly influence the process of extrapolating barrier performance from laboratory column data. Furthermore, as discussed below, there are significant differences in the treatment of these choices between low- and high-permeability systems.

The remainder of this chapter explores the above topics in greater detail, with specific examples from each class of sorbing barrier. In particular, the following issues are highlighted:

- Specification of the sorption term in Eq. 1;
- Design of column experiments for estimating sorption parameters;
- Specification of boundary conditions for interpreting column experiments; and
- Specification of boundary conditions for extrapolating column results to the field.

3. LOW-PERMEABILITY BARRIERS

Low-permeability vertical barriers such as soil/bentonite slurry walls are utilized extensively to restrict the migration of mobile subsurface contaminants. In light of the development of long-term remediation strategies such as funnel-and-gate systems, slurry walls may be expected to provide effective containment for time frames of decades or longer. However, several researchers have observed that significant contaminant breakthrough in slurry walls could occur due to molecular diffusion within a period of 5 to 30 years (e.g., Gray and Weber, 1984; Shackelford, 1989; Rabideau, 1996). In part, this limitation is attributed to the typically small sorption capacity associated with most native aquifer materials, which are the primary source of the backfill portion of a slurry wall.

A variety of sorbing additives have been proposed to restrict the transport of hydrophobic organic contaminants (HOCs) such as chlorinated solvents and petroleum compounds in slurry walls, as summarized in Table 1 (the authors are aware of only a single North American field installation of an amended slurry wall, as referenced by Gullick, 1998). Typically, candidate materials are characterized using batch tests and mathematical models are subsequently used to extrapolate the resulting improvement in barrier performance at the field scale. However, a number of factors can contribute to over-prediction of the time-to-breakthrough, including the interpretation of laboratory data and the formulation of the model used to simulate field conditions. These issues are examined in detail below.

Table 1. Studies of sorbing soil/bentonite additives.

Reference	Sorbing additive	Experimental Design
Mott and Weber (1992)	Flyash	Batch, diffusion cell
Bierck and Chang (1994)	Granular Activated Carbon (GAC)	Batch, column
Park (1997)	Shredded tires	Column
Khandelwal and Rabideau (1998; 2000)	Humus	Batch, column
Gullick (1998)	Organoclays, shale	Batch, diffusion cell

3.1 Sorption Model

In considering an organic-rich slurry wall additive, the governing sorption mechanism is often considered to be hydrophobic partitioning and the effects of solution chemistry (pH, ionic strength, etc.) are considered secondary. The specification of the appropriate sorption model requires several choices:

1) the form of the governing isotherm, 2) the choice between an equilibrium or kinetic model, and 3) the form of the kinetic model (if necessary). For a single nondecaying sorbing solute, the well-known equilibrium model is (subscripts removed for clarity):

$$\left[1+\frac{\rho_b}{n}\frac{\partial q}{\partial C}\right]\frac{\partial C}{\partial t}=-v\frac{\partial C}{\partial x}+D\frac{\partial^2 C}{\partial x^2} \tag{2}$$

where the bracketed term is designated the "retardation factor"(R_f), with the form dependent on the governing sorption isotherm. For a linear isotherm, the bracketed term reduces to a constant: $R_f = 1 + K_d\rho_b/n$ where K_d is the sorption distribution coefficient.

Most studies of contaminant transport in slurry walls have relied on the equilibrium assumption, either for convenience or because of the long residence times associated with diffusion-dominated transport. As an alternative, Rabideau and Khandelwal (1998a) have proposed the simple two-compartment mass transfer model to describe nonequilibrium sorption in soil/bentonite systems:

$$\frac{\partial C}{\partial t}=-v\frac{\partial C}{\partial x}+D\frac{\partial^2 C}{\partial x^2}-\frac{\rho_b\alpha}{n}\left(K_d C-q\right) \tag{3}$$

$$\frac{\partial q}{\partial t}=\alpha\left(K_d C-q\right) \tag{4}$$

where α is the sorption rate constant.

Eqs. 3-4 are amenable to semi-analytical solution techniques because of the linear form. The use of more complex kinetic models (e.g., intra-aggregate diffusion) has not been attempted, in part because the above models have proved adequate to describe the available data sets, and in part because of a limited understanding of the geometry of the soil/bentonite matrix (gel formation and the resulting diffusion geometry).

Because laboratory columns are often operated using flow rates higher than those associated with natural groundwater movement, the choice between an equilibrium and kinetic formulation is probably more significant for interpreting experimental data than for predicting barrier performance in the field. For example, we have observed kinetic effects in short-term (2 to 6 weeks) column experiments in which advection plays a significant (but not dominant) role. Similarly, other researchers have observed discrepancies between column results and modeling predictions based on batch isotherm tests, which could be partially due to kinetic effects (e.g., Mott and Weber, 1992; Bierck and Chang, 1994; Grathwohl, 1998). In general, neglecting kinetic effects in interpreting data from laboratory scale transport experiments may lead to mischaracterization of the barrier sorption capacity.

For the materials and concentration ranges we have studied, comparative batch tests have supported the use of linear isotherms (e.g., Khandelwal et al., 1998; Khandelwal and Rabideau, 2000), although the assumption of a linear isotherm may not be appropriate for all systems. In particular, Mott and Weber (1992) and Gullick (1998) have utilized nonlinear isotherms to describe the behavior of various sorbing additives, including flyash and organoclays. However, it is important to note that multi-parameter nonlinear isotherms are not easily calibrated from column data, particularly if a sorption rate constant must also be estimated.

3.2 Experimental Design

Both batch and column tests have been used to estimate the sorption capacity for barrier materials. Batch isotherm tests are easier to perform and are particularly useful for comparing the performance of alternative additives (e.g., Adu-Wusu et al., 1997; Bradl, 1997). However, isotherm tests may overestimate the sorption capacity applicable to a field setting due to kinetic effects and/or artifacts associated with the "solids effect" in batch experiments. Gullick (1998) provided a thorough review of possible explanations for differences in apparent sorption as a function of liquid /solid (L/S) ratio, and also conducted detailed batch experiments for different L/S conditions. The results indicated that a significant "solids effect" occurred for some, but not all, of the materials tested. Gullick recommended that batch experiments be performed at a L/S ratio as close to the anticipated field conditions as possible.

Alternatives to batch testing include the use of diffusion cells or flow-through columns. Diffusion cells are easier to operate, but are less representative of field conditions where some advection may occur. However, operation of columns at very low flow rates is difficult and subject to artifacts. To minimize possible wall effects associated with shrink/swell behavior of low-permeability clay materials, several researchers have utilized column devices that provide a confining pressure, such as flexible wall permeameters (e.g., Acar and Haider, 1990; Smith and Jaffe, 1994; Shackelford and Redmond, 1995; Khandelwal et al., 1998; Khandelwal and Rabideau, 2000).

A typical permeameter specimen is characterized by a diameter of approximately 5 cm and a length of 5 to 10 cm. To ensure uniform distribution of the influent across the cross-section, it is common to include "mixing zones" (MZs) of non-sorbing material at the column ends. As discussed by Khandelwal et al. (1998), our work has utilized glass wool MZs approximately 0.5 cm in thickness. At the end of a column run, the column specimen is placed in a rigid mold and hollow stainless steel tubes are inserted to extract several subsamples. The cylindrical subsamples are then extruded with a piston-like device and sectioned directly into a solvent extraction vial, followed by analysis to determine the contaminant spatial distribution. In general, this procedure results in smooth contaminant profiles with a minimum of volatilization losses (e.g., trichloroethylene mass balances typically range from 85 to 90 percent). An advantage of obtaining several column "core samples" is that the assumption of lateral homogeneity can be assessed. For example, we have found variations in extracted mass ranging from 2 to 10 percent across the column cross-section, with the variability increasing with the sorption capacity of the media. These variations are attributed to incomplete mixing in the influent zone and/or diffusion of contaminant across the outer membranes into the confining fluid (verified by sampling of the cell water).

Another important experimental consideration is the potential for nonequilibrium sorption in short experimental columns. In particular, a drawback of using spatial data is that model calibration of a sorption rate constant is more difficult. For example, Khandelwal and Rabideau (in press) showed that multiple combinations of K_d and α can be used to generate equivalent predictions for a single-time contaminant spatial distribution. As discussed below, this concern can be addressed by using multiple columns operated for different durations.

3.3 Boundary Conditions

Because diffusion dominates the transport of contaminants in barriers and columns constructed of low-permeability materials, model calibrations and predictions are extremely sensitive to the form of the specified boundary conditions. Two issues are of particular importance: 1) treatment of the entrance mixing zone in laboratory columns, and 2) specification of appropriate BCs to represent a slurry wall under field conditions.

Because considerable time may be required to flush the fluid initially present in the influent zones of low flow experimental columns, Rabideau and Khandelwal (1998b) proposed the following mixing zone (MZ) boundary condition for the column entrance:

$$C(0,t) = C_b + \frac{Q_i}{V_i}\int_0^t C_{in}(\tau)d\tau - \frac{A}{V_i}\int_0^t f(0,\tau)d\tau - \int_0^t \lambda_s C(0,\tau)d\tau \quad (4)$$

where C(0,t) is the concentration at the column entrance ($x = 0$), V_i is the volume of the influent MZ, Q_i is the flow rate into the column, A is the column area normal to x, C_{in} is the time-varying influent concentration, C_b is the initial concentration in the influent MZ (if applicable), λ_s is the contaminant decay rate for the influent MZ (if applicable), and $f(0,t)$ is the flux at $x = 0$, where the flux is given by:

$$f = n\left(vC - D\frac{\partial C}{\partial x}\right) \quad (5)$$

Eq. 4 is amenable to solution techniques based on the numerical inversion of Laplace-transformed equations; these calculations can be performed rapidly and are therefore suitable for calibration. In Figure 1, typical soil/bentonite column predictions are shown to highlight the effect of the influent mixing zone on the spatial contaminant distributions for low-flow systems. The simulation results, which were generated for column conditions described by Khandelwal et al. (1998), indicate that the mixing zone has a significant influence on the shape of the spatial contaminant distribution and, therefore should be considered explicitly in estimating sorption parameters from spatial column data.

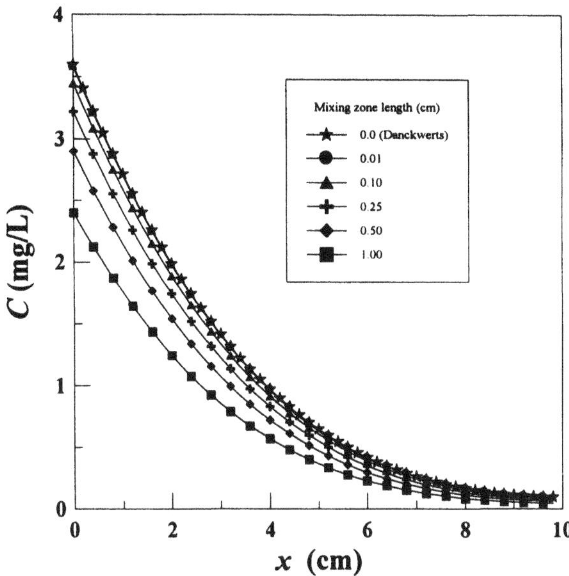

Figure 1. Effect of influent mixing zone size on predicted contaminant spatial profiles for a constant inflow concentration and zero initial concentration in the mixing zone (x is the distance from the interface between the influent mixing zone and barrier material; column velocity = 3.5×10^{-3} cm/d). Reprinted with permission from Rabideau and Khandelwal (1998b). Copyright 1998, American Society of Civil Engineers.

In extrapolating laboratory results to installed slurry wall of thickness L, the commonly applied idealized BCs may not provide an accurate representation of the transitions between an engineered barrier and native aquifer material in the field. To provide conservative predictions for design, Rabideau and Khandelwal (1998b) recommend the combination of a constant concentration entrance ($x = 0$) BC with a zero-concentration exit ($x = L$) BC. In particular, several commonly used BCs should be avoided because they distort the nature of the diffusive flux at the boundaries (e.g., the "Danckwerts" constant-flux entrance BC and the zero-gradient exit BC). These issues are discussed in greater detail by Rabideau and Khandelwal (1998b).

3.4 Example: Soil/bentonite Amended with Humus

In a comparative study of sorbing additives for soil/bentonite slurry walls, Adu-Wusu et al. (1997) identified a natural humus material as an attractive candidate in terms of cost and sorption capacity. Subsequently, column experiments were performed by Khandelwal and Rabideau (2000) using an SB mixture containing 5 percent humus. To account for possible nonequilibrium effects, three experiments of different durations were used. Key results from this work included:

- Calibrations performed using an equilibrium model indicated increasing K_d with time, which is consistent with kinetic effects (i.e., gradual approach to equilibrium). When the kinetic model was calibrated, good model fits were observed for all three columns using a calibrated K_d of 1.4 mL/g and first-order sorption rate constant of 0.15 day^{-1} (Figure 2).
- When extrapolated to a 1-meter barrier typical of field installations, predictions that incorporated kinetic sorption were essentially identical to those generated using an equilibrium model, due to the low hydraulic gradient and larger domain associated with field conditions.
- Model calculations indicated that the addition of 5 percent humus to a SB slurry wall would delay breakthrough (defined as 5 percent of the steady-state flux) from approximately 5 years to more than 100 years (Figure 3).
- The K_d calibrated from the column studies was approximately 58 percent of the value measured in batch isotherm tests using the same materials. The precise reasons for these differences is a subject of further investigation, but the result suggests that caution should be exercised in using the results of batch sorption tests to extrapolate the performance of amended slurry walls.

4. HIGH-PERMEABILITY BARRIERS

The application of high-permeability sorbing barriers is a relatively recent development, with only a few reported field installations (Table 3). To date, the most extensively studied systems include granular activated carbon (GAC) for removal of organic contaminants and natural zeolites designed to remove inorganic contaminants (notably, Sr-90). The development of a hybrid material for removal of both classes of compounds - surfactant modified zeolite - is discussed elsewhere in this book. Other sorbing materials receiving attention include activated alumina, bauxite, amorphous

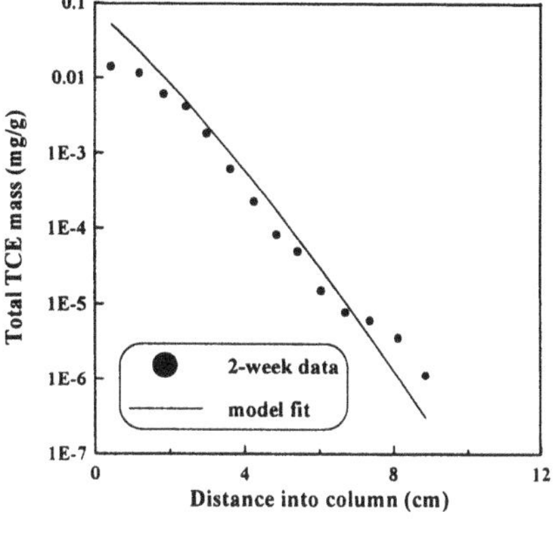

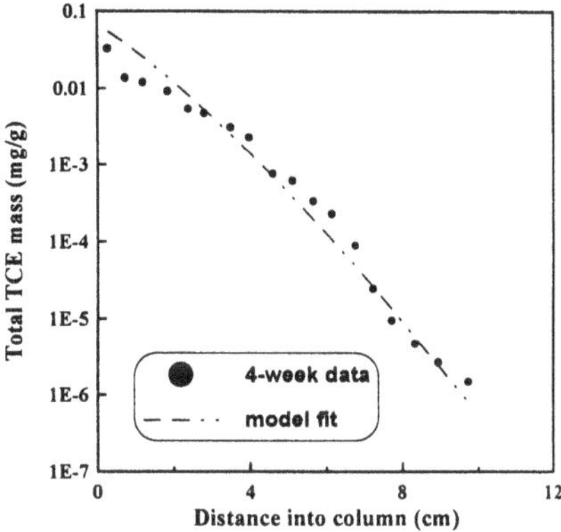

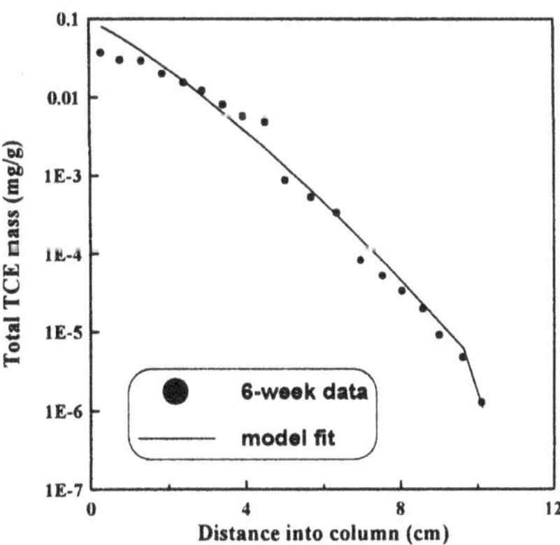

Figure 2. Spatial contaminant distributions and simultaneous model calibrations for column experiments using SB amended with 5 percent humus (column conditions summarized in Table 2). Reprinted with permission from Khandelwal and Rabideau (2000). Copyright 2000, Elsevier Science B. V.

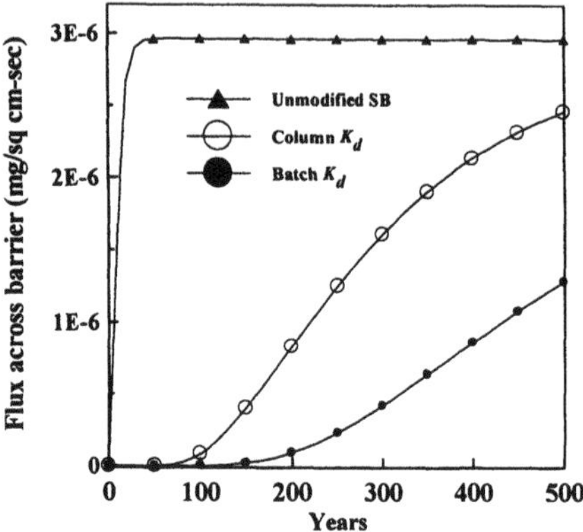

Figure 3. Simulated field performance of 1-meter humus-mofidied slurry wall (column conditions summarized in Table 2). Reprinted with permission from Khandelwal and Rabideau (2000). Copyright 2000, Elsevier Science B. V.

Table 2. Conditions for soil/bentonite experiments and model predictions

Parameters	Column Experiments (Figure 2)			Field Prediction
	2 week	4 week	6 week	(Figure 3)
Duration (h)	341	680	967	--
Domain length (cm)	11.1	12.2	11.7	100
Hydraulic conductivity (cm/s x 10^8)	3.1[a]	2.4[a]	2.7[a]	5
Hydraulic gradient	6.16[a]	5.01[a]	5.54[a]	0.1
Porosity	0.34[a]	0.34[a]	0.34[a]	0.34
Diffusion coefficient (cm^2/s x 10^6)	4.0[b]	4.0[b]	4.0[b]	4.0
Sorption rate constant (s^{-1} x 10^6)	1.7[b]	1.7[b]	1.7[b]	--
C_0 (mg/L)	480.0[a]	480.0[a]	480.0[a]	1.0
C_1 (mg/L-d)	-0.63[a]	-0.93[a]	-0.27[a]	0
C_2 (mg/L-d^2)	-0.049[a]	-0.009[a]	-0.052[a]	0

[a] = measured; [b] = calibration across all columns; influent concentration = $C_0 + C_1t + C_2t^2$

ferric oxyhydroxide, humate, phosphate, iron filings, peat moss, and various synthetic products (e.g., Baker et al., 1997; Watson et al., 1997; Whang et al., 1997).

Table 3. Selected field installations and studies of high-permeability sorbing barriers

Site	Sorbent	Reference	Comment
Chalk River, CN	Natural zeolite	Lee et al. (1998)	Constructed 1998
West Valley Demonstration Project, NY	Natural zeolite	Fuhrmann et al. (1995), Moore et al. (2000)	Pilot wall constructed 1999
Hanford, WA	Natural zeolite	Cantrell (1995)	Project discontinued due to stakeholder concerns
Marzone site, Atlanta, GA	GAC	Williamson (2000)	Constructed 1998; retrievable canisters
Karfsruhe, Germany	GAC	Schad and Gratwohl (1998)	Construction scheduled for 2000
Large Experimental Aquifer Program test site	Surfactant-modified zeolite	Bowman (this book)	Pilot test in 1997

The use of high-permeability materials enables greater flexibility in engineered measures to direct the target groundwater through the sorbing media. For example, the Chalk River Facility (Lee et al., 1998) and the Marzone Site (Williamson et al., 2000) include sumps and/or subsurface pumping that direct the groundwater through the reactive media. These "hybrid" systems retain the advantages of in situ treatment (low-flow operation, no aeration of processed groundwater) while providing an additional measure of hydraulic control. The remainder of this section discusses the use of natural zeolite to remove Sr-90 from groundwater, a technology that has recently been employed at two North American facilities.

4.1 Sorption model

4.1.1 Linear local equilibrium assumption

Natural and engineered zeolites have been used extensively in wastewater treatment (e.g., Clifford, 1990). For in situ barrier applications, the focus of study has been a natural clinoptilolite supplied by Teague Mineral Products (Adrian, Oregon), primarily because of its low cost and strong affinity for Sr.

In contrast to the removal of organic contaminants in organic-rich barriers, the removal of Sr-90 is considered to be a competitive cation-exchange process, with calcium serving as the primary competitor for zeolite exchange sites. However, performance assessment has generally been accomplished using the single-solute retardation factor approach (Eq. 2, e.g., Fuhrmann et al., 1995; Cantrell, 1996; Lee et al., 1998; Moore et al., 2000).

The use of the single-solute approach requires numerous assumptions, including: 1) the low flow rates associated with natural groundwater are conducive to the attainment of sorption equilibrium, 2) the Sr isotherm is linear over the concentration range of interest, 3) the sorption distribution coefficient (K_d) is estimated under experimental conditions that represent the strongest anticipated influence of competing solutes and other geochemical factors, and 4) the geochemical conditions will not change appreciably over the life of the barrier.

The assumption of equilibrium sorption has been supported by the long anticipated residence times in an installed barrier (e.g., ~ days for a barrier thickness of 1-2 m), and by batch kinetic data reported by Cantrell (1996) that indicate near-equilibrium is achieved on the order of one day. Similar assumptions have been applied to the analysis of GAC barriers (e.g., Schad and Gratwohl, 1998). Cantrell (1996) also observed linear isotherms for Sr concentrations below approximately 0.1 mg/L, although this result should be viewed as particular to the specific experimental conditions. Although these results lend support to the simplified modeling approach, more data are clearly needed to better evaluate the key assumption of linear equilibrium sorption.

Even if the linear equilibrium assumption is accepted, K_d values measured in laboratory experiments will be sensitive to the solution chemistry and experimental design. If conservative predictions of the barrier time-to-breakthrough are desired, the supporting laboratory experiments should be carefully designed, as discussed in the following section.

4.1.2 Multi-solute cation-exchange model

Although the simplicity of the retardation factor approach is appealing, the development of a more fundamental model is desirable for several reasons. First, a more mechanistic model could be used to explore the limiting behavior of the system and delineate the conditions under which the retardation factor approach is valid. Second, a properly designed "fundamental" model should yield parameters that are general in nature and could be used to extrapolate the behavior of a particular zeolite to other groundwater systems. Finally, since data from less strongly sorbing solutes are more readily obtained, a multi-species model may have advantages with respect to calibration, potentially enabling shorter experiments that do not depend solely on observations of the most strongly retarded species (Sr).

The effect of competitive ion-exchange is evident from Figure 4, which shows data from bench-testing performed for the West Valley Demonstration Project. For this 2-month experiment, the effluent concentrations of Na^{+}, Ca^{2+}, Mg^{2+}, and K^{+} have reached the approximate levels of the feed solution, but the Sr^{2+} effluent has not reached breakthrough (the effluent concentrations represent leaching from the natural zeolite). The known strong preference of clinoptilolite for Sr suggests that up to thousands of pore volumes would be required to achieve full effluent breakthrough, an undertaking that is probably infeasible for a typical bench-scale study performed under realistic flow rates. Taken alone, effluent data of this type are insufficient to parameterize a retardation factor for Sr, which, as described below, must be calibrated from spatial data.

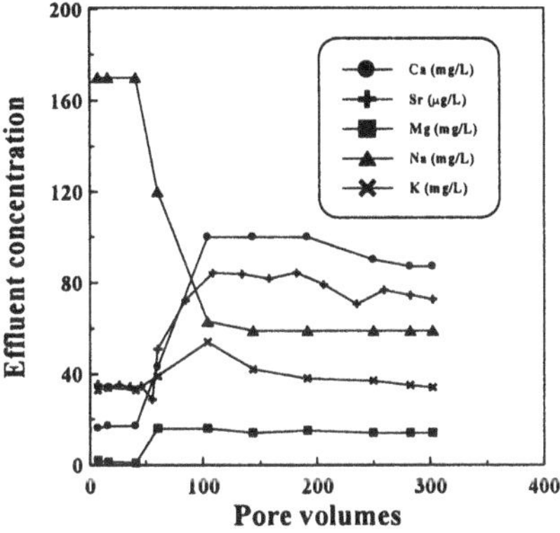

Figure 4. Effluent breakthrough curves for clinoptilolite laboratory columns (Table 4). The nonradioactive source groundwater was obtained from the West Valley Demonstration Project site and adjusted by increasing the Sr concentration from approximately 200 µg/L to 1 mg/L (for clarity, Sr units are µg/L, others are mg/L).

To provide a better mechanistic description of zeolite sorption dynamics, a competitive ion-exchange model has been developed, based loosely on similar models developed to analyze cation transport in soils and ion-exchange treatment systems (e.g, Perona, 1993; Cernik et al., 1994; Depaoli and Perona, 1996). In the current formulation, five solutes are considered to compete for cation exchange sites (Sr^{2+}, Ca^{2+}, Na^{+}, Mg^{2+}, K^{+}) resulting in 10 unknowns (5 sorbed and 5 aqueous species). The governing reaction equations are developed from five solute mass balances, a constraint on the total cation-exchange capacity, and four binary exchange equilibrium expressions. Several forms of the model equations are possible, depending on the convention adopted for the exchange reaction. To circumvent difficulties with estimating solid phase activities, the Gaines-Thomas expression was adopted for the binary equilibrium relationships (e.g., Valocchi et al., 1981; Cernik et al., 1994), illustrated here for sodium/strontium exchange:

$$K_{Na-Sr} = 2Q\left(\frac{q_{Na}}{\gamma_{Na}C_{Na}}\right)^2\left(\frac{\gamma_{Sr}q_{Sr}}{C_{Sr}}\right) \tag{6}$$

where C represents the aqueous phase concentration, q is the sorbed concentration expressed as an "equivalent fractions", $K_{Na\text{-}Sr}$ is the equilibrium constant, Q is the cation-exchange capacity, and γ represents a correction for activity.

Extension of the equilibrium model to column or field conditions requires coupling the ion-exchange equations with the transport equations for the 5 aqueous species (Eq. 1). To accomplish this coupling, we have adopted the split-operator approach (e.g., Miller and Rabideau, 1993), which provides considerable flexibility in adjusting the sorption submodel. In addition to the above conceptual model, we are pursuing more complex formulations that couple cation exchange with pore diffusion, surface diffusion, or combined pore/surface diffusion (e.g., Robinson et al., 1994; DePaoli and Perona, 1996; Ma et al., 1996). However, the currently available data are inadequate to parameterize such models, and the need for a kinetic formulation for the low-flow conditions expected for sorbing barriers has not been established. These issues will be addressed in a future publication.

In summary: while conceptually appealing, the application of complex multi-solute models for Sr sorption to zeolite is in the early stages of development. While preliminary results are encouraging, additional work is required to develop more efficient computational methods and develop an improved database for parameter estimation. The remainder of this section focuses on the simpler retardation factor approach.

4.2 Experimental Design

As discussed above for low-permeability systems, the analysis of spatial contaminant distributions is an appealing strategy for shortening the duration of column experiments. Again, the use of multiple columns operated over different time periods can facilitate identification of nonequilibrium sorption effects and lead to more robust parameter estimation.

If the retardation factor approach is adopted for performance assessment, the distribution coefficient should be measured under conditions that represent a "worst case" competition scenario. One approach to approximating "conservative" conditions is to pre-wash the zeolite in an effort to saturate the exchange sites with calcium, the chief competitor (e.g., Cantrell, 1996). Alternatively, it may suffice to operate the columns until full breakthrough of the competing solutes is observed, as suggested by Figure 4, followed by the analysis of spatial concentrations.

The primary difficulty with the spatial approach is the need to section and analyze column segments. In particular, saturated zeolite materials are difficult to extrude from a column. In our work, therefore, we have used a "scooping" procedure, in which a stainless steel spoon is used to remove material from one end of the column. This simple approach has yielded smooth contaminant mass profiles, with good reproducibility between replicate columns. A similar procedure was reported by Fuhrmann et al. (1995).

Another experimental issue concerns the choice of Sr compounds. For safety and logistical reasons, most researchers have used surrogates for Sr-90, including Sr-85, a gamma-emitter with a half-life of 65 days, and nonradioactive Sr (e.g., $SrCl_2$). The advantage of Sr-85 is that the sorbed activity can be easily measured to very low levels without the need for an extraction step. A disadvantage is that the rapid decay imposes a practical limit on the duration of the experiments and complicates the data interpretation. These difficulties are avoided by using nonradioactive Sr, but an additional extraction step is necessary to remove Sr from the sorbed phase. Our work utilized microwave-assisted digestion with concentrated nitric acid. Despite the advantages of nonradioactive Sr, two significant disadvantages remain: 1) the limited detection limits associated with conventional analytical procedures, and 2) the presence of Sr in natural zeolite, which complicates data interpretation and obscures the leading edge of the contaminant migration front. For these reasons, the source concentration and column duration should be carefully selected to ensure usable data.

4.3 Boundary Conditions

In interpreting the results of advection-dominated column experiments, it is essential that a mass-conserving entrance boundary condition be utilized (e.g., as discussed by Parker and van Genuchten, 1984):

$$C(0,t) = C_{in} + \frac{D}{v}\frac{\partial C}{\partial x}\bigg|_{x=0} \tag{7}$$

where C_{in} is the concentration in the feed solution.

Application of the above condition is especially important for studies of strongly sorbing materials because the reduced penetration of the contaminant results in sharp spatial concentration gradients over the entire duration of the study, which implies a significant dispersive flux at the column entrance. Thus, the calibration of the sorption coefficient is highly sensitive to the estimated dispersion coefficient. This sensitivity is reduced as the duration of the experiment increases, suggesting that the experiments should be performed for as long a period as reasonably possible. Also, an independent estimate for the dispersion coefficient (e.g., through tracer experiments) is needed for a more robust estimate of the sorption parameter(s).

For high-permeability barriers, predictions at the field scale are less sensitive to the form of the specified BCs than low-permeability systems. For most applications, a specified-concentration entrance condition should provide an adequate representation of the source. For the exit condition, the nature of the transition from "sorbing" to "nonsorbing" material suggests the use of a zero-derivative condition, although sensitivity analysis indicates that minimal error is introduced by the use of the common semi-infinite condition, which is more amenable to analytical solution.

4.4 Example: West Valley Demonstration Project

In the fall of 1999, a pilot zeolite barrier was installed at the West Valley Demonstration Project (WVDP) in Western New York. The WVDP is an environmental management project being conducted by the U.S. Department of Energy (DOE) with the cooperation of the New York State Energy Research and Development Authority. Details of the installation are reported by Moore et al. (2000). The clinoptilolite material used in the barrier had previously been studied by Cantrell (1996) for a proposed installation at the DOE Hanford Facility, by Fuhrmann et al. (1995) for use at the WVDP, and by Lee et al. (1998) for the 1998 installation at Chalk River, Ontario. The range of previously estimated distribution coefficients (K_d) was from 650 mL/g (Fuhrmann et al.) to 2600 mL/g (Cantrell). The variation across these studies is most likely attributable to differences in the source water and experimental conditions, although the data interpretation methodology may play a role. For example, Cantrell conducted batch tests using synthetic groundwater and zeolite that had been "prewashed" with a high-calcium solution, while Fuhrmann et al. utilized small columns and site groundwater from the WVDP. For the Fuhrmann et al. study, the unusually high reported dispersion coefficients suggest that the K_d estimation may have been influenced by the boundary effects described above.

In support of the WVDP, eight column tests were conducted at the University at Buffalo using WVDP groundwater spiked with nonradioactive Sr^{2+}, over four durations: 10, 20, 40, and 60 days. A single K_d of 2045 mL/g was calibrated from data from one of the 60-day columns, then used to successively predict the results for the other columns (Figure 5, 10-day data omitted for brevity). The importance of the specified boundary condition was highlighted by comparing results from various calibration schemes. For example, specification of a constant-concentration entrance boundary led to similar model fits but estimated K_d values that were 50% lower. Even when the recommended third-type BC was applied, efforts to simultaneously calibrate both the sorption and dispersion coefficient yielded similar fits for several combinations of parameters. Specification of the dispersion coefficient to a value obtained from an independent tracer test was necessary to obtain a robust estimate of the sorption coefficient.

The reasonable agreement between the K_d values from our study (2045 mL/g) and Cantrell's work (2600 mL/g) suggests that both studies similarly accounted for the effects of calcium competition in the experimental design. Also, the ability of a single K_d to describe data from several column durations (10 to 60 days) indicates that the equilibrium assumption is reasonable. However, the limited number of applicable studies suggests that more research is needed to confirm these conclusions.

Extrapolation of these results to the proposed field scale suggested a barrier life in excess of 25 years, based on the identified performance criteria (effluent Sr-90 < 1000 pCi/L). Work is ongoing to develop a more detailed model based on an explicit consideration of competitive ion-exchange effects using the general framework described above. One advantage of a multi-solute model is that less strongly sorbing solutes (e.g., Ca^{2+}, Mg^{2+}) could be monitored to develop improved models and predictive capability for the breakthrough of Sr-90.

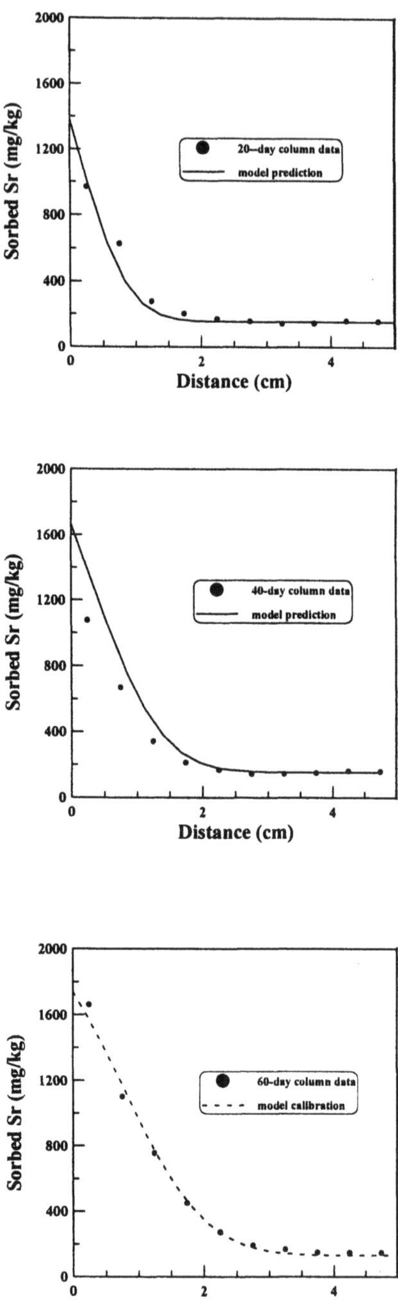

Figure 5. Data and model simulations for WVDP column tests (parameters summarized in Table 4).

Table 4. Conditions for zeolite column experiments

Parameters	Spatial data (Figure 5)			Breakthrough curve (Figure 4)
	20 day	40 day	60 day	
Zeolite length (cm)	11.1	12.2	11.7	13.5
Velocity (m/d)	0.43	0.45	0.45	0.42
Porosity	0.62	0.59	0.59	0.61
Dispersion coefficient (m^2/d)[a]	0.22	0.23	0.23	0.21
Bulk density (g/mL)	0.86	0.80	0.80	0.83
Influent Sr^{2+} (mg/L)	1.0	1.0	1.0	1.0
Influent Ca^{2+} (mg/L)	NA	NA	NA	90
Influent Mg^{2+} (mg/L)	NA	NA	NA	16
Influent K^{+} (mg/L)	NA	NA	NA	38
Influent Na^{+} (mg/L)	NA	NA	NA	60

[a] = estimate based on tracer test, all other parameters measured, NA = not analyzed

5. CLOSING REMARKS

Sorbing barriers represent an attractive alternative to pump-and-treat, provided that the lifetime of the system is sufficient to offset the replacement costs (if necessary). This chapter addresses some of the conceptual and practical issues associated with extrapolating the results of laboratory column tests to the field. In particular, attention to experimental design and mathematical model formulation is considered essential to accurate performance assessment.

In the case of low-permeability sorbing vertical barriers, the field applications lag the theory, at least in North America. At the time of this writing, the authors are aware of only a single installed slurry wall designed for enhanced sorption using flyash as the sorbing additive (Gullick, 1998). Two contributing factors are identified that may partially explain the lack of field implementation:

- For the most part, regulators are reluctant to embrace remediation strategies that rely only on containment (i.e., without some mass removal). The need to supplement the use of vertical barriers with other measures (e.g., pumping) detracts from the overall cost-effectiveness.
- Reliance upon sorbing additives necessitates more refined materials handling and Quality Control testing to ensure a reasonably uniform distribution of the sorbing additive within a slurry wall. This need for precision may complicate the current construction practice for slurry walls (e.g., as summarized by U.S. EPA, 1998) and will likely result in increased construction costs.

The prospects for deployment of high-permeability barriers are considerably greater, as indicated by the recent testing and deployment of zeolite treatment walls in North America. However, research is needed in several areas, including emplacement technology, monitoring techniques, and studies to enhance the current understanding of sorption dynamics, precipitation, and hydraulics over the extended anticipated life of these systems. The development of more sophisticated simulation models represents one promising vehicle for minimizing the need for costly and difficult site-specific experimentation to determine the barrier capacity.

BIBLIOGRAPHY

Acar, Y. B., and Haider, L. (1990). "Transport of low-concentration contaminants in saturated earthen barriers," *Journal of Geotechnical Engineering*, 116, 1031-1052.

Adu-Wusu, K., Whang, J., and McDevitt, M. F. (1997). "Modification of clay-based containment materials," *Proceedings of the 1997 International Containment Technology Conference and Exhibition*, St. Petersburg, FL, Feb. 9-12, 1997, 665-671.

Baker, M. J., Blowes, D. W., and Ptacek, C. J. (1997). "Phosphorus adsorption and precipitation in a permeable reactive wall," *Proceedings of the 1997 International Containment Technology Conference and Exhibition*, St. Petersburg, FL, Feb. 9-12, 1997, 697-703.

Bierck, B. R., and Chang, W. C. (1994). "Contaminant transport through soil-bentonite slurry walls: attenuation by activated carbon," *Proceedings, Water Environment Specialty Conference on Innovative Solutions for Contaminated Site Management*, Miami, FL, March 6-9, 1994, 461-472.

Bradl, H. B. (1997). "Vertical barriers with increased sorption capacities," *Proceedings of the 1997 International Containment Technology Conference and Exhibition*, St. Petersburg, FL, Feb. 9-12, 1997, 645-651.

Cantrell, K. (1996). A permeable reactive wall composed of clinoptilolite for containment of Sr-90 in Hanford groundwater, *Proceedings of the International Topical Meeting on Nuclear and Hazardous Waste Management*, American Nuclear Society, Inc., La Grange Park, IL, 1358-1365.

Cernik, M., Barmettler, K., Grolimund, D., Rohr, W., Borkovec, M., and Sticher, H. (1994). Cation transport in natural porous media on laboratory scale: multicomponent effects, *Journal of Contaminant Hydrology*, 16, 319-337.

Clifford, D. A. (1990). "Ion exchange and inorganic adsorption," in *Water quality and treatment*, McGraw Hill, New York, NY, 561-639.

DePaoli, S. M., and Perona, J. J. (1996). "Model for Sr-Cs-Ca-Mg-Na ion-exchange uptake kinetics on chabazite," *AICHe Journal*, 42(12), 3434-3441.

Furhmann, M., Aloysius, D., Zhou, H (1995). "Permeable subsurface sorbent barrier for Sr: laboratory studies of natural and synthetic materials," *Waste Management*, 15(7), 485-493.

Grathwohl, P. (1998). *Diffusion in natural porous media: contaminant transport, sorption/desorption and dissolution kinetics*, Kluwer Publishers, Boston, MA.

Gray, D. H., and Weber, W. J. (1984). "Diffusional transport of hazardous waste leachate across clay barriers," *Proceedings of the 7th Annual Waste Conference*, Sept 11-12, Univ. of Wisconsin, Madison, 373-389.

Gullick, R. W. (1998). "Effect of sorbent addition on the transport of inorganic and organic chemicals in soil-bentonite cutoff wall containment barriers," Ph.D. dissertation, University of Michigan, Ann Arbor, MI.

Khandelwal, A., Rabideau, A. J., and Shen, P. (1998). "Analysis of diffusion and sorption of volatile organic contaminants in soil/bentonite barrier materials," *Environmental Science & Technology*, 32(9), 1333-1339.

Khandelwal, A., and Rabideau, A. J. (1998). "Enhancement of soil/bentonite barrier performance by the addition of natural humus," *Proceedings of the 4th International Symposium on Environmental Geotechnology*, Boston, MA, 9-12 August, 1998.

Khandelwal, A., and Rabideau, A. J. (2000), "Amendment of soil/bentonite slurry walls with natural humus," *Journal of Contaminant Hydrology*, 25, 267-282.

Lee, D. R., Smith, D. J., Shikaze, S. G., Jowett, R., Hartwig, D. S., and Milloy, C. (1998). "Wall-and-curtain for passive collection/treatment of contaminant plumes," *Proceedings of the First International Conference on Remediation of Chlorinated and Recalcitrant Compounds*, Battelle Press, Columbus, OH.

Ma, Z., Whitley, R. D., and Wang, N. (1996). "Pore and surface diffusion in multicomponent adsorption and liquid chromatography systems," *AIChe Journal*, 42(5), 1244-1262.

Miller, C. T., and Rabideau, A. J. (1993). "Development of split-operator Petrov-Galerkin methods for simulating transport and diffusion problems," *Water Resources Research*, 29(7), 2227-2240.

Moore, H. R., Steiner, R. E., Fallon, E., Repp, C. L., Hemann, M. R., and Rabideau, A. J. (2000). "Permeable treatment wall pilot project at the West Valley Demonstration Project," to be presented at *Waste Management 2000*, 27 February - 2 March, 2000, Tucson, AZ.

Mott, H. V. and Weber, W. J., Jr. (1992). "Sorption of low molecular weight organic contaminants by fly ash: considerations for the enhancement of cutoff barrier performance," *Environmental Science & Technology*, 26, 1234-1241.

National Research Council (1994). *Alternatives for ground water cleanup*, National Academy Press, Washington, D. C.

Park, Jae K., Kim, Jae Y., Madsen, C. D., and Edil, T. B. (1997). "Retardation of volatile organic compounds movement by a soil-bentonite slurry wall amended with ground tires," *Water Environment Research*, 69(5), 1022-1031.

Parker, J. C., and van Genuchten, M. Th. (1984). "Flux-averaged and volume-averaged concentrations in continuum approaches to solute transport," *Water Resources Research*, 20(7), 866-872.

Perona, J. J. (1993). Model for Sr-Cs-Ca-Mg-Na ion exchange equilibria on chabazite, *AIChe Journal*, 39, 1716-1720.

Rabideau, A. J., and Khandelwal, A. (1998a). "Analysis of nonequilibrium sorption in soil/bentonite slurry walls," *ASCE Journal of Environmental Engineering*, 124(4), 329-335.

Rabideau, A. J., and Khandelwal, A. (1998b). "Boundary conditions for modeling contaminant transport in vertical barriers," *ASCE Journal of Environmental Engineering*, 124(11), 1135-1141.

Rabideau, A. J. (1996). "Contaminant transport modeling," in *Assessment of barrier containment technologies*, Rumer, R., and Mitchell, J., eds., NTIS #PB96-5083.

Robinson, S. M., Arnold, W. D., and Byers, C. H. (1994). "Mass transfer mechanisms for zeolite ion exchange in wastewater treatment," *AIChe Journal*, 40(12), 2045-2054.

Schad, H., and Gratwohl, P. (1998). "Funnel-and-gate systems for in situ treatment of contaminated groundwater at former manufactured gas sites," *Proceedings of NATO/CMS Phase III meeting*, 23-24 February, 1998, Vienna, Austria.

Shackelford, C. D. (1989). "Diffusion of contaminants through waste containment barriers," *Transportation Research Record*, 1219, 169-182.

Shackelford, C.D., and Redmond, P. L. (1995). "Solute breakthrough curves for processed kaolin at low flow rates," *J. Geotech. Eng.*, 121(1), 17-32.

Smith, J. A., and Jaffe, P. R. (1994). "Benzene transport through landfill liners containing organophilic bentonite," *Journal of Environmental Engineering*, 120, 6, 1559-1577.

U.S. Environmental Protection Agency (1998). "Evaluation of engineered subsurface barriers at waste sites," EPA 542-R-98-005, Office of Solid Waste and Emergency Response, Washington, DC.

Valocchi, A. J., Street, R. L., and Roberts, P. V. (1981). "Transport of ion-exchanging solutes in groundwater: chromatographic theory and field simulations," Water Resources Research, 17(5), 1517 – 1527.

Watson, D. Leavitt, M., Smith, C., Klasson, T., Bostick, B., Liang, L., and Moss, D. (1997). "Bear Creek valley characterization area mixed wastes passive in situ treatment technology demonstration project - status report," *Proceedings of the 1997 International Containment Technology Conference and Exhibition*, St. Petersburg, FL, Feb. 9-12, 1997, 730-736.

Whang, J. M., Adu-Wusu, K., Frampton, W, H., Staib, J. G. (1997). "In situ precipitation and sorption of arsenic from groundwater: laboratory and ex situ field tests," *Proceedings of the 1997 International Containment Technology Conference and Exhibition*, St. Petersburg, FL, Feb. 9-12, 1997, 737-743.

Williamson, D. F., Hoenke, K., Wyatt, J., and Anderson, J. (2000). "Construction of a funnel-and-gate treatment system for pesticide-contaminated groundwater," *2nd International Conference on Remediation of chlorinated and recalcitrant compounds*," 22-25 May, 2000, Monterey, CA.

Chapter Seven

Reductive Dechlorination Of Chlorinated Solvents On Zerovalent Iron Surfaces

BAOLIN DENG and SHAODONG HU

Department of Environmental Engineering
New Mexico Institute of Mining and Technology
Socorro, NM 87801

Key words: zerovalent iron; trichloroethylene; reactive barrier

Abstract: Optimal design of zerovalent iron-based permeable reactive barriers requires a complete understanding of dechlorination kinetics and mechanism. The effect of other ambient constituents, which may retard or enhance the dechlorination processes, should be considered. Literature has revealed that reduction of chlorinated compounds occurs on the iron surface and the reaction rate is limited by surface processes, rather than transport processes. Adsorption onto the surface can take place on both reactive sites that are responsible for the reductive dechlorination, and nonreactive sites that only sequester the contaminants. This chapter explores a model based on the assumptions that adsorption equilibrium on the two types of surface sites is always maintained, but the reduction rate is directly proportional to the amount sorbed onto reactive sites only. Numerical solutions are obtained to illustrate the effect of coadsorbates on the adsorption and reduction of chlorinated compounds under this mechanistic framework.

1. INTRODUCTION

Chlorinated compounds such as trichloroethylene (TCE) and perchloroethylene (PCE) have traditionally been used as solvents and degreasing agents in many industries. Intentional and inadvertent releases of chlorinated compounds have resulted in extensive soil and groundwater contamination. The prevailing method to restore groundwater contaminated with chlorinated solvents has been through pump-and-treat, i.e., the contaminated water is pumped out and treated at the surface. Although this approach is useful for some simple contamination sites, it is expensive and has failed to restore waste sites with complex hydrogeochemical settings and contamination history (Mackay and Cherry, 1989; NRC, 1994). Typically, contaminant concentrations in the water pumped out have a quick initial decline, then remain stable for an extended period of time (Cherry et al., 1996). In other instances, pumping may remove the dissolved contaminants, but contaminant concentrations can rebound in the groundwater once the pumping is stopped, partially due to the presence of denser-than-water non-aqueous phase liquids (DNAPLs) (NRC, 1994).

A timely "discovery" in 1990 by a group of researchers at the University of Waterloo helped to relieve some frustration with chlorinated solvent remediation (Reynolds et al., 1990). When testing materials for the construction of groundwater monitoring wells, the researchers at Waterloo found that halogenated compounds could be reductively transformed by metallic materials including stainless steel. Gillham's group recognized the significance of this zerovalent iron (ZVI)-based transformation process and proposed that a permeable reactive iron barrier could be built to remediate groundwater contaminated with chlorinated hydrocarbons (Gillham and O'Hannesin, 1994). The essence of the permeable barrier technology was that when contaminated groundwater passed through the reactive media, dissolved chlorinated compounds would be reduced by ZVI, forming non-toxic products. Although it was found later on that ZVI had been used earlier for treating industrial waste water with chlorinated compounds (Sweeny, 1981; Senzaki and Kumagai, 1988), the idea of building a wall against contaminants has created great excitement in environmental research communities in the past several years. By now, dozens of relevant papers have been published on the topic. The technology has also quickly gone from laboratory study to pilot testing and to full field installations. More than a dozen full-scale permeable reactive iron barriers have been installed and have successfully eliminated spreading of contamination plumes at those sites (Focht et al., 1996; EPA, 1998).

Our understanding of the iron reactive barrier is, however, still limited. Degradation rates of many chlorinated compounds have been reported, but the exact reaction kinetics and mechanism are much less understood. The long-term stability of the reactive barrier and the effects of other system constituents on the performance of the barrier should also be properly evaluated. This chapter will first review the kinetics and mechanism of reductive dechlorination reactions on ZVI, and then focus on the development of kinetic models for systems with varying complexities. A case study showing the utilities of a two-site model will be presented. The goal here is to provide a kinetic framework by which various factors influencing the reactive dechlorination reactions can be evaluated.

2. REDUCTIVE DECHLORINATION

Studies of reductive dechlorination by zerovalent iron has become active only very recently, but great progress has been made in the last several years as illustrated by works of Gillham and O'Hanesin (1994), Matheson and Tratnyek (1994), Burris et al. (1995), and Roberts et al. (1996).

2.1 Degradable Organic Contaminants by ZVI

ZVI and other metals (e.g., Zn, Cu, and Al) are strong reducing agents at ambient temperature, capable of reducing many halogenated methanes, ethanes, and ethenes and other halogenated organic compounds (Sweeny, 1979; Senzaki and Kumagai, 1988; Gillham and O'Hannesin, 1994; Matheson and Tratnyek, 1994; Burris et al., 1995; Roberts et al., 1996). To illustrate, some of the halogenated compounds degradable in the presence of ZVI are listed in Table 1. The list includes many compounds with significant environmental concerns, as indicated by the fact that eight of them are among the 25 most frequently detected ground water contaminants at hazardous waste sites (NRC, 1994).

The rates of reductive dechlorination reactions differ significantly among various compounds. Chlorinated ethanes and methanes are degraded very quickly, with the half-lives less than an hour for most of the compounds (Gavaskar et al., 1998). For chlorinated ethylenes, the half-lives of degradation normalized to 1 m^2 iron surface/ml solution ranges from less than an hour to several days (Johnson et al., 1996; Gavaskar et al., 1998). These degradation rates are orders of magnitude greater than the natural rates of dehalogenation (Vogel et al., 1987). Nevertheless, there are a few compounds that do not demonstrate noticeable degradation in the presence of ZVI, including chloromethane, dichloromethane, 1,2-dichloroethane, and 1,4-dichlorobenzene (EPA, 1998; Gavaskar et al., 1998).

Table 1. Degradability of halogenated organic compounds in the presence of zerovalent iron (Adapted from Gavaskar et al., 1998 and EPA, 1998). Compounds indicated in bold face are among the 25 most frequently detected groundwater contaminants at hazardous waste sites (NRC, 1994).

Methanes	Carbon Tetrachloride; **Trichloroethane**; Bromoform
Ethanes	Hexachloroethane; 1,1,2,2-Tetrachloroethane; 1,1,1,2-Tetrachloroethane; **1,1,1-Trichloroethane; 1,1-Dichloroethane**
Ethylenes	**Tetrachloroethylene (PCE); Trichloroethylene (TCE); 1,1-Dichloroethylene; trans-1,2-Dichloroethylene;** cis-1,2-Dichloroethylene; **Vinyl Chloride**
Propanes	1,2,3-Trichloropropane; 1,2-Dichloropropane; 1,3-Dichloropropane
Other Degradable Compounds	1,1,2-Trichlorotrifluoroethane; 1,2-Dibromo-3-chloropropane; 1,2-Dibromoethane; n-Nitrosodimethylamine (NDMA); Nitrobenzene

2.2 Dechlorination Pathways

In order to apply ZVI-based dechlorination for site remediation, we have to be sure that no toxic reaction products are produced in the end, and toxic intermediates, if there are any, are degradable. Product distribution is controlled by reaction mechanisms. For the reduction of chlorinated methanes such as carbon tetrachloride (CT), hydrogenolysis, where chlorines are sequentially replaced by hydrogen, takes place in the presence of ZVI (Vogel et al., 1987; Matheson and Tratnyek, 1994; Glod et al., 1997):

$$CCl_4 \xrightarrow{2e^- + H^+} CHCl_3 + Cl^- \xrightarrow{2e^- + H^+} CH_2Cl_2 + 2Cl^-$$

With each successive dechlorination, the reaction proceeds more slowly, and dichloromethane is not reduced by ZVI under ambient conditions (Matheson and Tratnyek, 1994).

During the reductive dechlorination of chloroethylenes such as PCE and TCE, a suite of reduction products have been identified, including *trans-* and *cis-* dichloroethylenes, vinyl chloride, methane, ethane, ethylene, acetylene, and small amounts of C_3-C_6 hydrocarbons (Burris et al., 1995; Orth and Gillham, 1996). Dichloroethylene isomers and vinyl chloride can be further reduced in the ZVI system, forming ethene and ethane (for dichlorothylene) and ethene (for vinyl chloride) as main products (Roberts et al., 1996; Allen-King et al., 1997; Deng et al., 1999). Less chlorinated ethylenes are reduced slower than the highly chlorinated ethylenes, similar to the reduction of chlorinated ethanes. The observed distribution of reaction intermediates and products with chloroethylenes suggests that the reduction proceed via more than one pathway: in addition to hydrogenolysis that has long been recognized, reductive β-elimination pathway is also important (Roberts et al., 1996). TCE reduction via both pathways is illustrated in Figure 1. The route of hydrogenolysis produces *trans-* and *cis*-dichloroethylenes as intermediates, and further reactions generate vinyl chloride and acetylene, and finally to ethene and ethane. The route of β-elimination will produce acetylene with chloroacetylene as an intermediate. Chloroacetylene has actually been detected by GC-MS, supporting the presence of the reductive elimination pathway (Campbell et al., 1997). Chloroacetylene is highly reactive (Arnold and Robert, 1998), so that it does not contribute significantly to the overall mass balance observed during TCE degradation.

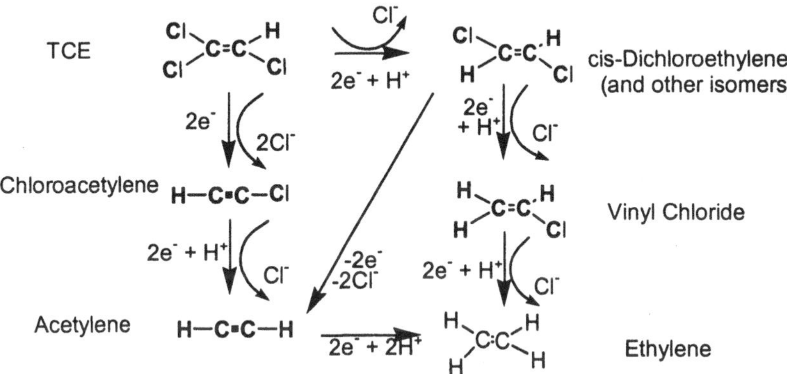

Figure 1. Reduction of trichloroethylene on ZVI surfaces through hydrogenolysis and reductive elimination pathways (adapted from Roberts et al., 1996).

2.3 Couplings of Iron Corrosion and Dechlorination

The reduction of chlorinated compounds is coupled with iron oxidation, or corrosion, as indicated by the reaction 1:

$$Fe^{o} + RCl + H_3O^{+} \Rightarrow Fe^{2+} + RH + Cl^{-} + H_2O \quad (1)$$

This dechlorination process, however, is just one of the corrosion reactions occurring in the ZVI-water system. The iron can react with water itself or with dissolved oxygen according to the following reaction stoichiometries:

$$Fe^{o} + 2H_2O \Rightarrow Fe^{2+} + H_2 + 2OH^{-} \quad (2)$$

$$2Fe^{o} + O_2 + 2H_2O => 2Fe^{2+} + 4OH^{-} \quad (3)$$

In the presence of excess dissolved oxygen, Fe^{2+} may further be oxidized to form iron oxides such as goethite and hematite. There are, therefore, at least three reductants in the ZVI-water system that could potentially reduce chlorinated compounds: (1) ZVI on the surface, (2) Fe^{2+} generated by iron corrosion, and (3) H_2 generated by iron corrosion (Matheson and Tratnyek, 1994). Direct reduction on ZVI surfaces is most likely to be the mechanism responsible for the dechlorination reaction (Matheson and Tratnyek, 1994). The debate on the role of Fe^{2+} for the dechlorination reaction is, however, still going on. The main question is whether Fe^{2+} adsorbed on ZVI can act as a reductant and thus serve as a catalyst for the reaction between chlorinated compounds and ZVI. Surface-bound Fe^{2+} is believed to be a much stronger reductant than the aqueous Fe^{2+} (Stumm and Sulzberger, 1992; Johnson et al., 1998). Reductive dechlorination by some iron sulfide minerals has been observed (Sivavec and Horney, 1997), suggesting that surface-bound Fe^{2+} is capable of reducing chlorinated compounds. However, batch experimental systems set up by mixing aqueous Fe^{2+} and a number of mineral surfaces (e.g., goethite, hematite, aluminum oxide, kaolinite) do not show any discernible TCE degradation in two weeks (Deng and Burris, unpublished data). In a recent experimental study on vinyl chloride reduction, additions of various Fe^{2+}/Fe^{3+} chelating reagents (1,10-phenanthroline, 2,2'-dipyridyl and nitrilotriacetic acid) do not alter the rate of vinyl chloride reduction by ZVI (Deng et al., 1999). These results suggest that the effect of Fe^{2+} on the reductive dechlorination reaction may not be significant. Identification of reactive phases and sites using microscopic (e.g., AFM, TEM) and spectroscopic (e.g., XPS) techniques may help to clarify the role of Fe^{2+}.

2.4 Adsorption as a Key Step for Dechlorination

Direct reduction of chlorinated compounds on ZVI requires the organic compounds to adsorb onto the surfaces before the electron transfer can take place. In this surface-mediated reaction, a sequence of physical and chemical steps are involved: (1) movement of reactant molecules into the interfacial region by convection and diffusion; (2) diffusion of reactant molecules within the interfacial region; (3) surface chemical reaction such as ligand exchange, electron and group transfers, addition or elimination reactions; and (4) outward movement of product molecules from the interfacial region to bulk solution (Stone and Morgan, 1990). The overall reaction rate depends upon one or more of these steps. Scherer et al. (1997) have examined the kinetics of carbon tetrachloride reduction at an oxide-free iron electrode and found the overall reaction is limited by the chemical process on the surface. Measuring activation energy of a surface reaction is another way to evaluate whether the reaction is controlled by mass transfer or surface reaction. Reactions that are controlled by diffusion in aqueous media normally have activation energies less than 20 kJ/mol (Lasaga, 1981). Reported activation energies are 42 kJ/mol for vinyl chloride (Deng et al., 1999) and 32.2 to 39.4 kJ/mol for trichloroethylene (Su and Puls, 1999), which indicate that surface processes, rather than mass transfer arriving and leaving the surface, control the overall rate of dechlorination.

Evaluating the adsorption of chlorinated compounds on ZVI is a challenging task because adsorption and reduction occur simultaneously. Burris et al. first attempted to measure sorption of PCE and TCE on ZVI (Burris et al., 1995). In that study, the amount adsorbed was estimated as the difference between aqueous contaminant concentration and total amount after acetonitrile extraction. Both PCE and TCE was found to exhibit nonlinear sorption behaviour and could be fitted by Langmuir adsorption isotherm. Competition for sorption sites between the two contaminants was observed, but not for the reactive sites of reductive dechlorination (Burris et al., 1995). The results suggest that surface sites for reductive dechlorination and sites for adsorption are not exactly the same, i.e., more than one kinds of surface sites exist. This is not surprising since solid surfaces are intrinsically heterogeneous, hosting a wide variety of surface sites with different affinities for adsorbates. For cast irons used in the experiments (Burris et al., 1995) and reactive barrier construction, carbon impurities including graphite can be as high as 3% (wt), which could be a very important phase for contaminant adsorption (Deng et al., 1997).

Sorption isotherms of less chlorinated ethylenes, *trans*- and *cis*-dichloroethylenes, are also nonlinear, but the sorbed amounts are much less than those of TCE and PCE (Allen-King et al., 1997). Vinyl chloride is the least chlorinated one in this series, thus its sorption onto ZVI is also the least (Burris et al., 1998; Deng et al., 1999).

2.5 Effect of Other Constituents on the Reductive Dechlorination

Studies on the reductive dechlorination on ZVI surfaces are mostly performed in systems containing only chlorinated compounds and iron in the aqueous phase. Processes actually occurring in reactive iron barriers, however, could be much more complex due to the presence of other chemical constituents in the system. The ambient constitutes may include anthropogenic contaminants that are co-disposed with chlorinated compounds and natural occurring constituents in soils and groundwater (e.g., natural organic matter, HCO_3^-, HS^-). These chemicals may compete with chlorinated solvents for the adsorption sites on ZVI and thus inhibit the reductive dechlorination, or facilitate the iron corrosion and thus enhance the rate of dechlorination. For example, the reduction rate of carbon tetrachloride is decreased by both redox active ligands (catechol and ascorbate) and non-redox active ligands (EDTA and acetate) (Johnson et al., 1998). Various natural organic matters also inhibit carbon chloride reduction (Tratnyek and Scherer, 1998). Site-specific hydrogeochemical information should be evaluated in the design of iron-based reactive barriers.

3. REDUCTION KINETICS: SINGLE-SITE VERSUS TWO-SITE MODEL

Reductive dechlorination by ZVI is often modelled by pseudo-first-order kinetics:

$$-\frac{d[X]_{aq}}{dt} = k_{obs}[X]_{aq} \tag{4}$$

where $[X]_{aq}$ is the contaminant concentration in the aqueous phase and k_{obs} is the observed first order rate constant. Such an approach can represent experimental data reasonably well and is simple to use for designing reactive barriers (Johnson et al., 1996). The model, however, is not a realistic representation of the processes occurring in the system, since the kinetics is expressed as being proportional to the aqueous reactant concentration, but the reaction does not actually occur in the aqueous phase. Recognizing that the dechlorination reactions occur on the iron surfaces, Johnson et al. (1996, 1998) proposed a model, assuming that the dechlorination of a substrate, X, involves reversible adsorption on iron surface (= Fe), followed by rate-limiting electron transfer to produce products:

$$X + \equiv Fe \underset{k_2}{\overset{k_1}{\longleftrightarrow}} \equiv Fe - X \tag{5}$$

$$\equiv Fe - X \xrightarrow{k_3} PRODUCTS \tag{6}$$

where =Fe-X is the adsorbed substrate on the surfaces. Probability of adsorption onto various surface sites is the same since all surface sites are assumed to be energetically identical. Adsorption processes are known to be faster than dechlorination steps (Burris et al., 1995; Scherer et al., 1997; Allen-King et al., 1997; Deng et al., 1999), thus the assumption of reversible adsorption is reasonable. The model of this type is one of a few theories commonly used to explain the saturation kinetics with respect to the substrate X (Wilkins, 1991). The effect of co-adsorbates on the dechlorination reaction has also been incorporated into the model by adding a competitive adsorption term (Johnson et al., 1998).

A general kinetic model should accommodate all chemical processes known to affect the dechlorination process. These include: (1) reductive dechlorination takes place on the iron surface, rather than in the aqueous phase, so adsorption must occur; (2) other components in the system may affect the dechlorination reaction by competing for the reaction sites; (3) surface sites for reduction and for sorption may not be the same, as for the system with PCE and TCE where dechlorination takes place on the reactive sites, but most of the adsorption is clearly on the nonreactive sites (Burris et al., 1995). In the following section we will first discuss a single-site model similar to the one used by Johnson et al. (1998), which has accounted for the first two observations, then develop a two-site model that will also take the third observation into consideration. We aim to illustrate how coadsorbates in the iron system will affect adsorption and reduction of chlorinated solvents. TCE will be used as an example since relevant adsorption and reduction data are available, from which the required parameters for simulation could be estimated.

3.1 Single-Site Model

A single site model assumes only one type of surface sites, i.e., the same type of sites is responsible for trichloroethylene adsorption and reduction. The kinetic model developed herein is based on the following reaction scheme:

$$S + X \longleftrightarrow SX \qquad \frac{k_1}{k_{-1}} = K_{SX} \tag{7}$$

$$S + I \longleftrightarrow SI \qquad \frac{k_2}{k_{2}} = K_{SI} \tag{8}$$

$$SX \xrightarrow{k_3} P + S \tag{9}$$

where S = surface sites on ZVI; X = chlorinated compound (e.g., TCE); I = coadsorbate (e.g., natural organic matter, S^{2-} species); SX = TCE sorbed on the surface; SI = coadsorbate sorbed on the surface; and P = products formed. It is assumed that adsorption reactions occur fast enough that adsorption equilibrium has established for both TCE and coadsorbate, with

K_{SX} and K_{SI} as the equilibrium constants, respectively. Irreversible reduction of TCE occurs, with the rate directly proportional to the amount adsorbed onto the surface (SX), while no coadsorbate reduction takes place. The model is more realistic than the empirical expression in Eq. 1 by representing the reaction based on surface species. Following this reaction scheme, concentration changes of various species as a function time can be represented by the differential equations below:

$$\frac{d[S]}{dt} = -k_1[S][X] + k_{-1}[SX] - k_2[S][I] + k_{-2}[SI] + k_3[SX] \tag{10}$$

$$\frac{d[X]}{dt} = -k_1[S][X] + k_{-1}[SX] \tag{11}$$

$$\frac{d[SX]}{dt} = k_1[S][X] - k_{-1}[SX] - k_3[SX] \tag{12}$$

$$\frac{d[I]}{dt} = -k_2[S][I] + k_{-2}[SI] \tag{13}$$

$$\frac{d[SI]}{dt} = k_2[S][I] - k_{-2}[SI] \tag{14}$$

$$\frac{d[P]}{dt} = k_3[SX] \tag{15}$$

Since analytical solutions for this set of coupled differential equations is not easily available in a general form, the adsorption and reduction of TCE are simulated numerically. A modelling software package, Scientist for Windows$^{(R)}$ (Micromath Scientific Software, Salt Lake City, UT, USA), has been used for the numerical simulation. Initial TCE concentration is set to be $1.0\text{x}10^{-3}$ M, which is close to the concentration used in the experiment by Burris et al. (1995), and the adsorption equilibrium constant (K_{SX}) is estimated according to the measured adsorption isotherm in the same study. The kinetic constant for the electron transfer step (k_3) is selected so that the simulated time period for TCE degradation is close to what has been observed in the study. In the system with coadsorbate, the initial concentration of coadsorbate and the adsorption equilibrium constant K_{SI} are set to be the same as for TCE. Adsorption equilibrium is modelled as a kinetic problem by setting the forward rate constants arbitrarily large and then calculating the backward rate constants based on the principles of detailed equilibrium (i.e., $k_{-i} = k_i/K_i$) (Prigogine, 1967). The modelling procedure using the software package involves creating a "model" file for the model to be used, a "parameter set" file containing modelling parameters, and a "spreadsheet" file for data output and plotting. Figure 2 illustrates some simulation results, as well as the parameters used for modelling. A "model' file for the single-site model with a coadsorbate (used to generate Figure 2b) is reproduced as Appendix A, to illustrate how a mechanistic model can be conveniently represented mathematically in this software package. The simulation results indicate that without coadsorbate competing for the surface sites (Figure 2a), TCE adsorbed accounts for a significant portion of the total TCE in the system. The adsorbed amount remains relatively constant until the aqueous TCE concentration has significantly decreased, then adsorbed TCE concentrations begin to decrease. In the presence of a coadsorbate (Fig. 2b), the amount of TCE sorbed on the surface is much less due to the competition from the coadsorbate. The overall rate of TCE reduction is also significantly decreased, due to the decreased TCE adsorption. With the single site model, the reductive dechlorination is directly proportional to the total amount adsorbed, so coadsorbate that competes strongly for adsorption sites will retard TCE reduction.

3.2 Two-Site Model

In order to explain the degradation kinetics of TCE and PCE, for which the adsorption onto the nonreactive sites is significant (Burris et al., 1995), a two-site model is developed. The basic assumption for the single-site model, i.e., pre-adsorption equilibrium followed by reductive dechlorination, is still valid here. In addition, the two-site model assumes that there are both reactive and nonreactive sites on the iron surface, and while the adsorption of TCE and coadsorbate can occur on both types of sites, reductive dechlorination of TCE only takes place on the reactive sites. Coadsorbate is not involved in redox reactions. The reaction scheme for this model is:

$$S + X \longleftrightarrow SX \qquad \frac{k_1}{k_{-1}} = K_{SX} \tag{16}$$

$$N + X \longleftrightarrow NX \qquad \frac{j_1}{j_{-1}} = K_{NX} \tag{17}$$

$$S + I \longleftrightarrow SI \qquad \frac{k_2}{k_{-2}} = K_{SI} \tag{18}$$

$$N + I \longleftrightarrow NI \qquad \frac{j_2}{j_{-2}} = K_{NI} \tag{19}$$

$$SX \xrightarrow{k_3} P + S \tag{20}$$

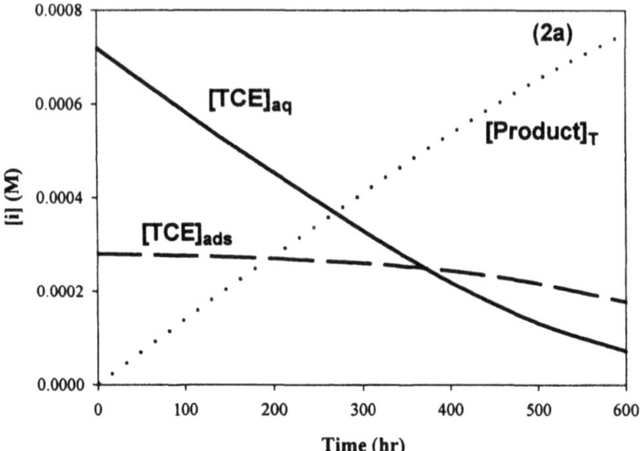

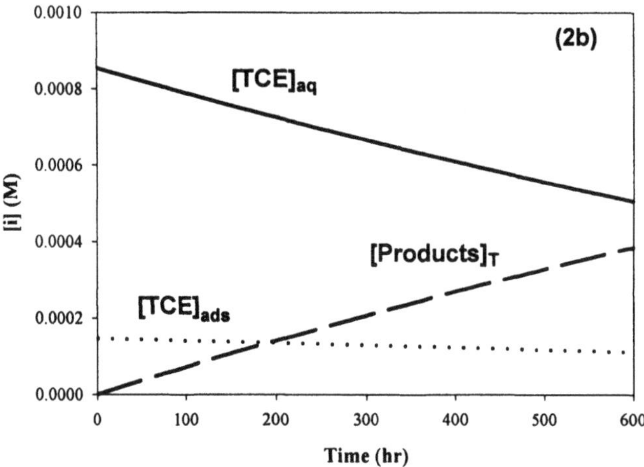

Figure 2. Single-site model simulation for the adsorption and reduction of TCE on zerovalent iron surfaces. Modelling parameters are:

without coadsorbate (2a)	**with coadsorbate (2b)**
$[TCE]_0 = 1\times10^{-3}$ M	$[TCE]_0 = [I]_0 = 1\times10^{-3}$ M
$K_{SX} = 2\times10^4$ l/mol	$K_{SX}=K_{SI} = 2\times10^4$ l/mol
$[S]_T = 3\times10^{-4}$ mol/l	$[S]_T = 3\times10^{-4}$ mol/l
$k_3 = 5\times10^{-3}$ hr^{-1}	$k_3 = 5\times10^{-3}$ hr^{-1}

where S = reactive sites; N = nonreactive sites; X = TCE; I = coadsorbate; SX = TCE adsorbed onto reactive sites; NX = TCE adsorbed onto nonreactive sites; SI = coadsorbate sorbed onto reactive sites; and NI = coadsorbate sorbed onto nonreactive sites. Again, adsorption reactions are assumed to occur fast so the equilibrium has established, and reductive dechlorination is directly proportional to the amount adsorbed on the reactive sites. Differential equations describing the kinetics as determined by this reaction scheme can be written as:

$$\frac{d[S]}{dt}=-k_1[S][X]+k_{-1}[SX]-k_2[S][I]+k_{-2}[SI]+k_3[SX] \tag{21}$$

$$\frac{d[N]}{dt}=-j_1[N][X]+j_{-1}[NX]-j_2[N][I]+j_{-2}[NI] \tag{22}$$

$$\frac{d[X]}{dt}=-k_1[S][X]+k_{-1}[SX]-j_1[N][X]+j_{-1}[NX] \tag{23}$$

$$\frac{d[SX]}{dt}=k_1[S][X]-k_{-1}[SX]-k_3[SX] \tag{24}$$

$$\frac{d[NX]}{dt}=j_1[N][X]-j_{-1}[NX] \tag{25}$$

$$\frac{d[I]}{dt}=-k_2[S][I]+k_{-2}[SI]-j_2[N][I]+j_{-2}[NI] \tag{26}$$

$$\frac{d[SI]}{dt}=k_2[S][I]-k_{-2}[SI] \tag{27}$$

$$\frac{d[NI]}{dt}=j_2[N][I]-j_{-2}[NI] \tag{28}$$

$$\frac{d[P]}{dt}=k_3[SX] \tag{29}$$

Apparently, analytical solution for these coupled equations is not feasible and we have to use numerical approaches. Model parameters are obtained similar to the single-site model. Additionally, since the adsorption experiments can only provide information on the total concentration of surface sites, we need to know the relative abundance of reactive versus nonreactive sites. That has not been determined unambiguously through experiments, so we simply assume that the reactive sites account for 10% of the total surface sites.

Simulations based on the two-site model, as well as parameters used for simulation, are illustrated in Figure 3. In the absence of coadsorbate, concentration v.s. time profiles similar to those from the single-site model can be obtained (Figure 3a), in spite of the fact that a dominant portion of adsorbed TCE is on nonreactive sites, and only a small portion goes to reactive sites. That is because the rate constant k_3 is adjusted according to the overall half-life of TCE degradation, and its higher value compensates for the lower adsorbed TCE concentration on the reactive sites. In the presence of a coadsorbate that might preferentially adsorb onto the reactive sites (see Figure 3b), reductive dechlorination could be blocked almost totally, while the total amount of adsorption remains unchanged in comparison to the single-site system. This result can't be derived from a single-site model. To illustrate the applicability of the two-site model, a case study on TCE degradation in the presence of a coadsorbate, cysteine, will be discussed below.

3.3 TCE Reduction in the Presence of Cysteine

Cysteine [$(NH_3^+)CH(COO^-)CH_2SH$] is an amino acid containing a thiol (-SH) functional group. The compound is not expected to be a major component in groundwater, but other reduced sulfide species like HS^- may demonstrate similar complexation and redox reactivities. Sulfide is present in many natural anaerobic environments. Microbial sulfate reduction is the main mechanism for sulfide production, which could also be the case in the highly-reducing zone created by iron barriers. A preliminary study has shown that among more than a dozen inorganic and organic chemicals tested for their effects on iron corrosion and TCE reduction, cysteine and adenine demonstrated the strongest inhibitory effect on iron corrosion and TCE reduction (Deng et al., 1998).

The effect of cysteine on TCE adsorption is illustrated in Figure 4. The experimental systems consisted of 5.00 g of iron filings (acid washed, Fisher) in 72 ml bottles, filled with solutions of TCE and various concentrations of cysteine. Each sample bottle was first shaken by hand for 10 min, followed by mixing on an incubator shaker (Innova 4335, New Brunswick Scientific) operated at 25 °C and a speed of 100 rpm. TCE concentrations were analysed at 60 min after setup of the system, which was long enough to establish a TCE sorption quasi-equilibrium and short enough to minimize significant TCE reduction. The difference of the aqueous TCE concentrations ($[TCE]_{aq}$) between the control and the samples was indicative of TCE adsorption. With a total initial TCE concentration of 336 μM, the results indicate a near constant $[TCE]_{aq}$ when different concentrations of cysteine are present. The data on the amount adsorbed are somewhat scattered, but this is reasonable considering the low fraction of TCE adsorption that is determined. It appears that TCE adsorption is not significantly affected by

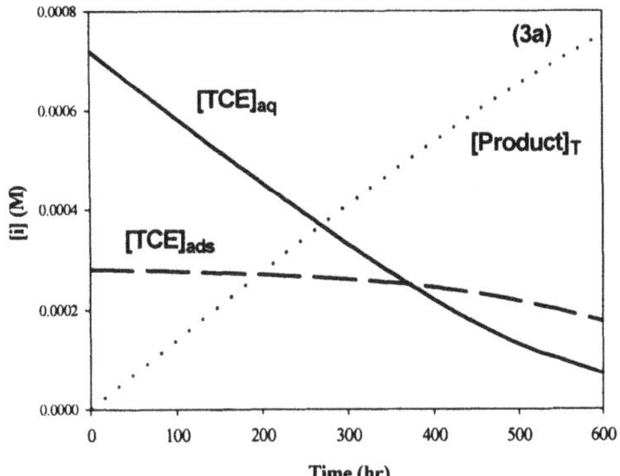

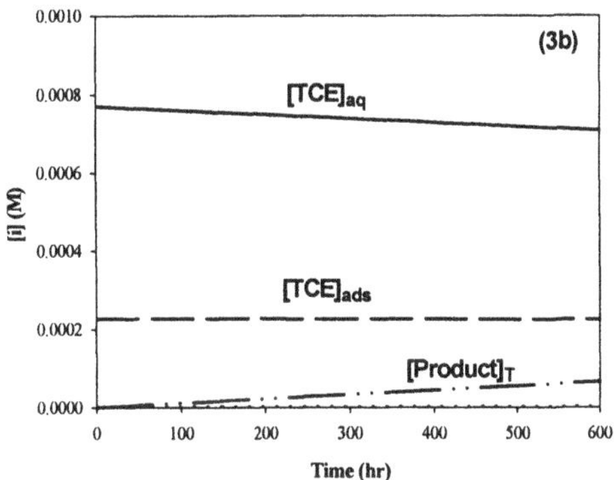

Figure 3. Simulation based on two-site model for the adsorption and reduction of TCE on zerovalent iron surfaces. Modelling parameters are:

without coadsorbate (3a)	**with coadsorbate (3b)**
$[TCE]_0 = 1\times10^{-3}$ M	$[TCE]_0 = [I]_0 = 1\times10^{-3}$ M
$K_{SX} = 2\times10^4$ l/mol	$K_{SX}=2\times10^4$ l/mol; $K_{SI}=2\times10^5$ l/mol;
$K_{NX} = 2\times10^4$ l/mol	$K_{NX}=2\times10^4$ l/mol; $K_{NI}=2\times10^3$ l/mol
$[S]_T = 0.3\times10^{-4}$ mol/l	$[S]_T = 0.3\times10^{-4}$ mol/l
$[N]_T = 2.7\times10^{-4}$ mol/l	$[N]_T = 2.7\times10^{-4}$ mol/l
$k_3 = 5\times10^{-2}$ hr^{-1}	$k_3 = 5\times10^{-2}$ hr^{-1}

cysteine up to a concentration of 1.0 mM. The calculated amount of TCE sorbed is approximately 40 μM or 0.53 μmol/m^2 of Fe surfaces.

The effect of cysteine on TCE degradation was examined in bottles with 35.0 ml aqueous phase and 37.0 ml headspace. The experiments were set up in an anaerobic glovebox (100% N_2). The reaction progress was monitored by headspace analysis of TCE and its degradation products on a SRI 8610C GC, which was equipped with a silicon capillary column and a flame ionization detector. TCE concentration in the aqueous phase ($[TCE]_{aq}$) was then determined by the Henry's law with K_H = 0.392 at 1 atm and 25 °C (Gossett, 1987). As illustrated by Figure 5, in the control with only aqueous TCE, its concentration is decreased from 328 μM to 230 μM in a period of 360 hrs due to TCE leaking from the system. In the presence of ZVI, TCE concentration is decreased, following a pseudo-first order kinetics with a half-life of 85 hrs.

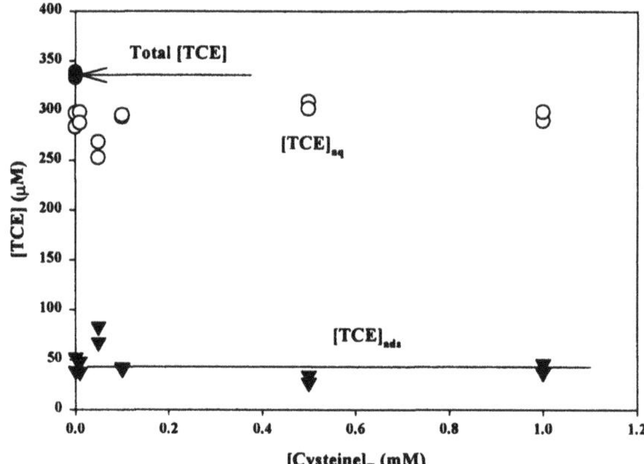

Figure 4. Effect of cysteine on TCE adsorption under an iron surface loading of 76 m^2/L.

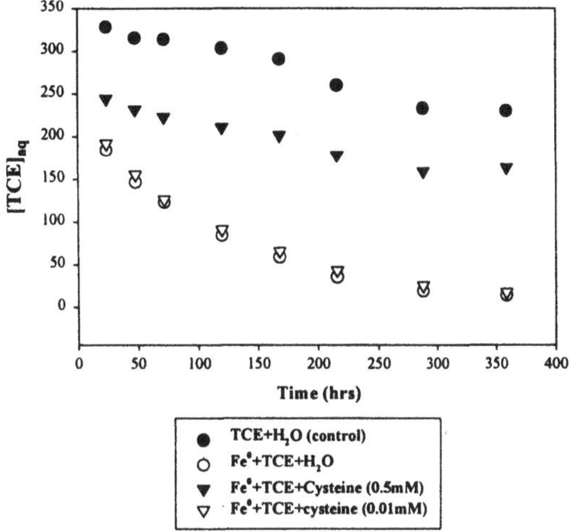

Figure 5. Effect of cysteine on the degradation of TCE

Cysteine, at a concentration of 0.50 mM, dramatically decreased the rate of TCE degradation as indicated by a concentration profile of TCE parallel to the control. Since 0.50 mM cysteine does not significantly affect the TCE sorption behavior but essentially shuts off its reduction, we conclude that there must be at least two types of surface sites, and cysteine can almost totally cover the reactive sites when its concentration is sufficiently high, so no significant TCE degradation takes place. Separate experiments have shown that cysteine adsorption is 0.35 mM when a total of 0.50 mM cysteine is added (Hu, 1999). This suggests that the reactive sites are equal or less than 0.35 mM. The maximum TCE adsorption is 4.3 mM under the same iron loading (Hu, 1999). Therefore the reaction sites represent only a minor fraction (ca. 8%) of the total sorption sites. On the other hand, in the presence of 0.010 mM cysteine, TCE degradation is not significantly decreased. The amount of cysteine adsorbed is 0.006 mM (or ca. 0.14% of the total surface sites) under these conditions, so the reactive sites must be higher than 0.006 mM.

Although reductive dechlorination is modelled either based on an empirical approach or a single site model, TCE dechlorination in the presence of cysteine apparently requires the use of a two-site model. The simulation as shown in Figure 3b indicates that quantitatively, the two-site model proposed here is sufficient for the interpretation of the experimental results. To enhance the predictive capability of the model, further research is needed to determine adsorption parameters independently.

4. CONCLUSIONS

Reductive dechlorination of chlorinated solvents in the ZVI system is a surface-mediated process. Adsorption of the chlorinated compounds takes place prior to the reduction, but the overall rate of reduction is limited by the electron transfer from the surface to the chlorinated compounds. The adsorption can occur on either reactive or nonreactive sites, while the reduction rate is directly proportional to the amount adsorbed onto the reactive sites. The proportion adsorbed onto reactive sites to the nonreactive sites is related to the nature of chlorinated compounds. Higher chlorinated ethylenes such as PCE and TCE are likely to have a larger portion going to the nonreactive sites compared to less chlorinated ethylenes like vinyl chloride. A two-site model incorporating the known observations related to the ZVI system has been developed and such a model can be applied to explain the adsorption and reduction of chlorinated solvents in the presence of competing coadsorbates.

ACKNOWLEDGEMENTS

We gratefully acknowledge two reviewers for their constructive comments. This work was supported in part by Waste Management-Education and Research Consortium (WERC) and by New Mexico Institute of Mining and Technology.

BIBLIOGRAPHY

Allen-King R. M., Halket R. M., and Burris D. R. (1997) Reductive transformation and sorption of cis- and trans-1,2-dichloroethylene in a metallic iron-water system. *Environmental Toxicology and Chemistry* **16**(3), 424-429.

Arnold W. A. and Robert A. L. (1998) Pathways of chlorinated ethylene and chlorinated acetylene reaction with Zn(0). *Environmental Science & Technology* **32**(19), 3017-3025.

Burris D. R., Allen-King R. M., Manoranjan V. S., Campbell T. J., Loraine G. A., and Deng B. (1998) Chlorinated ethene reduction by cast iron: sorption and mass transfer. *Journal of Environmental Engineering* **124**(10), 1012-1019.

Burris D. R., Campbell T. J., and Manoranjan V. S. (1995) Sorption of trichloroethylene and trichloroethylene in a batch reactive metallic iron-water system. *Environmental Science & Technology* **29**(11), 2850-2855.

Campbell T. J., Burris D. R., Roberts A. L., and Wells R. J. (1997) Trichloroethylene and tetrachloroethylene reduction in a metallic iron-water-vapor batch system. *Environmental Toxicology and Chemistry* **16**(4), 625-630.

Cherry J. A., Feenstra S., and Mackay D. M. (1996) Concepts for the remediation of sites contaminated with dense non-aqueous phase liquids (DNAPLs). In *Dense Chlorinated*

Solvents and Other DNAPLs in Groundwater (ed. J. F. Pankow and J. A. Cherry), pp. 475-506. Waterloo Press.

Deng B., Burris D. R., and Campbell T. J. (1999) Reduction of vinyl chloride in metallic iron-water systems. *Environmental Science & Technology* **33**(15), 2651-2656.

Deng B., Campbell T. J., and Burris D. R. (1997) Hydrocarbon Formation in Metallic Iron/Water Systems. *Environmental Science & Technology* **31**(4), 1185-1190.

Deng B., Hu S., and Burris D. R. (1998) Effect of iron corrosion inhibitors on trichloroethylene reduction. In *Physical, Chemical, and Thermal Technologies* (ed. G. B. Wickramanayake and R. R. Hinchee), pp. 341-346. Battelle Press.

EPA. (1998) Permeable Reactive Barrier Technologies for Contaminant Remediation. United States Environmental Protection Agency. EPA/600/R-98/125.

Focht R., Vogan J., and O'Hannesin S. (1996) Field application of reactive iron walls for in-site degradation of volatile organic compounds in groundwater. *Remediation* (Summer), 81-94.

Gavaskar A. R., Gupta N., Sass B. M., Janosy R. J., and O'Sullivan D. (1998) *Permeable Barriers for Groundwater Remediation*. Battelle Press.

Gillham R. W. and O'Hannesin S. F. (1994) Enhanced degradation of halogenated aliphatics by zero-valent iron. *Ground Water* **32**(6), 958-967.

Glod G., Angst W., Holliger C., and Schwarzenbach R. P. (1997) Corrinoid-mediated reduction of tetrachloroethene, trichloroethene, and trichlorofluoroethene in homogeneous aqueous solution: reaction kinetics and reaction mechanisms. *Environmental Science & Technology* **31**(1), 253-260.

Gossett, J. M. (1987) Measurement of Henry's Law constants for C_1 and C_2 chlorinated hydrocarbons. *Environ. Sci. Technol.* **21**, 202-208.

Hu S. (1999) *Reductive Dechlorination of Trichloroethylene by Metallic Iron: Effects of Iron Corrosion and Corrosion Inhibition*. Master Thesis, New Mexico Institute of Mining and Technology.

Johnson T. L., Fish W., Gorby Y. A., and Tratnyek P. G. (1998) Degradation of carbon tetrachloride by iron metal: Complexation effects on the oxide surface. *Journal of Contaminant Hydrology* **29**, 379-398.

Johnson T. L., Scherer M. M., and Tratnyek P. G. (1996) Kinetics of halogenated organic compound degradation by iron metal. *Environ. Sci. Technol.* **30**(8), 2634-2640.

Lasaga A. C. (1981) Transition state theory. In *Kinetics of Geochemical Processes*, Vol. 8 (ed. A. C. Lasaga and R. J. Kirkpatrick), pp. 135-169. Mineralogical Society of American.

Mackay D. M. and Cherry J. A. (1989) Groundwater contamination: pump and treat remediation. *Environmental Science & Technology* **23**, 630-636.

Matheson L. J. and Tratnyek P. G. (1994) Reductive dechlorination of chlorinated methanes by iron metal. *Environmental Science & Technology* **28**(12), 2045-2053.

NRC N. R. C. (1994) *Alternatives for Ground Water Cleanup*. National Academy Press.

Orth W. S. and Gillham R. W. (1996) Dechlorination of trichloroethylene in aqueous solution using Fe^0. *Environmental Science & Technology* **30**(1), 66-71.

Prigogine I. (1967) *Introduction to Thermodynamics of Irreversible Processes*. Wiley-Interscience.

Reynolds G. W., Hoff J. T., and Gillham R. W. (1990) Sampling bias caused by materials used to monitor halocarbons in groundwater. *Environmental Science & Technology* **24**(1), 135-142.

Roberts L. A., Totten L. A., Arnold W. A., Burris D. R., and Campbell T. J. (1996) Reductive elimination of chlorinated ethylenes by zero-valent metals. *Environmental Science & Technology* **30**(8), 2654-2659.

Scherer M. M, Westall J. C, ZiomekMoroz M, Tratnyek P. G, (1997) Kinetics of carbon tetrachloride reduction at an oxide-free iron electrode. *Environmental Science & Technology* 31(8) pp. 2385-2391.

Senzaki T. and Kumagai Y. (1988) Removal of chlorinated organic compounds from wastewater by reduction process: II. Treatment of trichloroethylene with iron powder. *Kogyo Yosui* **357**, 2-7.

Sivavec T. M. and Horney D. P. (1997) Reduction of chlorinated solvents by Fe(II) minerals, Preprints of ACS annual meeting-Division of Environmental Chemistry, San Francisco, CA. 115-117.

Stumm W. and Sulzberger B. (1992) The cycling of iron in natural environments: consideration based on laboratory studies of heterogeneous redox processes. *Geochimica et Cosmochimica Acta* **56**, 3233-3257.

Su C. and Puls R. W. (1999) Kinetics of trichloroethylene reduction by zerovalent iron and tin: pretreatment effect, apparent activation energy, and intermediate products. *Environmental Science & Technology* **33**(1), 163-168.

Sweeny K. H. (1979) Reductive degradation treatment of industrial and municipal wastewaters. *Water Reuse Symposium*, 1487-1497.

Sweeny K. H. (1981) Reductive treatment of industrial wastewaters. I. Process description. *AIChE Symp. Ser.*, 67-71.

Tratnyek P. G. (1996) Putting corrosion to use: remediating contaminated groundwater with zero-valent metals. *Chemistry & Industry*(1), 499-503.

Tratnyek P. G. and Scherer M. M. (1998) The effect of natural organic matter on reduction by zero-valent iron, Preprints of ACS annual meeting -Division of Environmental Chemistry, Boston, MA, 125-126.

Vogel T. M., Criddle C. S., and McCarty P. L. (1987) Transformations of halogenated aliphatic compounds. *Environmental Science & Technology* **21**(8), 722-736.

Wilkins R. G. (1991) *Kinetics and Mechanism of Reactions of Transition Metal Complexes*. VCH.

APPENDIX A

A "model" file used in Scientist@ software package to obtain the simulation of the single site model in the present of a coadsorbate (Figure 2b).

```
// MicroMath Scientist Model File
// TCE degradation on metallic iron: single site model with competitive adsorbate
// Based on adsorption equilibrium and assume that the reaction rate is directly
```

```
// proportional to  the amount of TCE sorbed.
// S=free site concentration; TS=total site conc.; SX=TCE adsorbed; SI= inhibitor
//  adsorbed; XT = total TCE conc.; X=TCE conc in aqueous phase; P=product conc.;
//  IT=total inhibitor conc.;  I = aqueous inhibitor conc.; SI= inhibitor adsorbed. k1, k2,
//  k3 – rate constants
IndVars: t
DepVars: S, X, SX, I, SI, P
Params: k1, k2, k3
KX=2.0*10^4
KI=2.0*10^4
S'=-k1*S*X+(k1/KX)*SX-k2*S*I+(k2/KI)*SI +K3*SX
X'=-k1*S*X+(k1/KX)*SX
SX'=k1*S*X-(k1/KX)*SX-k3*SX
I'=-k2*S*I+(k2/KI)*SI
SI'=k2*S*I-(k2/KI)*SI
P'=k3*SX
// Initial conditions
t=0.0
S=3*10^(-4)
X=1*10^(-3)
SX=0
I=1.0*10^(-3)
SI=0.0
P=0.0
***
```

Chapter Eight

Pilot Test of a Surfactant-Modified Zeolite Permeable Barrier for Groundwater Remediation

ROBERT S. BOWMAN[1], ZHAOHUI LI[1], STEPHEN J. ROY[1], TODD BURT[1], TIMOTHY L. JOHNSON[2], and RICHARD L. JOHNSON[2]
[1]*Department of Earth and Environmental Science, New Mexico Institute of Mining and Technology, Socorro, NM 87801*
[2]*Department of Environmental Science and Engineering, Oregon Graduate Institute of Science and Technology, Beaverton, OR 97006*

Key words: chromate, perchloroethylene, sorption, transport

Abstract: Two pilot-scale tests of surfactant-modified zeolite (SMZ) permeable barriers were conducted at the Large Experimental Aquifer Facility of the Oregon Graduate Institute. The tests were performed in an 8.5-m-wide, 8.5-m-long, 3-m-deep concrete tank. The SMZ was installed in a 1-m-wide, 6-m-long, 2-m-deep barrier frame in the center of the tank. The rest of the tank was filled with sand to form a simulated aquifer. A three-dimensional sampling array consisting of 405 sampling points was installed in the tank. Controlled water flow across the tank was maintained using ten upgradient injection wells and ten downgradient withdrawal wells. A specific discharge of 0.17 m day^{-1} was imposed, resulting in an average linear groundwater velocity of approximately 0.5 m day^{-1} in the sand. The upgradient wells allowed injection of a three-dimensional contaminant plume composed of 10 mg L^{-1} (0.19 mmol L^{-1}) Cr, in the form of chromate, and 1.8 mg L^{-1} (0.011 mmol L^{-1}) perchloroethylene (PCE).

In the first pilot test, 12 metric tons of 14-40 mesh (1.4-0.4 mm) SMZ, manufactured at a cost of about $460 per metric ton, was used as the barrier material. Intensive sampling showed that much of the contaminant plume was being deflected under and around the SMZ barrier. Hydraulic testing failed to conclusively isolate the cause(s) of the flow restriction but suggested that a partially plugged barrier frame, along with a possible decrease in SMZ permeability, were responsible.

For the second pilot test, the 14-40 SMZ was excavated from the frame, a nylon screen on the barrier frame was removed, and two sections of the frame were refilled with 8-14 mesh (2.4-1.4 mm) SMZ. The remaining one-third of the frame was filled with iron/SMZ pellets as part of another project. After steady water flow was reestablished, chromate and PCE were injected over a period of eight weeks. No plume deflection occurred in the test with the 8-14 SMZ. The SMZ fully intercepted the contaminant plume and prevented migration of contaminants downgradient of the barrier. Near the end of the test, chromate and PCE were detected in samplers installed in the upgradient portion of the SMZ. The estimated retardation factors for chromate and PCE in the pilot test were 44 and 39, respectively. These retardation factors are very close to the values of 42 and 29 for chromate and PCE predicted from laboratory sorption isotherm experiments.

The pilot test results demonstrate that contaminant retardation by an SMZ permeable barrier can be well predicted from laboratory characterization of the SMZ. Furthermore, the engineered water control, sampling, and containment system developed for this project serves as a general model for testing permeable barrier performance.

1. INTRODUCTION

Groundwater cleanup at industrial, government, and military facilities continues to be a high priority. Waste disposal or leakage at many of these sites has resulted in contaminants that are distributed in shallow, broad areas. Many sites have contamination that is spreading and needs containment. The contaminants are often complex mixtures that may include chlorinated and nonchlorinated organic compounds as well as highly oxidized metal species such as arsenate, chromate, and selenate. These metal oxyanions are mobile in most aquifers. Sorbent/reactive materials emplaced in permeable subsurface barriers are promising tools for dealing with such complex and expensive groundwater contamination problems. Barrier materials that retain organic compounds, radionuclides, and other hazardous contaminants while allowing the passage of groundwater are needed to prevent plume migration from near-surface waste sites. Such barriers will allow concentration of contaminants in a narrow zone, increasing the efficiency and lowering the costs of other *in situ* treatment methods such as enhanced biodegradation or air stripping.

The use of surfactant-modified zeolite (SMZ) as a permeable barrier sorbent may offer several unique advantages when dealing with mixed contaminant plumes. Zeolites are hydrated aluminosilicate minerals characterized by cage-like structures, high internal and external surface areas, and high cation exchange capacities. Both natural and synthetic zeolites find use in industry as sorbents, soil amendments, ion exchangers, and molecular sieves. Clinoptilolite is the most abundant naturally occurring zeolite. It has a two-dimensional 8-ring and 10-ring channel structure with the largest cavity dimension measuring 4.4 by 7.2Å (Newsam 1986). The unit-cell formula is $(Ca, Na_2, K_2)_3\ [Al_6Si_{30}O_{72}]\bullet24H_2O$. The low cost of natural zeolites ($45-$60 ton^{-1}) makes their use attractive in water treatment applications.

Zeolite surface chemistry resembles that of smectite clays. In contrast to clays, however, natural zeolites can occur as millimeter- or greater-sized particles and are free of shrink-swell behavior. As a result, zeolites exhibit superior hydraulic characteristics and are suitable for use in filtration systems (Breck 1974) and as permeable barriers to dissolved chemical migration. Internal and external surface areas up to 800 $m^2\ g^{-1}$ have been measured. Total cation exchange capacities in natural zeolites vary from 250 to 3000 meq kg^{-1} (Ming and Mumpton 1989). External cation exchange capacities have been determined for a few natural zeolites and typically range from 10 to 50 percent of the total cation exchange capacity (Bowman et al. 1995).

Due to their large specific surface areas and high cation exchange capacities, natural zeolites have a high affinity for cationic heavy metals such as Pb^{2+} and Cd^{2+} (Colella et al. 1995). Zeolites have been used commercially to remove Pb^{2+} and NH_4^+ from waste waters (Groffman et al. 1992; Mumpton and Fishman 1977). However, natural zeolites have little affinity for inorganic anions such as chromate (CrO_4^{2-}) or for dissolved nonpolar organics.

Treatment of natural zeolites with cationic surfactants dramatically alters their surface chemistry. The large organic cations exchange essentially irreversibly with native cations such as Na^+, K^+, or Ca^{2+} (Bowman et al. 1995; Li et al. 1998). Surfactant modification of zeolites enables them to sorb nonpolar organics such as benzene and chlorinated hydrocarbons, including perchloroethylene (PCE) and 1,1,1-trichloroethane, while retaining their ability to sorb heavy metal cations (Li and Bowman 1998; Neel and Bowman 1992). The sorption of target organics is little affected by the presence of other nonpolar organics in aqueous solution (Neel and Bowman 1992). It has also been shown that strongly hydrolyzed, anionic metals such as arsenic, chromium, and selenium are selectively removed by SMZ (Haggerty and Bowman 1994). While sorption of target anions is reduced in the presence of competing anionic species, the SMZ is selective for sorption of multivalent oxyanions.

Although a variety of surfactants can be used for alteration of zeolite surface chemistry (Bowman et al. 1995), hexadecyltrimethylammonium (HDTMA) is preferred due to its ready availability and low cost. The SMZ prepared using HDTMA is stable under a wide range of pH and Eh conditions and in organic solvents and is resistant to microbial degradation (Li et al. 1998). After the SMZ is saturated with an anion or volatile organic contaminant, it can be regenerated with little loss of sorption capacity (Bowman 1996).

The work described above shows that SMZ is a physically and chemically stable sorbent that can simultaneously remove organics, inorganic cations, and inorganic anions from contaminated water. Because of the broad sorptive capabilities of SMZ, its superior hydraulic characteristics, its low unit cost, and the historical use of zeolites in water treatment facilities, this material is very promising as a sorbent for *in situ* treatment of contaminated groundwater.

The research described in this chapter is part of an effort to develop and test a zeolite-based permeable barrier system for containing and remediating contaminated groundwater. The specific goals of this work were to scale up the production of SMZ, install the SMZ in a pilot-scale permeable barrier, and compare barrier performance to predictions based on laboratory characterization of the SMZ.

2. FACILITIES, METHODS, AND MATERIALS

2.1 Pilot-Test Tank Design

The pilot test was conducted at the Large Experimental Aquifer Program (LEAP) site at the Oregon Graduate Institute (OGI). The LEAP tank is an 8.5-m-wide, 8.5-m-long, 3-m-deep concrete pool. The permeable barrier was 1-m wide, 6-m long, and 2-m deep, and designed to simulate a real-world application where the barrier was not keyed in to a low-permeability layer. A schematic diagram of the pilot-test facility is shown in Figure 1.

A steel liner reinforced with a steel I-beam baffle was installed in the LEAP tank for secondary containment of the experiment. The primary liner was a 0.64-cm thick welded seam high-density polyethylene (HDPE) liner installed within an outer steel liner. Baffles in the steel liner were backfilled with pea gravel and contained eight fully screened 5-cm-inner-diameter (ID) wells. These wells could be used for leak detection as well as mitigation by inducing a hydraulic gradient into the tank. The finished interior dimensions of the tank were 8.5-m wide, 8.5-m long, and 3-m deep.

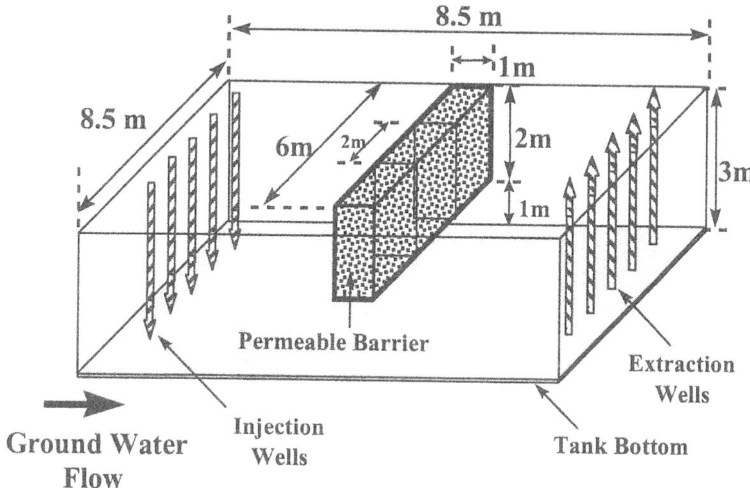

Figure 1. Schematic diagram of the pilot-test tank at OGI's LEAP facility.

2.2 Barrier Construction and Installation

The permeable barrier was composed of a steel frame that was constructed to hold the SMZ and to allow for media replacement. The frame was constructed of 5-cm steel angle iron and 2.5-cm and 7.6-cm square steel tube. The frame had solid floor and end walls (1.3-cm-thick steel plates) to divide it into three distinct cells. Each cell had perforated metal walls (0.16-cm thick perforated steel sheets with 0.64-cm holes covering 50% of the surface area) transverse to the direction of flow. The perforated metal was installed on both the inside and the outside of the steel tube skeleton, resulting in a 7.6-cm-wide annulus between the inner and outer walls of the frame. The entire frame assembly was professionally painted with high-quality, rust-resistant paint. The barrier frame was placed in the pilot-test tank in three sections on top of a 1-m depth of aquifer sand that had been previously added to the tank in lifts. The physical and chemical properties of the sand are described later in this chapter. The three frame sections were bolted together after applying a silicone caulk (Sika-Flex™) for sealing. The end of the barrier in contact with the side of the tank was sealed to the

HDPE liner with Sika-Flex™ and a silicone-based glue. To prevent sand from flowing into the barrier when it was empty, the interior and exterior perforated metal walls of the frame were covered with 100-mesh nylon screen attached with silicone-based glue.

After the barrier frame was in place, pipes for the injection/extraction wells, piezometers, and the sampling network were suspended in the tank from cables. Sand was then added in lifts to fill the remaining aquifer portion of the tank. After the aquifer material was in place the SMZ was packed into the permeable barrier frame. The pipes were buried in place as the tank and frame were filled.

2.3 Water Injection/Extraction and Monitoring Systems

Ten injection and ten extraction wells were used to control water flow within the tank and to inject contaminants. The wells were bundled in pairs, with five pairs at the upgradient end of the tank and five pairs at the downgradient end. In each pair, one well controlled water flow in the lower half of the tank while the other controlled flow in the upper half. All wells were constructed of 5-cm-ID schedule 40 polyvinyl chloride (PVC) tubing. The lower wells were 3-m long and screened over the bottom 1.5 m while the upper wells were 1.5-m long and screened over the entire length. A moveable packer in the upper injection wells allowed control of the contaminant injection interval. Eight piezometers were installed to monitor water pressures within and outside the barrier. Piezometers were constructed of 1-in- (2.5-cm-) ID schedule 40 PVC of 1-m length screened over the bottom 0.17 m.

2.4 Sampling System

The sampling system consisted of a 9 x 9 grid of multilevel samplers with five depths each (Figure 2), for a total of 405 sampling locations. The sampler construction consisted of 1.25-cm PVC stock to which was attached HDPE tubing terminating at successive 0.5-m depths below the surface. The ends of all of the sample tubes were wrapped in 100-mesh nylon screen attached with an HDPE ziptie. The 18 sample nests within the SMZ barrier had four of the sampling levels inside the barrier and one sampling level penetrating the solid base to a depth of 0.5 m below the barrier. A small hole was drilled into the bottom of the barrier during construction and the appropriate sample tubes were inserted through the holes and sealed into place with Sika-Flex™. The samplers were labeled using a letter (column), number (row), and color (depth) system. For example, sampler designation

A3B translates to column A, row 3, depth black. The depth roughly corresponded to the color spectrum, beginning with the red (deepest), followed by yellow, green, black, and white (shallowest). Color coding the samplers and bottles increased accuracy significantly in collecting and analyzing large numbers of samples.

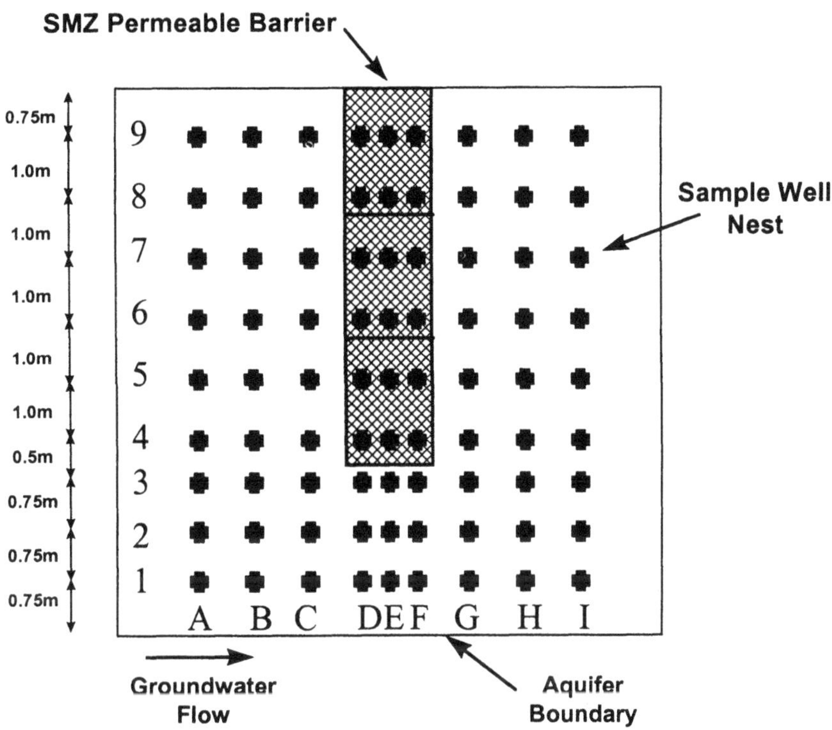

Figure 2. Locations of the sampler nests in the pilot-test tank.

2.5 Flow Control and Contaminant Injection Systems

Figure 3 is a schematic diagram of the pilot-test flow control system. The flow in the pilot-test tank was stratified into upper and lower zones created by the five upper injection/extraction wells and the five lower injection/extraction wells. The feed water was dispensed through a flow totalizer into the injection wells via a manifold and a series of ten controllable flow meters. The lower injection wells were fed directly with Beaverton city water. The upper injection wells were fed from three 6800-L tanks. Using three tanks allowed one tank to be actively supplying feed water, one to be full with the appropriate input solution, and one to be receiving tank effluent. Water for the upper injection wells was supplied under gravity pressure. Water was removed from the extraction wells using five Cole-Parmer Masterflex® L/S™ Variable-Speed Standard Console 1/10 hp drives, each capable of operating four peristaltic pump heads from 1 to 100 rpm. Injection and extraction rates were controlled manually using valves. These rates, as well as the water levels in the feed tanks and pilot-test tank, were monitored continuously using HPVee™ software. Redundant float switches connected to automatic cutoff valves protected against fluid escapes from the system.

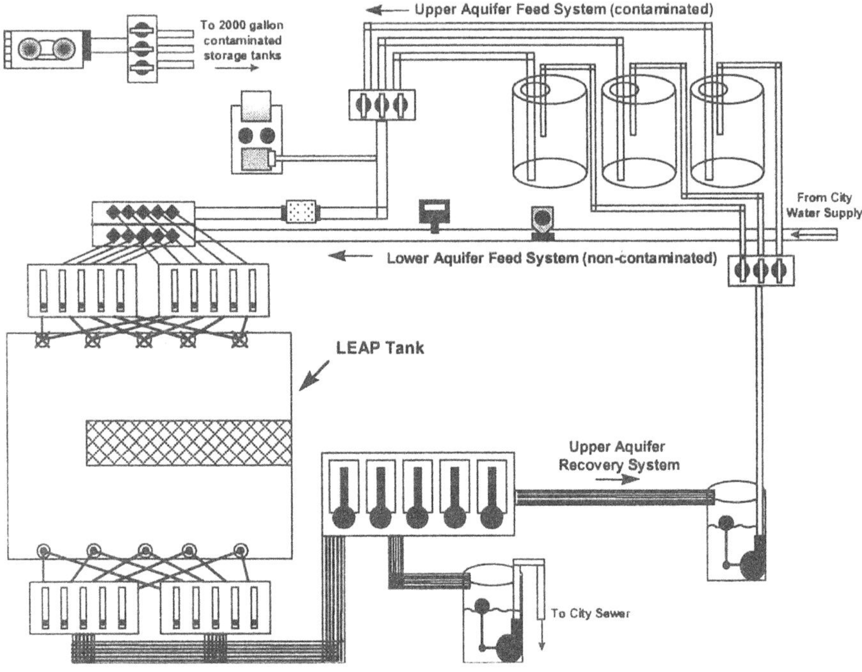

Figure 3. Schematic diagram of the pilot-test flow control system.

During pilot-test tank saturation and flushing, the feed tank contained Beaverton city water. During contaminant injection, the feed tank also contained the target input concentration of chromate. A 9:1 methanol: PCE mixture was injected into the feed line via a high-performance-liquid-chromatography (HPLC) pump to yield the appropriate PCE input concentration. The target contaminant concentrations in the injection wells were 10 mg L^{-1} (0.19 mmol L^{-1}) Cr(VI) as chromate and 1.8 mg L^{-1} (0.011 mmol L^{-1}) PCE. Effluent from the extraction wells was checked for chromate and PCE concentrations and discharged to the municipal sewage system if it met regulatory standards. Effluent not meeting standards was directed to a receiving tank where it was air-stripped of PCE. Chromate was then added as necessary to establish the correct input concentration. This tank was now ready to serve as a feed tank. In this manner each of the three tanks served alternately as a feed, standby, or receiving tank.

2.6 Sample Collection

Samples were collected in 40-mL volatile-organics analysis (VOA) vials using dedicated tubing and a five-head peristaltic pump. To collect a representative sample and to prevent volatilization losses of PCE, several fluid volumes were flushed through the vials prior to sealing with zero headspace. Synoptic samples were taken within different subsets of the 405 samplers depending upon the stage of the experiment. The frequency of sampling ranged from semi-weekly to monthly, again depending upon the stage of the experiment.

2.7 Chemical Analyses

Samples from the 40-mL VOA vials were split for chromate and PCE analyses and typically analyzed within 48 hours of collection. Chromate concentration was determined via an HPLC method using a Gilson Model 116 UV detector set at 365 nm and a 2- by 150-mm Waters Nova-Pak® C18 60A HPLC column packed with 4-μm particles. The mobile phase consisted of 5-mM tert-butylammonium hydrogen sulfate buffered to pH 4.4 with NaOH with 10% acetonitrile (v/v) as a modifier. The eluent flow rate was 0.8 mL min^{-1}. Samples were filtered through a 0.45-μm filter as they were injected by an Alcott 708 autosampler with a 0.1-mL sample loop. The typical run time was 4 min with a calibration range of 0.05 to 20 mg L^{-1} (0.001 - 0.38 mmol L^{-1}) Cr as chromate.

The PCE concentration was determined by a headspace method using an Hewlett Packard 5890 Series II gas chromatograph (GC) with an HP 7694 autosampler. A sample volume of 750 μL was injected onto a 30-m by 0.53-mm ID DB-1 column at an oven temperature of 140 °C. Detection was by electron capture with a calibration range of 1 to 2500 μg L^{-1} (6.0×10^{-6} - 1.5×10^{-2} mmol L^{-1}) and a 3-min run time.

2.8 Characterization of Aquifer Sand

The aquifer material was a silica beach sand from the Columbia River basin in Oregon. The well-sorted sand had less than 1% by weight of organic matter and less than 1% by weight of iron-oxide-cemented aggregates. The hydraulic conductivity of the sand was determined at New Mexico Tech (NMT) using a constant-head permeameter and at OGI using a falling-head permeameter (Freeze and Cherry 1979). To test the interaction of the contaminants with the aquifer sand, sorption isotherms for both chromate and PCE were prepared using methods described earlier (Li and Bowman 1997; Li and Bowman 1998).

2.9 Preparation of SMZ

We used a natural clinoptilolite-rich zeolitic tuff ("zeolite") from the St. Cloud deposit near Winston, New Mexico, for the pilot test. The fundamental zeolite crystals are of micron to sub-micron size, cemented together by finer-grained material. The bulk material can be crushed and ground to any desired aggregate size. The zeolite used for our tests was ground and screened at the St. Cloud mine to the appropriate size, either 14-40 mesh (1.4 to 0.4 mm), or 8-14 mesh (2.4 to1.4 mm). The mineral content of the zeolite, based on internal standard XRD analysis (Chipera and Bish 1995; Sullivan et al. 1997), was 74% clinoptilolite, 5% smectite, 10% quartz plus cristobalite, 10% feldspar, and 1% illite. The zeolite had an internal (zeolitic) cation exchange capacity (CEC) of 800 meq kg^{-1} and an external (nonzeolitic) cation exchange capacity (ECEC) of 100 meq kg^{-1}, as determined using a method modified from that of Ming and Dixon (Li and Bowman 1997; Ming and Dixon 1987). The external surface area using nitrogen adsorption was 15.7 m^2 g^{-1} (Sullivan et al. 1997). When packed to a dry bulk density of 1 kg L^{-1}, the porosity of the zeolite was 0.6.

The SMZ was bulk-produced in an existing batch plant at the St. Cloud mine. The sequence of steps used to produce the SMZ is illustrated in Figure 4. Raw zeolite (A) was metered (B) into a mixing chamber (F) where HDTMA solution (C), fed by pump (D) and monitored with a flow meter (E), was added. The mixture was further homogenized and fed by a screw-auger (G) into a rotating dryer (H) heated by a propane burner (I). The mixture exited the dryer onto a conveyor belt (J) leading to a 1-m^3 "super sack" storage container (K).

Two separate large-scale batches of SMZ were manufactured. In February 1997, 20 cubic meters (about 20 metric tons) of 14-40 mesh SMZ was prepared at the St. Cloud facility. In May 1998, nine cubic meters (nine metric tons) of 8-14 mesh SMZ was prepared. In each case, SMZ was manufactured at a rate of 3-4 m^3 h^{-1}. A 30% (by weight) aqueous HDTMA-Cl solution from Lonza Chemical Company was used as the feed solution for both batches. The target HDTMA loading for the 14-40 batch was 180 mmol kg^{-1}, while the target HDTMA loading for the 8-14 batch was 140 mmol kg^{-1}.

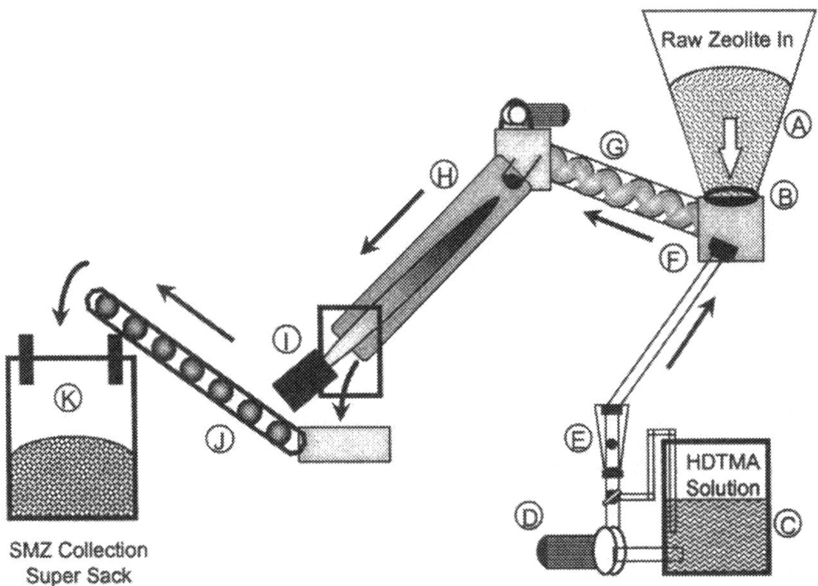

Figure 4. Schematic diagram of the process for bulk production of SMZ.

2.10 Physical and Chemical Characterization of SMZ

The grain size distributions of the two batches of SMZ were determined by sieve analysis. Chromate and PCE sorption isotherms for each batch of SMZ were prepared using methods described earlier (Li and Bowman 1997; Li and Bowman 1998).

2.11 Installation of Sand and SMZ in the Pilot-Test Tank

After manufacture and characterization, the SMZ was shipped to OGI on flatbed trucks in 1-m^3 super sacks. As described above, the sand and SMZ were packed into the pilot-test tank in lifts around the previously installed instrumentation. The aquifer sand was placed into a hopper with a small end-loader, then traveled up a conveyor and down a large tube to the approximate location where it was needed in the tank. The sand was then further spread and leveled by hand. The SMZ was added to the barrier in a similar manner. The final depth of aquifer sand was 3.0 m while the depth of the SMZ was 2.0 m.

3. RESULTS AND DISCUSSION

3.1 Physical and Chemical Properties of Aquifer Sand and SMZ

The mean hydraulic conductivities for the aquifer sand determined using the two laboratory methods (constant-head and falling-head permeameters) were $5*10^{-4}$ m sec^{-1} and $2*10^{-4}$ m sec^{-1}, respectively. The isotherm results showed that the sand had negligible sorption capacity for either chromate or PCE.

The grain size distributions for the two batches of zeolite are shown in Figures 5 and 6. In both cases the grain size distribution of the SMZ is almost the same as that of the raw zeolite, demonstrating that the SMZ manufacturing process caused little breakdown of zeolite aggregates. This preservation of size distribution is important for manufacturing SMZ with specific hydraulic characteristics. Using the constant-head permeameter we determined a hydraulic conductivity of $20*10^{-4}$ m sec^{-1} for the 14-40 SMZ and a much higher value (difficult to measure accurately with this apparatus) for the 8-14 SMZ. Using the falling-head permeameter we determined a conductivity of $10*10^{-4}$ m sec^{-1} for the 14-40 SMZ.

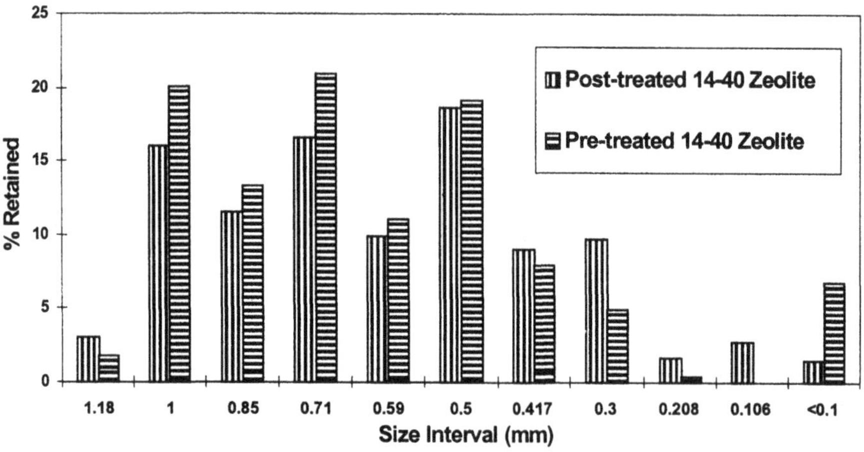

Figure 5. Grain size distribution of 14-40 zeolite before and after surfactant treatment.

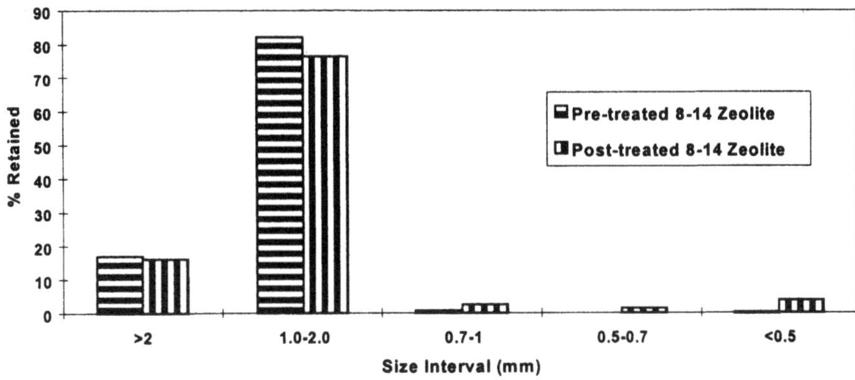

Figure 6. Grain size distribution of 8-14 zeolite before and after surfactant treatment.

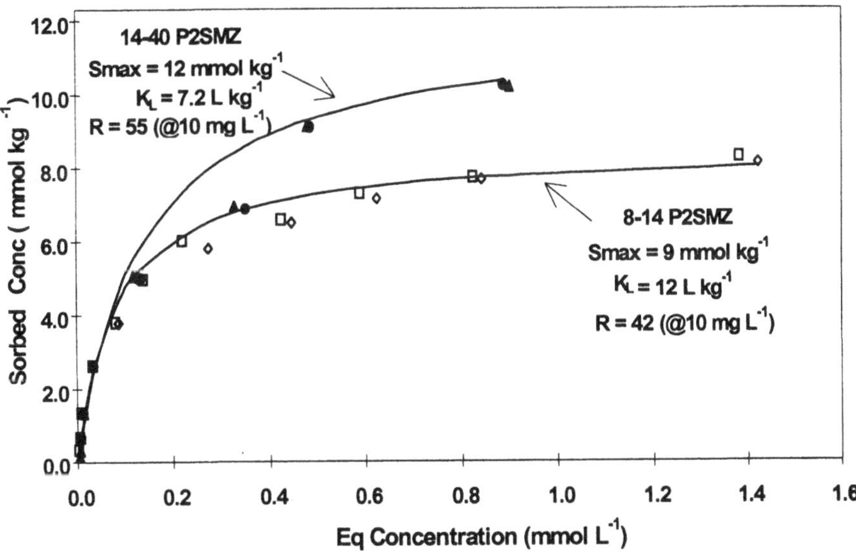

Figure 7. Sorption isotherms for chromate on 14-40 and 8-14 SMZ.

The chromate sorption isotherms for the 14-40 and the 8-14 SMZ are shown in Figure 7. The sorption of chromate on SMZ was Langmuirian with sorption maxima of 12 mmol kg^{-1} for the 14-40 SMZ and 9 mmol kg^{-1} for the 8-14 SMZ. These values are similar to those measured for SMZ prepared in the laboratory (Bowman 1996; Li and Bowman 1997).

The retardation factor R for a chemical undergoing Langmuir-type sorption is:

$$R = 1 + \frac{\rho K_L \beta}{\theta (1 + K_L C)^2} \tag{1}$$

where ρ is the bulk density (M L^{-3}), β is the sorption maximum (M/M), θ is the porosity (L^3 L^{-3}), C is the aqueous concentration (M L^{-3}), and K_L is the Langmuir sorption coefficient (L^3 M^{-1}). The chromate K_L values were 7.2 L $mmol^{-1}$ for the 14-40 SMZ and 12 L $mmol^{-1}$ for the 8-14 SMZ.

The R defined by Equation 1 is a function of concentration and will change during solute migration through a sorbing medium. Nonetheless, it is illustrative to calculate R for a specified concentration. Given the measured SMZ properties of 1 kg L^{-1} for ρ and 0.6 for θ, and assuming a chromate concentration of 10 mg L^{-1} (equivalent to 4.5 mg L^{-1} Cr, about half the input Cr concentration in the pilot tests), the estimated R values for chromate are 55 for the 14-40 SMZ and 42 for the 8-14 SMZ.

The PCE sorption isotherms for the 14-40 and the 8-14 SMZ are shown in Figure 8. The sorption of PCE for both samples was well described by linear sorption isotherms. The linear sorption coefficient (K_d) was 20 L kg^{-1} for the 14-40 SMZ and 17 L kg^{-1} for the 8-14 SMZ. Again, these values are similar to PCE sorption coefficients for SMZ prepared in the laboratory (Bowman 1996; Li and Bowman 1998).

The retardation factor for a chemical undergoing linear-type sorption is:

$$R = 1 + \frac{\rho K_d}{\theta} \tag{2}$$

Again assuming that ρ equals 1 kg L^{-1} and θ equals 0.6, the estimated R values for PCE are 34 for the 14-40 SMZ and 29 for the 8-14 SMZ, regardless of PCE input concentration.

The retention of both chromate and PCE is similar for the two size fractions of SMZ, with the 14-40 material showing about 20-30% greater sorption and retardation. This greater retention by the 14-40 material is consistent with its 30% greater loading of HDTMA.

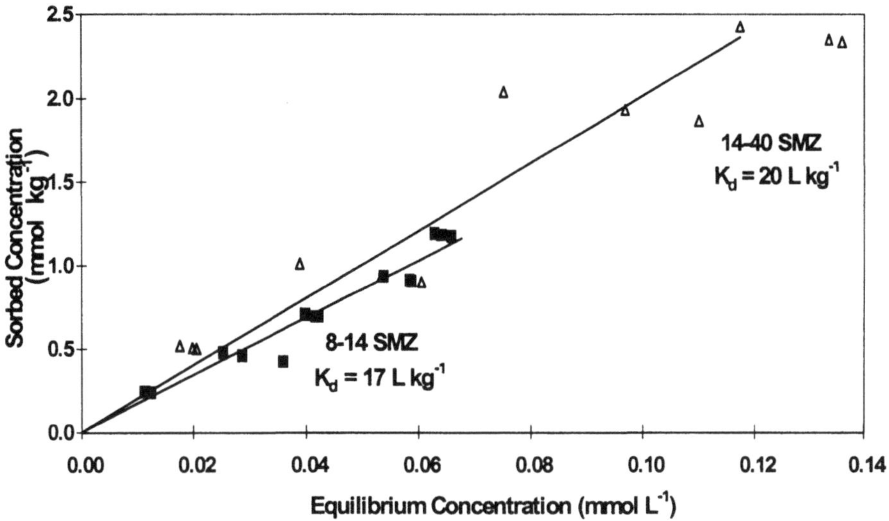

Figure 8. Sorption isotherms for PCE on 14-40 and 8-14 SMZ.

3.2 Pilot-Scale Testing and Analysis

Two separate pilot tests were performed, one using the 14-40 SMZ and a second using the 8-14 SMZ. The original intent was to perform a single, long-term test using the 14-40 SMZ only. However, due to the hydraulic and operational problems described below, the decision was made to remove the 14-40 material and replace it with the coarser 8-14 SMZ.

3.2.1 Pilot Test with 14-40 SMZ

3.2.1.1 *In Situ* Hydraulic Testing

Soon after saturation of the pilot-test tank and establishment of the flow regime, slug tests were performed to estimate the *in situ* hydraulic conductivities of the aquifer sand and the SMZ. The slug tests were performed using piezometers (2.5-cm ID schedule 40 PVC, 1 deep, with a 0.17-m screened interval at the bottom) installed during tank filling. Hydraulic conductivities were determined from the slug test data using the Hvorslev method (Freeze and Cherry 1979). Tests were performed at different locations within the sand and the SMZ. The average results from these tests, along with the laboratory-measured hydraulic conductivity values, are shown in Table 1. Whereas the laboratory tests showed a desired hydraulic conductivity contrast of about 5:1, the *in situ* tests showed essentially identical hydraulic conductivities for the SMZ and the sand. The major change was a much lower conductivity of the SMZ in the barrier than what had been determined in the laboratory. The explanation for this decreased conductivity is not certain. At the time, it was thought that entrapped air, which would slowly dissolve under sustained water flow, might be responsible for the decreased conductivity of the SMZ. Another potential explanation was that some compaction of the SMZ had occurred during barrier filling. The decision was made to proceed with the pilot test even though the sand/SMZ permeability contrast was lower than desired.

Table 1. Hydraulic conductivity of the 14-40 SMZ measured in the laboratory and after installation in the pilot-test tank.

	Hydraulic Conductivity (m/sec * 10^{-3})		
Material	Constant-Head Permeameter	Falling-Head Permeameter	Slug Test
Sand	0.5	0.2	0.2
SMZ	2.0	1.0	0.2

3.2.1.2 Pilot Test Operation and Contaminant Sampling

A steady flux rate of 0.17 m day^{-1}, resulting in an average linear groundwater velocity of approximately 0.5 m day^{-1}, was established in the tank. Beginning on 27 December 1997, 10 mg L^{-1} (0.19 mmol L^{-1}) Cr (in the form of chromate) and 1.8 mg L^{-1} (0.011 mmol L^{-1}) of PCE were injected through all five upper injection wells. The same flux of contaminant-free water was provided to the bottom half of the aquifer. Contaminant injection continued through 12 February 1998. The cumulative volume of contaminant solution injected was about $5*10^4$ L. At approximately weekly intervals, samples were collected from transects (rows) parallel to the water flow direction to monitor the performance of the SMZ barrier. As standard practice Transects 2, 5, 7, and 9 were sampled. This scheme provided one transect within the barrier-free portion of the tank and one transect through each of the three barrier cells. Depending upon the previous week's results, additional locations were sampled to provide a detailed picture of selected flow regions. Eight sampling rounds were performed over a period of six weeks. During this period approximately 3000 samples were collected and analyzed for PCE and chromate.

3.2.1.3 Results for 14-40 SMZ

The chromate/PCE plume contacted the barrier during the second week of contaminant injection. The barrier-free portion of the aquifer (the control section) saw free migration of both contaminants across the tank. By the beginning of the third week, contaminants had migrated completely through the control section and were being recovered from the extraction wells at the downgradient end of the tank. Conversely, by late January 1998, neither chromate nor PCE had migrated through the SMZ barrier. At the end of the fourth week, PCE had been detected in the upgradient sampling nests in the barrier for the first time at a concentration of 20 μg L^{-1}, while chromate still had not been detected in the barrier.

Within two weeks following contaminant injection, however, both chromate and PCE were detected below the barrier. As the experiment progressed, it became clear that the contaminants (and hence the flow field) were being deflected beneath and around the barrier. Figure 9, which shows the chromate and PCE distributions after about five weeks of injection, illustrates the deflection of the plume. The contaminant distributions indicated that the barrier frame or the SMZ or both were causing a hydraulic restriction resulting in the plume deflection. Since evaluation of the SMZ performance depended upon accurate knowledge of the amount of chromate and PCE entering the barrier, contaminant injection was terminated. Attempts were then made to ascertain and ameliorate the cause(s) of the hydraulic restriction.

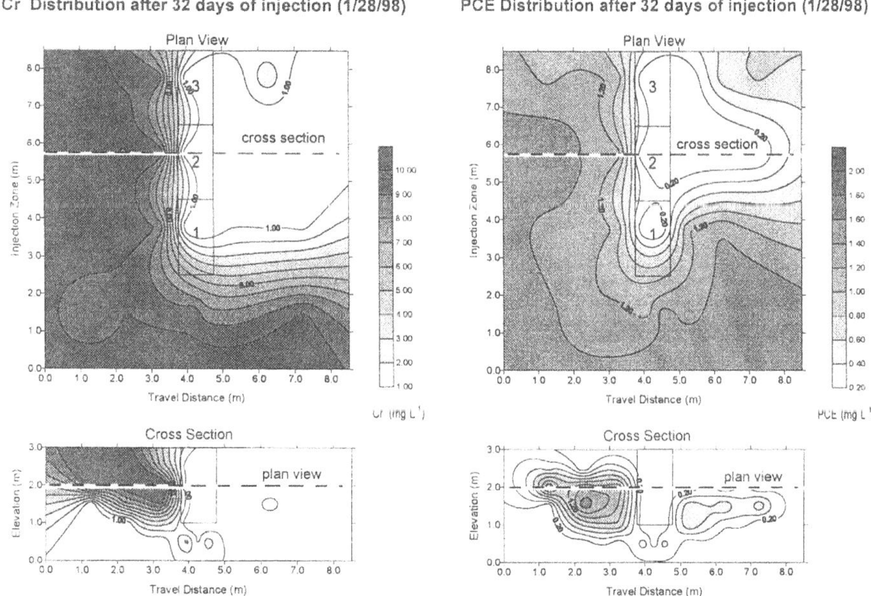

Figure 9. Chrome and PCE distributions following 32 days of contaminant injection, 14-40 experiment.

Substantial effort was spent evaluating the hydraulics of the tank/barrier system. The most likely causes of plume deflection were considered to be changes in the conductivity of the aquifer sand or SMZ (due to compaction or particle breakdown) or plugging of the barrier frame's nylon screen (by mobilized fines, precipitates, or biological growth). New slug tests in the SMZ and the aquifer material resulted in values similar to those obtained prior to contaminant injection. Localized pumping tests seemed to indicate a restriction at the upgradient face of the barrier frame. Several tracer tests with fluorescein were conducted in an attempt to determine whether the low barrier conductivity was localized or consistent across the barrier. Two-meter long laboratory columns were packed with SMZ and subjected to water flows several times greater than used in the pilot test. These column tests showed a decreasing hydraulic conductivity of the SMZ bed over time, due to compaction and/or collection of fines at the screened column end.

Based on the assumption that much of the flow restriction was due to plugging of the upgradient barrier screen, the sand in the barrier frame annulus was removed and an attempt made to flush the screen with high-pressure water jets. Additional tracer tests conducted after the jetting showed that a hydraulic restriction was still present. Since the cause of the hydraulic restriction could not be unambiguously determined or ameliorated, the decision was made to replace the 14-40 SMZ with the higher hydraulic conductivity 8-14 SMZ, and at the same time to remove the 100-mesh screen from the barrier frame.

Although the plume deflection during this phase of the experiment was a setback and a disappointment, it illustrated the value of an intensive sampling array in assessing barrier performance. With less spatial resolution of the contaminant plume, we may have reached the erroneous conclusion that contaminants were passing through, rather than around, the barrier.

3.2.2 SMZ Replacement and Barrier Retrofit

3.2.2.1 Removal of 14-40 SMZ

Clean water was flushed through the tank for several weeks in order to reduce contaminant concentrations. Water was then pumped from the tank to lower the water table to the 1-m depth in preparation for excavation of the 14-40 SMZ. The 14-40 SMZ was manually excavated from the barrier cells starting with Cell 1 (farthest away from the tank wall). Two methods of emptying the cells were used. The first method involved using a small trash pump capable of pumping slurries. This method was somewhat effective and could be used successfully in larger-scale applications where full-size pumps could be employed. The effluent from the trash pump was placed in a super sack, which acted as a filter to remove the SMZ from the water stream. The second, and more effective, method of SMZ removal was manual excavation from the cell into super sacks for disposal.

3.2.2.2 Modification of Barrier Frame

Once a cell was empty, the interior 100-mesh nylon screen was removed. The inner perforated metal was then removed by cutting or breaking off the bolts holding it to the barrier frame. The perforated metal on the outer barrier frame was thus left exposed, with the outer nylon screen accessible through the perforations. This outer nylon screen, directly adjacent to the aquifer sand, was removed by burning it off through the holes in the perforations using a propane torch. The method was quite effective and left only a small amount of charred material. The upper 30 cm of sand adjacent to the frame exterior was excavated to examine the burn results and to manually remove the remaining screen.

3.2.2.3 Barrier Refilling

Cells 1 and 2 of the barrier frame were filled with 8-14 SMZ while Cell 3, adjacent to the pilot-test tank wall, was filled with iron/surfactant-modified zeolite pellets (Fe/SMZ pellets, see below). During refilling, sheets of plywood were temporarily placed against the inner faces of the barrier frame to retain the fill material. The SMZ was transferred using a conveyor belt that ran from the outside the pilot-test tank to the appropriate cell. While the SMZ was loaded in the barrier frame the samplers were reinstalled in their original positions. The annular space between the plywood and the outer perforated metal of the frame was filled with aquifer sand. The plywood was then pulled out of the cell using a jack and appropriate blocking.

The refilling of the barrier provided the opportunity to test Fe/SMZ pellets along with the SMZ. The Fe/SMZ allows both sorption and chemical reduction of PCE and chromate and shows promise for enhancing contaminant removal and reducing the thickness of permeable barriers (Burt et al. 1999; Jones et al. 1998; Li et al. 1997; Li et al. 1999). Due to limited pellet availability, an Fe/SMZ barrier thickness of 0.5 m, rather than 1.0 m, was used. For Cell 3, the plywood retainers were used to split the cell into equal halves. The Fe/SMZ pellets were placed in the upgradient half while the downgradient 0.5-m width was simultaneously filled with aquifer sand. Samplers were reinstalled in their original locations, resulting in four sampling locations in the Fe/SMZ and two sampling locations in the downgradient sand of Cell 3.

3.2.3 Pilot Test with 8-14 SMZ

3.2.3.1 Pilot Test Operation and Contaminant Sampling

Following the barrier retrofit and replacement of the reactive media, the pilot-test tank was slowly resaturated and the flow regime reestablished. The same injection and extraction rates were used as in the 14-40 experiment, resulting in the same volumetric water flux and linear velocity. After several days of steady flow, contaminant injection began. In this case, only the three upper injection wells directly upgradient from the barrier (Wells 3, 4, and 5), received chromate and PCE, while upper injection Wells 1 and 2 as well as all the lower injection wells received contaminant-free water. This configuration resulted in a plume that could be completely captured by the barrier. Contaminant injection began on 10 July 1998 and ended on 11 September 1998. Due to budgetary and schedule constraints, a longer injection period could not be sustained. The cumulative volume of contaminant solution injected was again about $5*10^4$ L.

Following cessation of chromate and PCE injection, contaminant-free water was injected at the same rate through all wells for an additional six weeks to flush residual contaminants from the aquifer sand. Sampling continued during this period, allowing further characterization of the SMZ's ability to retard the chromate and PCE. Samples were collected approximately weekly during the first several weeks, with a decreasing frequency later in the experiment. Transects 3, 5, 7, 8, and 9 were sampled. This scheme provided one transect within the barrier-free portion of the tank, one transect through each of the two SMZ-filled barrier cells, and two transects through the Fe/SMZ cell. Eleven sampling rounds were conducted over a period of 16 weeks. During this period approximately 1500 samples were collected and analyzed for PCE and chromate.

3.2.3.2 Results and Discussion

The migration of the chromate and PCE plumes early and late in the test are shown in Figures 10 and 11. The chromate/PCE plume contacted the barrier during the second week of contaminant injection. In contrast to the earlier experiment with the 14-40 SMZ, no deflection of the contaminant plume occurred. All of the chromate and PCE entered the barrier and both were retarded in their movement relative to water flow. After 56 days (equivalent to 20 pore volumes of contaminated water passing through the barrier), low concentrations of chromate and PCE were detected in the in-barrier samplers, but no contamination was detected downgradient of the SMZ. Chromate broke through the Fe/SMZ section of the barrier after about 20 days of contaminant injection, while PCE began to break through after about 56 days. The results for the Fe/SMZ pellets are described in detail in a separate report (Bowman et al. 1999).

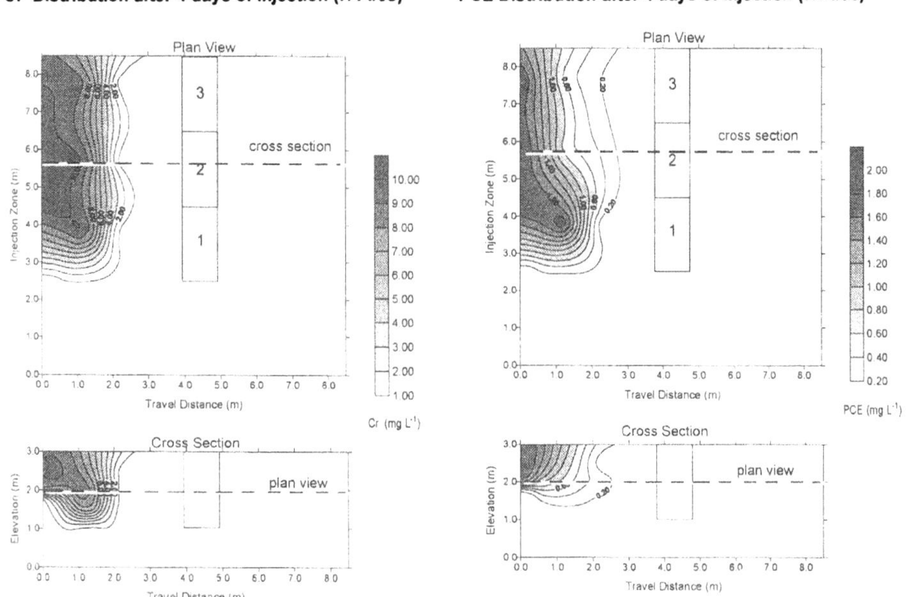

Figure 10. Chrome and PCE distributions following 4 days of contaminant injection, 8-14 experiment.

Quantitative evaluation of barrier performance would have required several additional months of contaminant injection in order to collect complete contaminant breakthrough profiles for the in-barrier and downgradient samplers. Budgetary and time constraints prevented such longer-term monitoring. Nonetheless, a semi-quantitative evaluation of the SMZ performance was made by comparing the velocities of the contaminant contours in the aquifer sand upgradient from the barrier with the arrival times of specific contaminant concentrations at the in-barrier samplers. Using this approach, the estimated velocity of contours for both contaminants in the aquifer sand was about 0.35 m d^{-1}. The velocity of chromate within the barrier, obtained by tracking concentration contours, was about 0.008 m d^{-1}. The PCE velocity within the barrier was about 0.009 m d^{-1}.

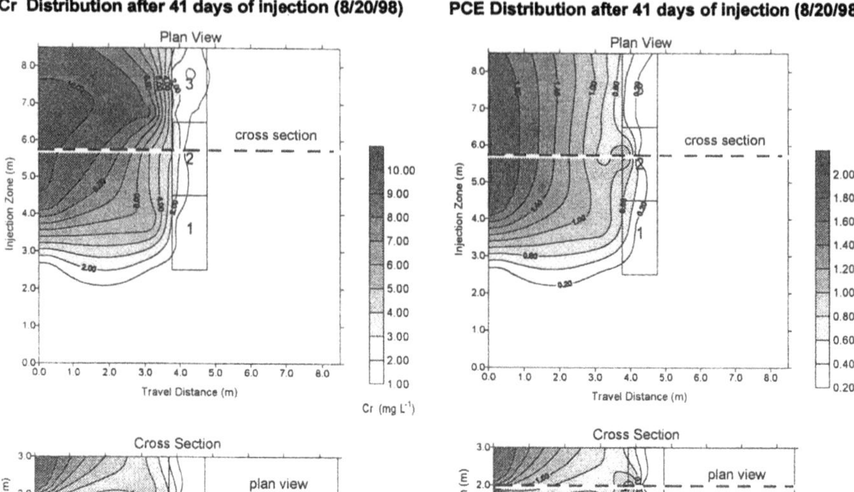

Figure 11. Chrome and PCE distributions following 41 days of contaminant injection, 8-14 experiment.

The retardation factors of the contaminants in the SMZ are simply the ratios of the velocities in the aquifer sand and the SMZ:

$$R = V_{sand} / V_{SMZ} \tag{3}$$

Using the above equation we calculated retardation factors of 44 for chromate and 39 for PCE. These pilot-test retardation factors are very close to the estimates of 42 and 29 calculated in section 3.1 for chromate and PCE based upon their laboratory sorption isotherms. The pilot test results thus confirm that the field barrier performance of SMZ may be predicted reasonably well from laboratory characterization of contaminant interactions.

4. SMZ COST COMPARISONS

The preparation of 30 metric tons of SMZ provided solid information on the cost of bulk production of the material. The major cost of the final product was associated with the price of the raw surfactant. Using 14-40 SMZ (with about 5% HDTMA-Cl by weight) as an example, the production cost was about \$50 for raw zeolite, \$335 for HDTMA, and \$75 for processing, for a total cost of about $460 per metric ton. Since the dry bulk density of SMZ is about 1 g cm^{-3}, the cost on a unit volume basis was about $460 per cubic meter.

In this project, we used high-purity HDTMA to manufacture the SMZ, since all of our laboratory testing had been done using this surfactant. Clearly, the bulk cost of SMZ could be greatly reduced if a less expensive surfactant were used. Surfactant mixtures which contain lower molecular-weight cationic surfactants, in addition to HDTMA, are available at lower cost. Limited experimentation with these lower-purity formulations indicated they produced SMZ that was less stable chemically and had a lower contaminant sorption capacity than SMZ prepared with pure HDTMA. The effectiveness of SMZ prepared from these alternative surfactant formulations warrants further examination, however.

Table 2 compares the cost of SMZ to other materials currently used or proposed for use in permeable reactive barriers. The most effective material for a particular barrier application will depend upon the material's unit cost and upon the contaminants to be controlled. Installation costs should be similar for each material.

Table 2. Costs of potential permeable reactive barrier materials.

Medium	Cost per Metric Ton	Cost per Cubic Meter
Granular Activated Carbon	$1450-5500	$700-2100
Ion Exchange Resins	$3500-6500	$3200-5000
Iron Metal	$350-450	$1100-1200
Organoclay	$1700-4500	$1100-3200
SMZ	$450-500	$450-500

In determining the material cost for a barrier of a given geometry, it is the cost per unit volume, rather than per unit weight, that is important. One clear advantage of SMZ is that its cost per unit volume is one-half to one-tenth that of the other materials listed in Table 2.

5. SUMMARY AND CONCLUSIONS

An SMZ permeable barrier was successfully deployed under field-like conditions while providing hydraulic containment. A highly automated water and chemical delivery system was designed to provide uniform water flows and contaminant plumes of defined geometries, while minimizing contaminant discharges to the environment. The research showed that intensive sampling can and should be performed when evaluating prospective permeable barrier systems. Without an extensive sampling array and close monitoring of contaminant plumes, barrier performance will be difficult to evaluate. The results also showed that extreme care must be taken to prevent hydraulic restrictions at barrier/aquifer interfaces.

Surfactant-modified zeolite can be manufactured in multi-ton quantities at a cost of about $460 per cubic meter ($460 per metric ton). The bulk-produced SMZ has physical and chemical properties essentially identical to SMZ prepared in the laboratory. In particular, the contaminant (chromate and PCE) sorption characteristics of bulk- and laboratory-produced SMZ are the same.

Compaction of SMZ under the loading conditions of a permeable barrier is a potential problem. Since the hydraulic conductivity of SMZ can be tailored by varying the particle size, SMZ with a laboratory conductivity significantly greater than that of the aquifer material should be used in permeable barriers. Based upon the pilot-test data collected, it appears that contaminant retention by SMZ in a permeable barrier can be well-predicted from laboratory sorption measurements.

ACKNOWLEDGMENTS

This work was conducted for the DOE Office of Environmental Management at the Federal Energy Technology Center under contract DE-AR21-95MC32108. The authors thank Kirk Jones and Shawn Williams of New Mexico Tech and Matt Perrott, Ameer Tavakoli, and Emily Keene of the Oregon Graduate Institute for their technical assistance.

BIBLIOGRAPHY

Bowman, R. S. (1996). "Surface-altered zeolites as permeable barriers for in situ treatment of contaminated groundwater." *Phase I Topical Report*, U.S. Dept. of Energy, Morgantown, WV.

Bowman, R. S., Haggerty, G. M., Huddleston, R. G., Neel, D., and Flynn, M. M. (1995). "Sorption of nonpolar organic compounds, inorganic cations, and inorganic anions by surfactant-modified zeolites." Surfactant-enhanced subsurface remediation, D. A. Sabatini, R. C. Knox, and J. H. Harwell, eds., American Chemical Society, Washington, DC, 54-64.

Bowman, R. S., Li, Z., Burt, T. A., Johnson, T. L., Johnson, R. L., and Helferich, R. (1999). "Pilot test of an iron/zeolite permeable barrier for destruction of groundwater contaminants." *Technical Completion Report*, Waste-management Education and Research Consortium, Las Cruces, NM.

Breck, D. W. (1974). *Zeolite molecular sieves: structure, chemistry, and use*, John Wiley and Sons, New York.

Burt, T. A., Jones, H. K., Li, Z., Bowman, R. S., and Helferich, R. "Perchloroethylene and chromate reduction using a surfactant-modified zeolite/zero-valent iron pellet." *1999 WERC Conf. on the Environment*, Albuquerque, NM, (in press).

Chipera, S. J., and Bish, D. L. (1995). "Multireflection RIR and intensity normalizations for quantitative analyses: Applications to feldspars and zeolites." *Powder Diffract.*, 10, 47-55.

Colella, C., DeGennarao, M., Langella, A., and Pansini, M. (1995). "Cadmium removal from wastewaters using chabazite and phillipsite." Natural zeolites '93: occurrence, properties, and use, D. W. Ming and F. A. Mumpton, eds., International Committee on Natural Zeolites, Brockport, NY, 377-384.

Freeze, R. A., and Cherry, J. A. (1979). *Groundwater*, Prentice-Hall, Inc., Englewood Cliffs, NJ.

Groffman, A., Peterson, S., and Brookins, D. (1992). "Removing lead from wastewater using zeolite." *Water Environ. Tech.*, 5, 54.

Haggerty, G. M., and Bowman, R. S. (1994). "Sorption of chromate and other inorganic anions by organo-zeolite." *Environ. Sci. Technol.*, 28, 452-458.

Jones, H. K., Li, Z., and Bowman, R. S. "Reduction of dissolved perchloroethylene by pelletized zero valent iron/surfactant-modified zeolite." *WERC-WRHSRC-NMHWMS 1998 Joint Conf. on the Environment*, Albuquerque, NM, 3-7.

Li, Z., and Bowman, R. S. (1997). "Counterion effects on the sorption of cationic surfactant and chromate on natural clinoptilolite." *Environ. Sci. Technol.*, 31, 2407-2412.

Li, Z., and Bowman, R. S. (1998). "Sorption of perchloroethylene by surfactant-modified zeolite as controlled by surfactant loading." *Environ. Sci. Technol.*, 32, 2278-2282.

Li, Z., Jones, H. K., Bowman, R. S., and Helferich, R. "Incorporation of zero valent iron into pelletized surfactant-modified zeolite for groundwater remediation." *WERC/HSRC '97 Joint Conference on the Environment*, Albuquerque, NM., 463-467.

Li, Z., Jones, H. K., Bowman, R. S., and Helferich, R. (1999). "Enhanced degradation of chromate and PCE by pelletized surfactant-modified zeolite/zero valent iron." *Environ. Sci. Technol.*, 33, (in press).

Li, Z., Roy, S. J., Zou, Y., and Bowman, R. S. (1998). "Long-term chemical and biological stability of surfactant-modified zeolite." *Envrion. Sci. Technol.*, 32, 2628-2632.

Ming, D. L., and Mumpton, F. A. (1989). "Zeolites in soil." Minerals in soil environments, J. B. Dixon and S. B. Weed, eds., Soil Science Society of America, Madison, WI, 873-911.

Ming, D. W., and Dixon, D. B. (1987). "Quantitative determination of clinoptilolite in soils by a cation-exchange capacity method." *Clays Clay Min.*, 35, 463-468.

Mumpton, F. A., and Fishman, P. H. (1977). "The application of natural zeolites in animal science and aquaculture." *J. Anim. Sci.*, 45, 1188-1203.

Neel, D., and Bowman, R. S. "Sorption of organics to surface-altered zeolites." *36th Annual New Mexico Water Conf.*, Las Cruces, NM, 57-61.

Newsam, J. M. (1986). "The zeolite cage structure." *Science*, 231, 1093-1099.

Sullivan, E. J., Hunter, D. B., and Bowman, R. S. (1997). "Topological and thermal properties of surfactant-modified clinoptilolite studied by tapping-mode atomic force microscopy and high-resolution thermogravimetric analysis." *Clays Clay Miner.*, 45(1), 42-53.

Chapter Nine

Effects of Surfactant Sorption on the Equilibrium Distribution of Organic Pollutants in Contaminated Subsurface Environments

SEOK-OH KO[1], MARK A. SCHLAUTMAN[2,3], AND ELIZABETH R. CARRAWAY[3]

[1] *Environmental Research Team, Daewoo Institute of Construction Technology, Seoul, Korea*
[2] *Department of Agricultural and Biological Engineering, Clemson University, Clemson, SC 29634-0357*
[3] *Department of Environmental Toxicology and the Clemson Institute of Environmental Toxicology, Clemson University, Pendleton, SC 29670*

Key words: surfactant-enhanced aquifer remediation; surfactant sorption; distribution coefficients; surfactant micelles; soil solution chemistry

Abstract: Partitioning of two hydrophobic organic contaminants (HOCs), phenanthrene and naphthalene, to surfactant micelles, kaolinite and sorbed surfactants was studied to provide further insight on (1) the effectiveness of using sorbed surfactants to remove HOCs from water and (2) the feasibility of using surfactant-enhanced aquifer remediation (SEAR) for contaminated subsurface systems. Sorbed surfactant partition coefficients (K_{ss}) showed a strong dependence on the surfactant sorption isotherms. K_{ss} values for sodium dodecyl sulfate (SDS) were always larger than the corresponding micellar partition coefficient (K_{mic}) values. For Tween 80, however, K_{ss} values were higher than K_{mic} values only at the lower sorbed surfactant concentrations. HOC distributions between immobile and mobile compartments varied with surfactant dose and solution chemistry, and were primarily dependent on the competition between sorbed and micellar surfactants for HOC partitioning. Overall results of this study demonstrate that surfactant sorption to the solid phase can lead to *increases* in HOC retardation, a desirable effect when the treatment objective is to immobilize HOCs by removing them from water but an undesirable effect in SEAR applications. Therefore, appropriate consideration must be given to surfactant sorption and HOC partitioning to immobile versus mobile phases when using surfactants to remediate contaminated subsurface systems.

1. INTRODUCTION

The widespread occurrence of hydrophobic organic contaminants (HOCs) in soils, sediments, and aquifers has led to intensive studies of the mobility and fate of these compounds in subsurface environments. Because of their low solubilities and slow dissolution/desorption rates, many HOCs are associated with the solid phase or exist as nonaqueous phase liquids (NAPLs). Depending on the particular site conditions, two complementary remediation alternatives are often considered: (1) HOC sorption to an immobile phase that subsequently decreases HOC mobility, or (2) HOC partitioning to a mobile phase that results in an increase in HOC mobility (and apparent solubility) in water. For the first approach, use of organoclays (e.g., Wagner et al., 1994; Gullick et al., 1994; Xu and Boyd, 1995) or organooxides (e.g., Holsen et al., 1991; Kibbey and Hayes, 1993; Sun and Jaffé, 1996) to remove HOCs from water has received much attention. For the second alternative, in-situ surfactant-enhanced aquifer remediation (SEAR) has been suggested as an economically and technically feasible remediation approach (e.g., Kile and Chiou, 1989; Abdul et al., 1990; Pennell et al., 1993). To date, most field-scale SEAR applications have focused on the removal of NAPLs from contaminated aquifers by utilizing NAPL mobilization (i.e., brought about by lowering the NAPL-water interfacial tension) as well as enhanced solubilization of NAPL components in surfactant micelles. In contrast to NAPLs, however, we are aware of only one field-scale SEAR application that has been used to remove HOCs existing in a sorbed state for the purpose of aquifer restoration or risk mitigation (Smith et al., 1997; Sahoo et al., 1998).

Enhanced HOC solubility in surfactant systems generally has been quantified by a distribution coefficient that only considers HOC partitioning to surfactant micelles that exist above the critical micelle concentration (CMC). Although surfactants can form a mobile micellar pseudophase that leads to the facilitated transport of solubilized HOCs, they also can be adsorbed by the solid matrix and thereby lead to HOC partitioning to immobile sorbed surfactants and, thus, enhanced HOC retardation. Therefore, the effectiveness of a remediation scheme utilizing surfactants depends on the distribution of an HOC between immobile compartments (e.g., subsurface solids, sorbed surfactants) and mobile compartments (e.g., water, micelles).

Although the partitioning of HOCs to surfactant micelles has been well studied, HOC partitioning to sorbed surfactants (e.g., hemimicelles, admicelles) has received much less attention. Holsen et al. (1991) examined the sorption of several HOCs on ferrihydrite coated with sodium dodecyl sulfate (SDS) and found that the HOCs with the lowest water solubility showed the highest sorption. Holsen et al. also found a linear relationship between HOC sorption and the amount of SDS on the ferrihydrite, suggesting that the SDS coating was primarily responsible for HOC sorption. Sun and Jaffé (1996) investigated the partitioning of phenanthrene to dianionic monomers and micelles and to the same surfactants sorbed on alumina, and found the sorbed surfactants to be generally 5 to 7 times more effective as a partitioning medium than the aqueous-phase micelles. Likewise, Nayyar et al. (1994) reported partition coefficients for several organic contaminants to SDS sorbed on alumina that were higher in value than the corresponding micellar partition coefficients. Similar experiments were performed by Sun et al. (1995) to obtain partition coefficients for three chlorinated HOCs to a silt loam with a sorbed nonionic surfactant (Triton X-100). Sun et al. observed that the sorbed surfactant increased HOC partitioning relative to the untreated soil; however, when the aqueous surfactant concentration was greater than the CMC, the micelles competed against the sorbed surfactant for HOC partitioning and led to an overall decrease in HOC distribution coefficients.

Many previous studies of HOC partitioning to sorbed surfactants examined conditions favorable for the formation of surfactant bilayers resulting from high adsorption densities; in these studies, a single partition coefficient was often observed. However, for many expected surfactant remediation applications, a surfactant solution would likely be pumped into or near the contaminated subsurface environment, and thus aqueous surfactant concentrations would vary spatially and temporally from zero to the applied concentration. Also, because soil-surfactant and surfactant-surfactant interactions lead to highly nonlinear sorption isotherms, the transport of surfactant monomers and micelles would exhibit very complex behavior. Correspondingly, the HOC distribution between immobile and mobile phases would also be expected to show complex behavior depending on the surfactant mass in each phase. Finally, it is likely that the varying soil solution chemistries found in different subsurface environments will affect surfactant sorption to solid phases and subsequent HOC partitioning to micelles and sorbed surfactants. Therefore, a quantitative evaluation of any potential surfactant remediation approach must consider the distribution of surfactant and subsequent HOC partitioning to each phase as a function of solution chemistry to maximize efficiency and minimize remediation costs.

The objectives of this work were to (1) study the equilibrium sorption characteristics of an anionic surfactant (SDS) and a nonionic surfactant (Tween 80) to kaolinite, a common soil mineral, as a function of solution chemistry; (2) examine the equilibrium partitioning of two HOCs (phenanthrene and naphthalene) to the surfactant micelles and sorbed surfactants for varying solution chemistry conditions; and (3) quantify overall HOC distribution coefficients that consider sorbed surfactant amounts and the presence of micelles as a function of surfactant dose and aqueous chemistry. Results from this investigation can be used to elucidate the role of sorbed surfactants in HOC partitioning in contaminated subsurface environments and to provide a framework for evaluating HOC removal efficiencies for alternative surfactant remediation applications.

2. EXPERIMENTAL METHODS

Detailed experimental procedures have been previously reported (Ko, 1998; Ko et al., 1998a,b); therefore, they are only briefly described here. Phenanthrene (Aldrich, 99.5+%), naphthalene (Aldrich, 99+%), SDS (Sigma, 99.5+%), and Tween 80 (Aldrich, no purity reported) were used as received; selected physicochemical properties for these compounds are shown in Table 1. Kaolinite, a nonswelling 1:1 layer phyllosilicate clay and common constituent of many subsurface environments, was used as received from Sigma. Solution pH and ionic strength were adjusted as necessary with 0.5 M HCl and/or 0.5 M NaOH and NaCl, respectively. Aqueous phenanthrene and naphthalene concentrations were quantified by fluorescence (PTI, Inc.) at the excitation/emission wavelengths of 250/364 and 278/322 nm, respectively. A total organic carbon (TOC) analyzer (Shimadzu Model 5050) was used to determine aqueous SDS concentrations and Tween 80 concentrations were determined by UV absorbance at 234 nm.

Table 1. Selected HOC and surfactant properties. [a]

compound	formula	MW	solubility in distilled water	log K_{ow}	CMC [b]
phenanthrene	$C_{14}H_{10}$	178.23	7.2 μM	4.57	-
naphthalene	$C_{10}H_8$	128.17	240 μM	3.36	-
SDS	$C_{12}H_{25}SO_4Na$	288.38	complete	-	1.45 mM [c]
Tween 80	$C_{18}S_6E_{20}$ [d]	1310	complete	-	9.92 μM [e]

[a] HOC data from Karcher (1988). Surfactant data from supplier unless noted. [b] 0.1 M NaCl. [c] Israelachvili (1992). [d] Jafvert et al. (1994). E_n is n repeating $-CH_2CH_2O-$ groups; S_6 is a sorbitan ring. [e] Pennell et al. (1993).

Surfactant equilibrium isotherms and sorption envelopes on kaolinite were determined in triplicate batch experiments for the appropriate solution chemistry conditions. After equilibration, the solids were separated by centrifugation at 7000 rpm for 30 min and aliquots of the supernatant were taken for analysis. Residual SDS and Tween 80 concentrations (S_{surf}, mM) were determined after taking into account dilution factors and system losses, and the sorbed surfactant concentrations (S_{sorb}, μmol/g) were calculated by mass balance.

Fluorescence techniques (e.g., Miyagishi et al., 1987, 1994) were used to determine surfactant CMC and micellar partition coefficient (K_{mic}) values under various solution chemistry conditions (Ko et al., 1998a,b). For some conditions, solubility enhancement experiments were also conducted to compare K_{mic} values obtained at HOC saturation. To quantify HOC partitioning to sorbed surfactants, two different types of experiments were conducted. In the *varying surfactant dose* experiments, samples containing kaolinite and the appropriate surfactant were first prepared using the procedure described above for the surfactant sorption tests (i.e., same pH, ionic strength, and surfactant concentrations). After an initial equilibration, HOCs were then added such that the total phenanthrene and naphthalene concentrations were 4.49 and 78 μM, respectively. After a second equilibration, the samples were centrifuged at 7000 rpm for 30 min to separate the aqueous and solid phases. HOC concentrations in the aqueous phase were determined by fluorescence using external standards in appropriate surfactant solutions, and HOC amounts partitioned to the sorbed surfactants were calculated by mass balance. In addition to the varying surfactant dose experiments, *fixed surfactant dose* (i.e., varying HOC concentration) experiments were conducted to test the linearity of HOC sorption isotherms for various sorbed surfactant concentrations and to determine whether surfactants sorbed to the same level but under different solution chemistry conditions would have similar HOC partitioning capabilities.

3. RESULTS AND DISCUSSION

3.1 Surfactant Sorption on Kaolinite

SDS sorption on kaolinite was relatively quick (~6 h; Ko et al., 1998b), and no significant solids effects (i.e., decreasing distribution coefficients with increasing sorbent concentrations) were observed for solid-to-liquid ratios of 1:10 to 1:4 (i.e., 100 to 250 g/L kaolinite; Fig. 1). SDS isotherms exhibited the classical S-shaped curves as previously reported (e.g., Fuerstenau and Wakamatsu, 1975; Holsen et al., 1991; Chandar et al., 1987); those studies generally used positively-charged mineral surfaces (ferrihydrite or alumina), and their sorption isotherms clearly showed three to four distinct regions, indicative of the varying importance of sorption interactions between solid surfaces and surfactant molecules. SDS sorption onto kaolinite, however, appeared to show only three distinct regions for the conditions examined here. The lower amount of SDS sorption in Region I of Fig. 1 generally results because of the electrostatic repulsion between the anionic head group (sulfate ion) of SDS and the overall negatively-charged kaolinite surface; any sorption occurring in Region I can be attributed to hydrophobic interactions between SDS tails and the kaolinite basal plane and/or to anion exchange at the small number of positively charged sites that exist. In Region II, the sharp rise in the isotherm indicates associations between SDS molecules at the surface, presumably through lateral interactions of their hydrophobic tails, and the formation of hemimicelles and/or admicelles. The sorption plateau that occurs in Region III corresponds to either an increase in electrostatic repulsion between the anionic head groups, complete surface coverage, and/or the attainment of a constant surfactant monomer concentration in the aqueous phase, possibly due to incorporation of any additional surfactant oligomers into micelles. In Fig. 1, it is noteworthy that the sorption of SDS begins to level out near its CMC value of 1.5 mM but that the isotherm becomes flat only above the CMC.

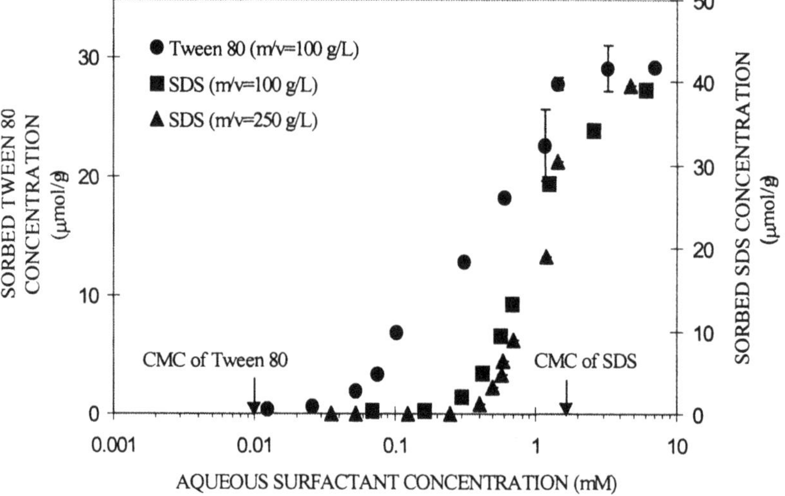

Figure 1. Surfactant sorption isotherms on kaolinite (pH 4.6, I.S. = 0.1 M). Error bars for some data points are smaller than the symbols. Kaolinite concentrations were 100 g/L (SDS, Tween 80) or 250 g/L (SDS). Adapted from Ko et al. (1998b).

Tween 80 sorption to kaolinite also shows a high degree of nonlinearity and an S-shaped curve; however, most of its sorption occurs above the CMC (Fig. 1). This observation is consistent with some previous studies (e.g., Pennell et al., 1993; Edwards et al., 1994; Sharma, 1995) that found nonionic surfactant sorption occurring well above the CMC but contrasts with other studies (e.g., Liu et al., 1992; Brownawell et al., 1997) that found the sorption of nonionic surfactants to plateau near their CMC values. It generally has been thought that surfactant sorption should reach a limiting maximum value at the CMC if the sorbing species are surfactant monomers because the concentration of monomers is constant above the CMC. Although no satisfactory explanation for the contrasting observations above is yet available, a comparison of the quantity of native organic matter present in the sorbents is noteworthy: those studies using sorbents with relatively low organic carbon mass fractions (f_{oc} < 0.06%; Pennell et al., 1993; Edwards et al., 1994; Ko et al., 1998a,b) observed sorption above the CMC, whereas the studies using sorbents with much larger amounts of native organic matter (0.76% < f_{oc} < 3.04%; Liu et al., 1992; Brownawell et al., 1997) did not.

The influence of ionic strength on surfactant sorption is shown in Fig. 2. In general, SDS sorption at 0.1 M NaCl was greater than for no added NaCl, consistent with previous observations (Xu and Boyd, 1995). Increased SDS sorption at the higher ionic strength can be explained by a decrease in the electrostatic repulsion between sorbed SDS molecules as well as between

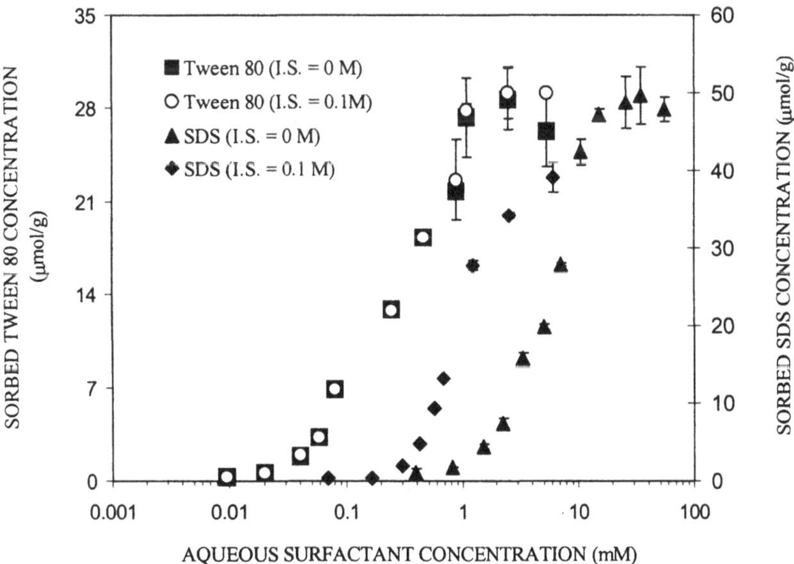

Figure 2. Surfactant sorption on kaolinite for two ionic strength conditions (as added NaCl) at pH 4.6. Error bars for some data points are smaller than the symbols. Kaolinite concentration was 100 g/L. Adapted from Ko et al. (1998a).

SDS and kaolinite (i.e., at pH 4.6 both the kaolinite surface and SDS have net negative charges). The initial enhancement of sorption that occurs with increasing NaCl at low SDS concentrations most likely results from a screening effect between SDS and kaolinite that allows SDS molecules to first sorb; enhancement at higher SDS concentrations most likely results from decreasing repulsions between sorbed SDS headgroups when hydrophobic forces become more important. For the nonionic Tween 80 surfactant, isotherms for 0 and 0.1 M NaCl show that differences in sorption are minor for these conditions, consistent with results from Brownawell et al. (1997).

pH effects on SDS sorption to kaolinite are summarized in Fig. 3. A wide range of pH values at fixed ionic strength were investigated for both sorption versus dose experiments (Fig. 3a) and sorption envelope tests (Fig. 3b). It is clear that SDS sorption decreases with increasing pH over the entire dose region (Fig. 3a), an observation that is even more dramatic when shown as a sorption envelope (Fig. 3b). This observation is consistent with the idea that a decrease in pH leads to a decrease in the negative charge density on the kaolinite surface; this, in turn, reduces the repulsive force between the surface and the negative head group of SDS molecules, thereby leading to increased SDS sorption. When the solution pH is below the point of zero charge (PZC) of kaolinite (pH 4.1 to 4.3; Ko et al., 1998a), the net surface charge of kaolinite becomes positive and more SDS molecules can be sorbed due to the electrostatic attraction between the surface and SDS head groups.

A Tween 80 sorption envelope is also shown in Fig. 3b. Tween 80 sorption generally decreases with increasing pH because of the corresponding increase of net negative surface charge on kaolinite above its PZC (i.e., pH-dependent surface sites). For example, hydrogen bonding has been suggested as the mechanism responsible for the sorption of nonionic surfactants on mineral surfaces (Cummins et al., 1992; Brownawell et al., 1997). Depending on the solution pH relative to the PZC, surface hydroxyl groups of kaolinite can be either protonated to positively-charged species or deprotonated to negatively-charged species. Therefore, as the solution pH decreases, favorable hydrogen bonding between Tween 80 oxyethylene groups and sorbed protons on the surface can occur. Conversely, increasing the solution pH results in deprotonation of surface hydroxyl groups, and therefore less hydrogen bonding between Tween 80 and kaolinite and decreased sorption.

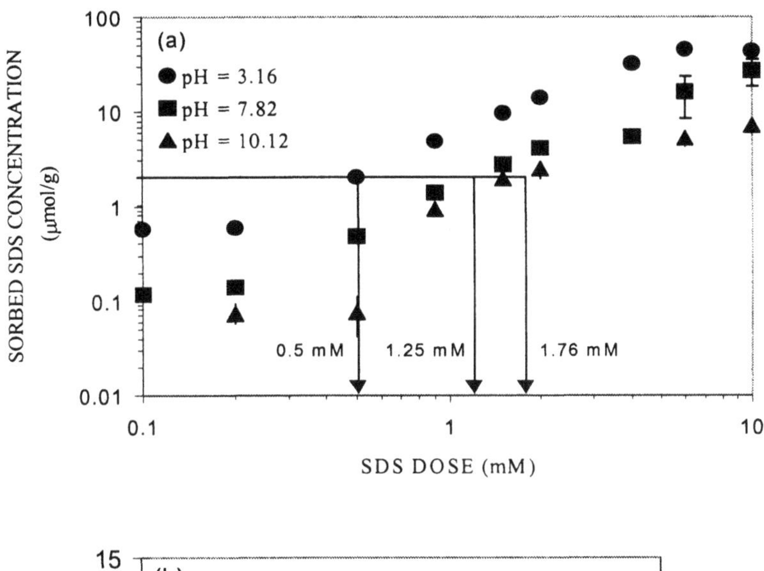

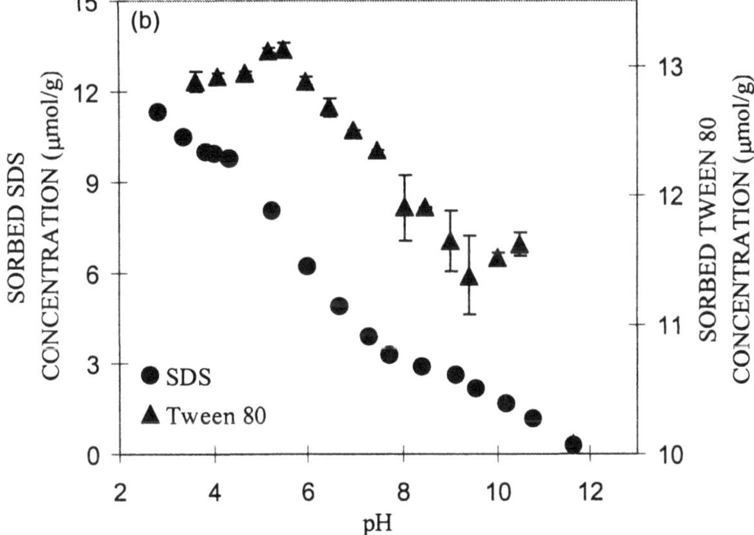

Figure 3. Surfactant sorption on kaolinite for 0.1 M NaCl. Error bars for some data points are smaller than the symbols. Kaolinite concentration was 100 g/L. (a) SDS sorption curves for three pH values. The solid lines identify pH-dependent doses necessary to obtain equivalent SDS loadings (see text). (b) Sorption envelopes. Surfactant doses were 2 and 1.67 mM for SDS and Tween 80, respectively. Adapted from Ko et al. (1998a).

3.2 HOC Micellar Solubilization

Surfactant titrations of aqueous solutions containing a hydrophobic fluorescent probe result in two distinct fluorescence regions that can provide relatively well-defined surfactant CMC values (Ko et al., 1998b). The influence of ionic strength (as added NaCl) on the CMC of SDS was dramatic (Table 2); as explained by Ko et al. (1998a) and references therein, increases in solution ionic strength lead to stronger bonding energies between SDS molecules and, therefore, lower CMC values. In contrast to ionic strength, no pH effect on the CMC of SDS was evident over the pH range 4 to 10; the latter result is consistent with the low pK_a (~2.3) of SDS. For Tween 80, CMC values showed negligible differences under varying ionic strength and pH conditions (Table 2).

Micellar partition coefficient (K_{mic}) values for phenanthrene and naphthalene below their aqueous solubility limits were determined from experimental fluorescence measurements using nonlinear regression analysis of the following equation:

$$F_t = \frac{F_w + F_{mic} \cdot K_{mic} \cdot (S_{surf} - CMC)}{1 + K_{mic} \cdot (S_{surf} - CMC)} \tag{1}$$

where S_{surf} is the total aqueous surfactant concentration, and F_t, F_w, and F_{mic} are the HOC-normalized fluorescence intensities in the total system, aqueous phase, and micellar phase, respectively (Ko et al., 1998b). Using the appropriate CMC values for SDS and Tween 80, K_{mic} values for phenanthrene and naphthalene were then determined for varying solution chemistry conditions (Tables 2 and 3). The general results show that the more hydrophobic compound, phenanthrene, has a larger partition coefficient than naphthalene and that the nonionic Tween 80 surfactant has larger K_{mic} values than does the anionic SDS surfactant, agreeing with observations from previous studies (Kile and Chiou, 1989; Nayyar et al., 1994; Park and Jaffé, 1993). As shown in Table 3, K_{mic} values obtained by fluorescence generally decreased with increasing HOC concentration. Although the deviations were not large and some values had more relative uncertainty associated with them, the decrease appears to be significant at the 95% confidence level.

An increase in the phenanthrene partition coefficient for SDS micelles is observed with increasing ionic strength at a fixed pH of 6 (Table 2). A conceptual model has been proposed to describe the effects of electrolyte addition on the partitioning of nonpolar compounds such as phenanthrene into the core (or deep region within the palisade layer) of ionic surfactant

Table 2. Surfactant critical micelle concentrations (CMC) and micellar partition coefficients (K_{mic}) for phenanthrene under various solution chemisty conditions. [a]

	SDS			Tween 80		
NaCl (M)	CMC (mM)	$K_{mic} \times 10^{-3}$ (M^{-1})	K_{oc} (L/g)	CMC (μM)	$K_{mic} \times 10^{-3}$ (M^{-1})	K_{oc} (L/g)
0	8.1	1.43 ± 0.12	10.28 ± 0.83	9.92	47.4 ± 2.15	66.02 ± 2.99
0.001	8.0	1.67 ± 0.15	11.59 ± 1.04	nd [b]	nd	nd
0.01	5.6	2.12 ± 0.26	14.72 ± 1.81	9.92	54.6 ± 2.23	76.04 ± 3.11
0.1 [c]	1.5	2.23 ± 0.41	15.83 ± 2.84	9.92	55.0 ± 1.33	76.60 ± 1.85
0.1	1.5	2.17 ± 0.17	15.07 ± 1.18	9.92	48.7 ± 3.43	67.83 ± 4.78
0.1 [d]	1.5	2.23 ± 0.14	15.49 ± 0.97	9.92	51.3 ± 7.55	71.45 ± 10.52

[a] Obtained by nonlinear regression analysis of fluorescence data using procedures described by Ko et al. (1998b). Values for K_{mic} and K_{oc} are ± SD. The pH value was 6 unless otherwise noted. [b] nd, not determined. [c] pH 4. [d] pH 10.

Table 3. HOC partition coefficients to surfactant micelles (K_{mic}) and sorbed surfactants (K_{ss}). [a]

		SDS						Tween 80					
HOC	C_t (μM)	K_{mic} [b] (M^{-1})	R^2	N[c]	K_{ss} [d] (M^{-1})	R^2	N	K_{mic} [b] (M^{-1})	R^2	N	K_{ss} [d] (M^{-1})	R^2	N
phenanthrene	1.68	2693 ± 42	0.993	15	nd [e]			63495 ± 1982	0.999	15	nd		
	2.81	2282 ± 610	0.998	15	nd			55011 ± 1337	0.996	15	nd		
	4.49	1635 ± 72	0.998	15	13804 ± 621	0.990	27	51507 ± 2810	0.999	15	58606 ± 384	0.971	30
	6.73	3467 ± 156 [f]	0.998	27	nd			35481 ± 807 [f]	0.999	15	nd		
naphthalene	39	381 ± 5.1	0.999	24	nd			nd			nd		
	117	315 ± 22	0.998	24	525 ± 24 [g]	0.959	27	2754 ± 408	0.996	18	8323 ± 579 [g]	0.960	30
	195	152 ± 27	0.999	24	nd			nd			nd		
	241	245 ± 5.1 [f]	0.998	24	nd			2239 ± 289 [f]	0.986	15	nd		

[a] From experiments using fixed HOC concentrations (C_t) and varying surfactant concentrations. Values are ± SD. Ionic Strength = 0.1 M NaCl and pH 4.0 (K_{mic}) or 4.6 (K_{ss}). [b] Determined by nonlinear regression of eq 1 except as noted. [c] N, number of data points. [d] Determined by nonlinear regression of eq 2. [e] nd, not determined. [f] Obtained by linear regression of solubility enhancement data. [g] Naphthalene concentration was 78 μM.

micelles (Attwood and Florence, 1983). For example, displacement of the CMC of ionic surfactants to a lower value as a result of electrolyte addition leads to increased partitioning overall because of the increased fraction of surfactant molecules existing in the micellar form (Attwood and Florence, 1983; Israelachvili, 1992); however, this effect does not explain the increase in K_{mic} observed here because the values have already been normalized for the concentration of micelles present. "Salting out" effects can lead to increases in HOC partition coefficients; however, using a Setschenow constant of 0.28 M^{-1} for phenanthrene (Schwarzenbach et al., 1993), calculations show that the K_{mic} value should have increased only ~6% for the 0.1 M NaCl solution, a value much smaller than the ~47% relative difference observed here. Therefore, it appears that the differences in K_{mic} values for phenanthrene are caused primarily by changes in SDS micellar properties with ionic strength. No significant effects of solution pH were observed for SDS solubilization of phenanthrene (Table 2). Again, this can be attributed to the fact that SDS molecules have strong dissociation characteristics (i.e., low pK_a of ~2.3). For Tween 80, little to no effects on phenanthrene solubilization were observed with changing solution chemistry conditions. A few studies, however, have reported increases in the solubilization of organic compounds by nonionic micelles at ionic strength values much higher than the range used here (Attwood and Florence, 1983).

3.3 Equilibrium Partitioning of HOCs to Sorbed Surfactants

Figure 4 shows phenanthrene and naphthalene sorption isotherms to kaolinite covered with varying levels of sorbed surfactant; these levels of surfactant coverage correspond to the different regions existing in the surfactant sorption isotherms discussed earlier (Fig. 1). The linearity of each isotherm was evaluated using Freundlich and linear sorption models. It is apparent from Fig. 4 and Table 4 that HOC partitioning to kaolinite with and without adsorbed surfactants results in linear or near-linear isotherms. As the amount of surfactant adsorbed on the kaolinite surface increased, the sorption of phenanthrene and naphthalene to the solid phase also increased. However, upon normalizing by the amount of sorbed surfactant present, the sorbed surfactant partition coefficient (K_{ss}) decreased with increasing sorbed surfactant amounts (Table 4).

The effectiveness of a treatment/remediation scheme utilizing surfactants will depend on the distribution of an HOC between immobile and mobile phases, which is commonly quantified by a distribution coefficient (K_D). Representative results for K_D as a function of aqueous surfactant concentration are shown in Fig. 5. These experiments were conducted by holding total HOC concentrations constant and varying the surfactant doses; K_D values were then determined directly after centrifuging to remove solids. At low aqueous SDS concentrations, K_D values increased with increasing surfactant concentration (Fig. 5a,b) because the amount of SDS adsorbed to kaolinite increases rapidly in this region (Fig. 1) and because the sorbed SDS is very effective for HOC partitioning (Holsen et al., 1991; Jafvert, 1991). When the aqueous phase SDS concentration reaches its CMC, SDS sorption to kaolinite plateaus and micelles begin competing for HOC molecules, thereby causing a decrease in K_D. For Tween 80, all but one data point is above the CMC; thus, the HOCs are partitioning to both surfactant phases (sorbed surfactant and micelles) over the majority of the concentration range examined (Fig. 5c,d). K_D values initially increase at the lower concentrations near the CMC because the affinity of sorbed Tween 80 for the two HOCs is greater than that of the micellar Tween 80 in this region. As the amount of micellar Tween 80 becomes larger relative to the sorbed Tween 80, K_D values begin to decrease. Note that this decrease in K_D occurs even though the Tween 80 sorption to kaolinite has not yet reached its maximum sorption plateau (Fig. 1).

Our results for HOC partitioning in the presence of sorbed surfactant and micelles demonstrate that large differences can exist in the HOC sorption capacity of surfactant aggregates in micellar versus sorbed forms. This can be seen quite readily by calculating K_{ss} values as a function of surfactant dose from the experimental K_D values. The distribution coefficient defines the HOC mass balance and can be expressed as:

$$K_D = \frac{C_{immob}}{C_{mob}} = \frac{S_{sorb} \cdot K_{ss} + K_{min}}{1 + S_{mic} \cdot K_{mic}} \qquad (2)$$

where C_{immob} (mol/g-kaolinite) and C_{mob} (mol/L) are the immobile and mobile HOC concentrations, respectively, S_{sorb} is the sorbed surfactant concentration (mole/g-kaolinite), and K_{min} (L/g-kaolinite) is the HOC sorption constant to the bare kaolinite surface. From the previously determined values for micellar solubilization (K_{mic}), surfactant distribution (S_{sorb} and S_{mic}) and HOC sorption to kaolinite (K_{min}), K_{ss} values can be calculated for each distribution data point using eq 2. Alternatively, an overall average K_{ss} value can be calculated by fitting eq 2 to the distribution data using nonlinear regression analysis.

Average K_{ss} values determined by nonlinear regression of eq 2 are shown in Table 3. As expected, these values are intermediate of the ones calculated for each individual data point (Fig. 5). In all cases, K_{ss} values were larger for

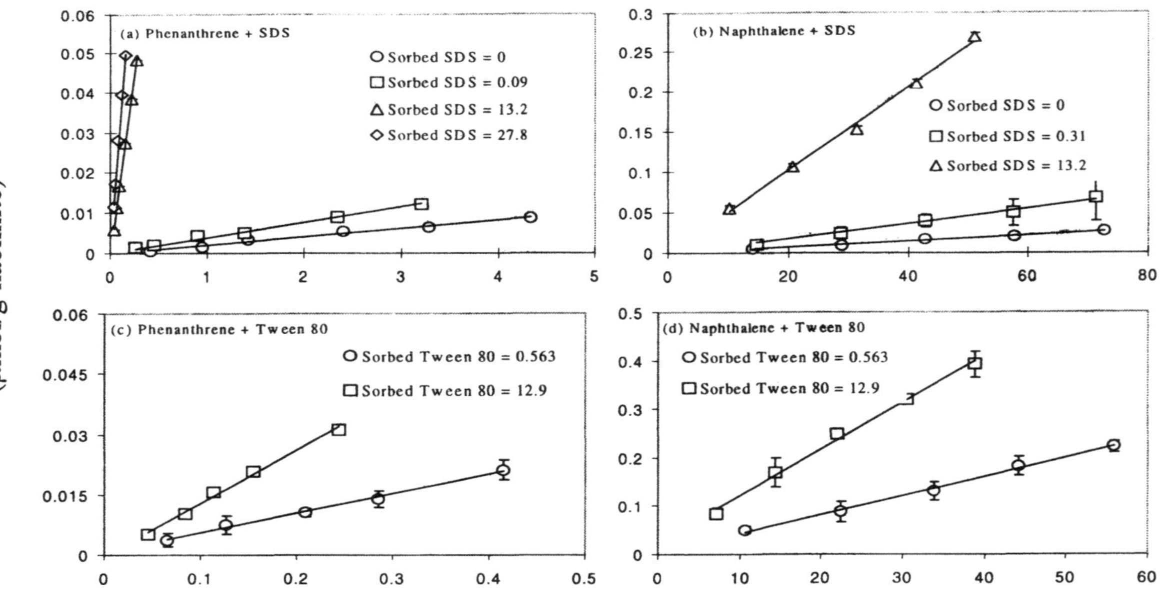

Figure 4. HOC sorption isotherms on kaolinite with varying concentrations of sorbed surfactant (units of μmol/g-kaolinite). Error bars for some data points are smaller than the symbols. Kaolinite concentration was 100 g/L, except for naphthalene sorption to bare kaolinite (i.e., sorbed SDS = 0), where it was 300 g/L. Adapted from Ko et al. (1998b).

Table 4. Parameters of the HOC sorption isotherms for sorbed surfactants. [a]

		S_{sorb} [b] (µmol/g)	N[e]	Freundlich Model [c]			Linear Model [d]	
				K_{ss} (M^{-1})	n	R^2	K_{ss} (M^{-1})	R^2
phenanthrene								
	SDS	0.0	18	0.002 ± 2E-4 [f]	0.996 ± 0.107	0.994	0.002 ± 5E-5 [f]	0.986
		0.09	21	51503 ± 2540	0.972 ± 0.067	0.995	51111 ± 641	0.991
		13.2	18	13149 ± 1134	0.992 ± 0.029	0.999	13497 ± 584	0.999
		27.8	15	11480 ± 1229	0.992 ± 0.026	0.992	11934 ± 231	0.999
	Tween 80	0.563	15	123027 ± 3844	1.083 ± 0.035	0.994	162885 ± 7994	0.992
		12.9	15	85113 ± 2201	0.932 ± 0.011	0.995	65228 ± 1477	0.994
naphthalene								
	SDS	0.0	15	3.04E-4 ± 8.3E-6 [f]	1.091 ± 0.111	0.996	3.64E-4 ± 1.9E-5 [f]	0.992
		0.31	15	2319 ± 229	1.121 ± 0.096	0.992	2932 ± 338	0.985
		13.2	15	367 ± 44	1.033 ± 0.015	0.995	385 ± 8.7	0.996
	Tween 80	0.563	15	8511 ± 841	1.091 ± 0.209	0.989	7304 ± 772	0.993
		12.9	15	1122 ± 104	1.107 ± 0.125	0.991	977 ± 68	0.982

[a] From experiments using varying HOC concentrations and fixed sorbed surfactant concentrations (S_{sorb}). Values for K_{ss} and n are ± SD. Ionic strength = 0.1 M NaCl and pH 4.6. [b] Maximum sorption plateaus for SDS and Tween 80 were approximately 40 and 29 µmol/g, respectively. [c] Model parameters determined from nonlinear regression analysis of $q_{HOC} = K_{ss} C_{HOC}^{n}$. [d] Model parameters determined from linear regression analysis of $q_{HOC} = K_{ss} C_{HOC}$. [e] N, number of data points. [f] Values are K_{min} (L/g-kaolinite).

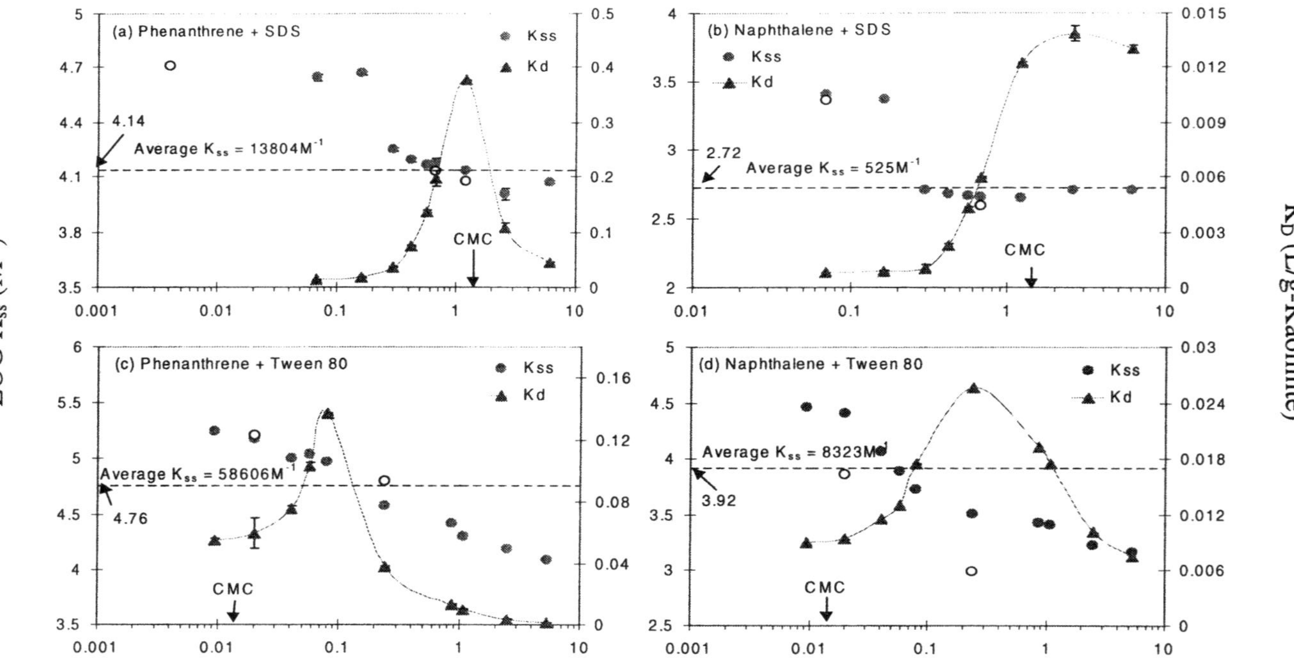

Figure 5. HOC distribution (K_D) and sorbed surfactant partition (K_{ss}) coefficients. Kaolinite concentration was 100 g/L. Individual K_{ss} values (filled circles) were determined from the K_D values using eq 2 and the micellar partition (K_{mic}) and kaolinite sorption (K_{min}) constants below. Isotherm K_{ss} values (open circles) are from Table 4 (linear values). (a) $K_{mic} = 1635\ M^{-1}$. $K_{min} = 0.002$ L/g. (b) $K_{mic} = 280\ M^{-1}$. $K_{min} = 0.0003$ L/g. (c) $K_{mic} = 51507\ M^{-1}$. $K_{min} = 0.002$ L/g. (d) $K_{mic} = 2496\ M^{-1}$. $K_{min} = 0.0003$ L/g. Adapted from Ko et al. (1998b).

the more hydrophobic HOC (phenanthrene) and for the nonionic surfactant (Tween 80). For both SDS and Tween 80, the average K_{ss} values calculated for phenanthrene and naphthalene were always larger than the K_{mic} values at equivalent HOC concentrations (Table 3). When all K_{mic} values in Table 3 are considered, only one (i.e., Tween 80 for a phenanthrene concentration of 1.68 μM) is larger than the corresponding K_{ss}. Previous studies have also reported sorbed surfactant partition coefficients that were generally larger than the micellar partition coefficients (Nayyar et al., 1994; Mukerjee et al., 1995; O'Haver and Harwell, 1995; Sharma, 1995; Sun and Jaffé, 1996). No convincing explanation for this observation has yet been advanced; presumably it results from geometric differences between sorbed and micellar surfactant aggregate structures.

Individual K_{ss} values for SDS calculated directly from each HOC distribution data point clearly show two distinct ranges that generally correspond well with the locations of the different surfactant sorption regions (i.e., compare the filled circles in Fig. 5a,b with Fig. 1). For both phenanthrene and naphthalene, as the amount of surfactant sorbed to kaolinite increased, the respective K_{ss} values decreased. In addition, the individual K_{ss} values showed excellent agreement with the isotherm K_{ss} values (filled vs. open circles, respectively, in Fig. 5). For phenanthrene, all individual and isotherm K_{ss} values obtained for SDS were much larger than any of the K_{mic} values observed (Fig. 5a and Table 3). Although all K_{ss} values for naphthalene and SDS were also larger than the K_{mic} values, the difference was much smaller; in fact, K_{ss} values approached K_{mic} as sorbed SDS levels increased.

The dependence of K_{ss} on sorbed SDS levels appears to be qualitatively consistent with proposed surfactant structures at mineral surfaces (Fuerstenau and Wakamatsu, 1975; Holsen et al., 1991; Chandar et al., 1987). For example, the configuration of adsorbed SDS in Region I is expected to be different from that in Regions II and III, where micelle-like structures are thought to exist at these relatively higher sorption densities. Apparently, these differences in sorbed surfactant structure that result from regional sorption mechanisms and sorption densities lead to regional differences in K_{ss} values.

In contrast to SDS, calculated individual K_{ss} values for Tween 80 did not show distinct ranges but instead decreased monotonically over the surfactant concentration range examined (Fig. 5c,d). In addition, although the decreasing trend exhibited by individual K_{ss} values (filled circles) agreed qualitatively with that observed for the isotherm K_{ss} values (open circles), there was not the same good agreement in values as was obtained for SDS. This is most likely attributable to the fact that Tween 80 micelles existed throughout the surfactant concentration range studied (e.g., see CMC arrows in Fig. 5); as can be seen in eq 2, calculation of individual K_{ss} values for Tween 80 thus required the additional input values (and associated uncertainties) of K_{mic} and S_{mic}. Conversely, the close agreement in values for SDS resulted primarily because, with the exception of Region III, HOC partitioning to sorbed SDS did not have to compete with partitioning to SDS micelles over the majority of the SDS sorption regions examined (Fig. 5a,b). Although average K_{ss} values for Tween 80 were larger than corresponding K_{mic} values, some of the individual and isotherm K_{ss} values actually fell below K_{mic}. Whether this proves to be a general observation for nonionic surfactants or is merely a result of the complications from working with Tween 80 will need to be investigated in future studies.

The influence of ionic strength (as added NaCl) on phenanthrene partitioning to sorbed SDS was investigated at pH 6.5 (Table 5 and Fig. 6a). As discussed previously, SDS sorption increases with increasing ionic strength; therefore, one would logically expect phenanthrene sorption to the solid phase (which includes sorbed SDS aggregates) to correspondingly increase with ionic strength. To determine whether SDS molecules sorbed under different ionic strength conditions exhibit different partitioning characteristics, carbon-normalized partition coefficients (K_{oc}) were calculated from the linear distribution coefficients (K_D). The organic carbon fraction (f_{oc}, % mass of carbon sorbed per sorbent mass) was calculated from SDS sorption results for the same pH and dose (i.e., 2 mM). K_{oc} values for sorbed SDS were approximately 10 times greater than those for SDS micelles (Tables 2 and 5), indicating a higher affinity of the sorbed SDS molecules for phenanthrene. Sorbed SDS K_{oc} values increased with increasing ionic strength; although the percentage increase in phenanthrene partitioning was slightly less than that observed for SDS micelles (i.e., 20% versus 30% for a NaCl increase from 0.001 to 0.1 M), the increase was still larger than the estimated 6% increase expected due to salting out.

Phenanthrene sorption isotherms for varying pH conditions at 0.1 M NaCl were observed to be linear (Fig. 6b); therefore, distribution coefficients were determined using linear regression analysis and K_{oc} values were calculated as before (Table 5). K_D values decreased with increasing pH, which can be attributed primarily to decreased SDS sorption at high pH (Holsen et al., 1991). Interestingly, Fig. 7 and Table 5 show that sorbed SDS K_{oc} values can be divided into two distinct regions depending on whether the solution pH is above or below the PZC of kaolinite; K_{oc} values below the PZC (greater SDS sorption) are much greater than those above the PZC (less SDS sorption). This observation contrasts with the results presented earlier where phenanthrene K_{oc} values decreased as the amount of sorbed SDS increased under fixed solution chemistry conditions when the pH was above the kaolinite PZC (Fig. 5 and Table 4). However, those results showed that

Table 5. Phenanthrene distribution (K_D) and organic carbon normalized partition (K_{oc}) coefficients to sorbed surfactants on kaolinite for varying solution chemistry conditions. [a]

	pH	S_{sorb} (μmol/g)	K_D [b] (L/g)	R^2	N [c]	f_{oc} [d] (%)	K_{oc} (L/g)
SDS	3.38	10.51	0.556 ± 0.012	0.976	18	0.1509	368 ± 7.6
	3.85	9.99	0.383 ± 0.010	0.992	15	0.1434	264 ± 5.9
	4.32	9.80	0.222 ± 0.005	0.988	18	0.1406	155 ± 3.5
	5.25	8.07	0.100 ± 0.020	0.999	15	0.1156	86.5 ± 1.4
	5.87	7.79	0.098 ± 0.008	0.991	18	0.1122	87.6 ± 7.4
	6.50						
	I.S. = 1×10^{-3} M	0.84	0.0144 ± 0.002	0.996	18	0.0121	119 ± 14.1
	I.S. = 1×10^{-2} M	0.87	0.0168 ± 0.001	0.999	15	0.0125	135 ± 9.6
	I.S. = 1×10^{-1} M	3.51	0.0727 ± 0.003	0.983	18	0.0505	144 ± 5.9
	7.85	3.26	0.068 ± 0.004	0.992	15	0.0470	145 ± 10.4
	8.40	2.91	0.055 ± 0.005	0.997	18	0.0410	131 ± 10.4
	9.11	2.62	0.051 ± 0.002	0.991	15	0.0369	138 ± 5.6
Tween 80	4.60	12.90	0.1301 ± 0.0015	0.994	18	0.926	90.9 ± 2.1
	6.25	12.13	0.0289 ± 0.0008	0.987	18	0.871	71.2 ± 3.4
	7.85	9.47	0.0157 ± 0.0005	0.973	15	0.679	89.6 ± 2.7

[a] Values for K_D and K_{oc} are ± SD. Ionic strength = 0.1 M unless noted. Dose was 2 and 1.527 mM for SDS and Tween 80, respectively. [b] Obtained from linear regression analysis of experimental isotherms. [c] Number of data points. [d] For SDS, f_{oc} was calculated from the theoretical carbon content specified by the chemical formula. For Tween 80, the carbon content was determined by TOC measurements.

higher partition coefficients for sorbed SDS are obtained in Region I of the SDS sorption isotherm whereas the sorbed SDS here exists as surface aggregates such as admicelles or hemicelles for the particular dose used (i.e., 2 mM SDS, which corresponds to sorption in Region II and possibly III as shown in Figs. 2 and 3a).

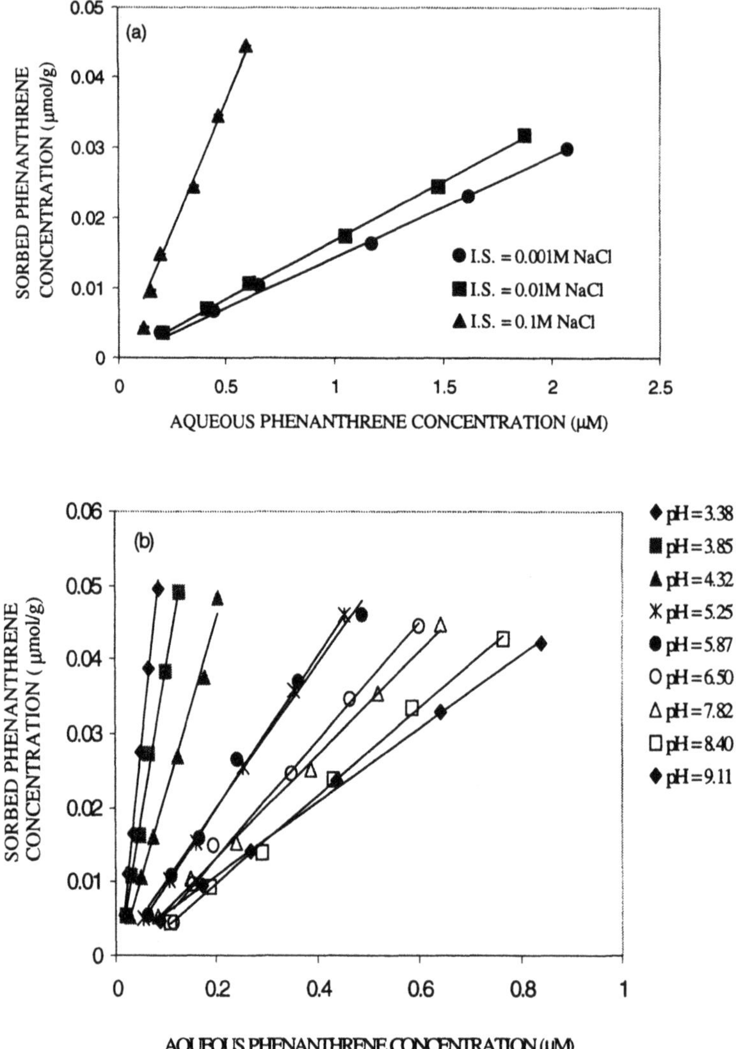

Figure 6. Phenanthrene sorption isotherms on kaolinite with sorbed SDS. Error bars for some data points are smaller than the symbols. Kaolinite concentration was 100 g/L and the SDS dose was 2 mM. (a) Ionic strength dependence at pH 6.5. (b) pH dependence at 0.1 M NaCl.

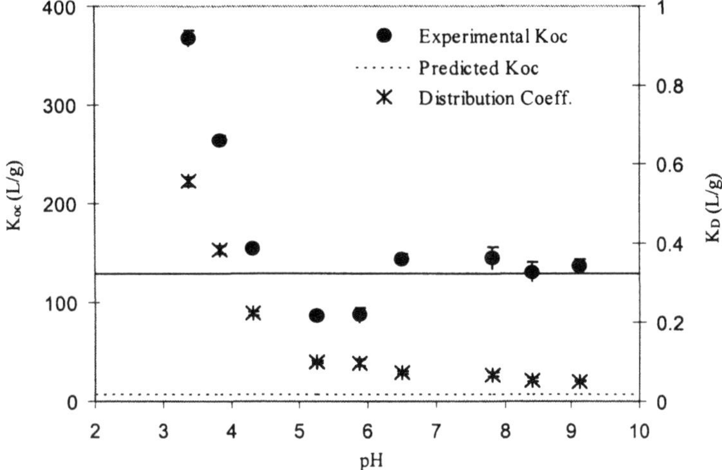

Figure 7. Effect of pH on phenanthrene distribution (K_D) and organic carbon normalized partition (K_{oc}) coefficients for sorbed SDS. The ionic strength was 0.1 M NaCl, and the kaolinite concentration was 100 g/L. Error bars for some data points are smaller than the symbols. The solid line is the average K_{oc} value for pH values above the PZC of kaolinite. The predicted K_{oc} value is based on an equation from Schwarzenbach et al. (1993). Adapted from Ko et al. (1998a).

To further examine differences in partitioning to sorbed SDS aggregates, phenanthrene sorption tests were conducted on kaolinite loaded with the same amount of SDS under different pH conditions (i.e., q = 1.95 μmol/g as shown in Fig. 3a). Ideally, phenanthrene isotherms and distribution coefficients should have been identical regardless of pH because of the constant f_{oc} value. However, it is obvious from Fig. 8 that the K_D value for the isotherm below the PZC of kaolinite (i.e., the one conducted at pH 3.2) is much greater than those two conducted above the PZC (i.e., pH 7.8 and 10.1). K_{oc} values calculated from the K_D values and sorbed SDS amounts were 435 ± 9.5, 122 ± 2.1, and 150 ± 3.0 L/g for pH 3.2, 7.8, and 10.1, respectively. These values agree well with those obtained using a constant SDS dose of 2 mM (Table 5); additionally, the values fall into the same two regions described above. These results suggest that sorbed SDS aggregates formed at pH values below and above the PZC of a mineral surface will have different HOC partitioning characteristics. A variety of interactions between SDS and kaolinite can occur depending on the surface charge of kaolinite, thus resulting in different sorbed surfactant structures (e.g., SDS monolayers versus bilayers) (Ko et al., 1998a). Consequently, sorbed surfactant partitioning results can thus be interpreted as HOC molecules partitioning to these different surface structures (Ko et al., 1998a).

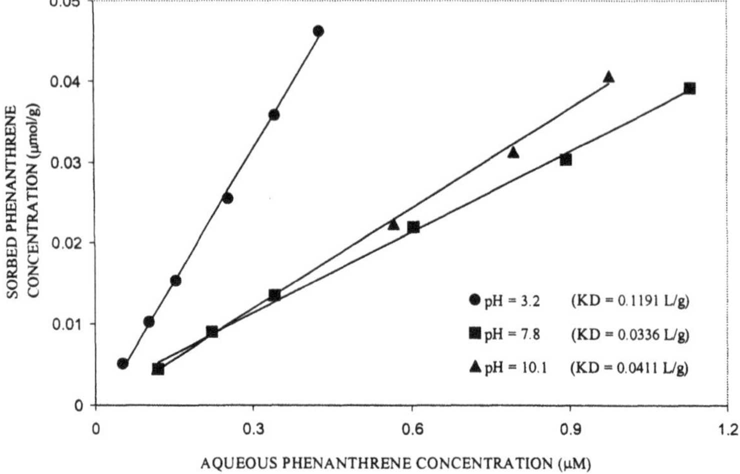

Figure 8. Phenanthrene sorption isotherms on kaolinite loaded with the same amount (1.95 μmol/g) of sorbed SDS under different pH conditions. The ionic strength was 0.1 M NaCl, and the kaolinite concentration was 100 g/L. Error bars are smaller than the symbols. Solid lines are the best fits from linear regression and represent the K_D values. Adapted from Ko et al. (1998a).

The influence of pH on phenanthrene partitioning to sorbed Tween 80 is summarized in Fig. 9 and Table 5. The Tween 80 dose for these experiments was 1.67 mM, which is well above the CMC value; therefore, Tween 80 micelles contributed to the overall phenanthrene distribution. As can be seen, K_D values decrease with increasing pH. This trend was expected because of the decrease in Tween 80 sorption with increasing pH (Fig. 3b) and the concurrent increase in the number of Tween 80 micelles. Similarly to sorbed SDS, the affinity of sorbed Tween 80 for phenanthrene can be evaluated by calculating K_{oc} values. Because of the presence of micelles, however, a mass balance that includes micelles and sorbed surfactant is required to calculate the partitioning capacity of sorbed Tween 80, expressed previously as K_{ss} in eq 2. K_{ss} values calculated with this equation were then converted to K_{oc} values using the average carbon fraction of Tween 80 (i.e., 0.548 as determined by TOC analysis). As shown in Table 5, K_{oc} values for sorbed Tween 80 are similar in magnitude to those reported earlier for Tween 80 micelles (Table 2). K_{oc} values for the pH conditions investigated here showed no significant differences, implying that the sorbed Tween 80 aggregates had similar structures and/or physicochemical characteristics.

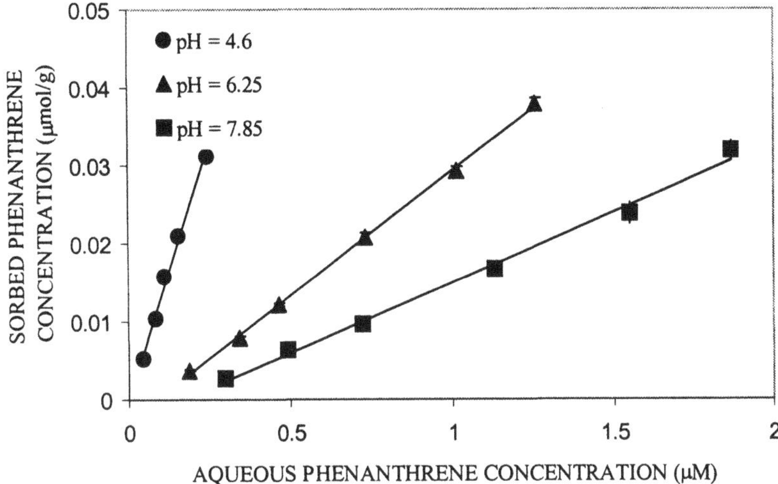

Figure 9. Phenanthrene sorption isotherms on kaolinite with sorbed Tween 80 for three pH values at 0.1 M NaCl. Error bars for some data points are smaller than the symbols. Kaolinite concentration was 100 g/L and the Tween 80 dose was 1.67 mM.

4. IMPLICATIONS FOR SUBSURFACE REMEDIATION ALTERNATIVES USING SURFACTANTS

Depending on the desired treatment methodology and goals, addition of surfactants to a subsurface system should either increase HOC distribution coefficients (i.e., immobilization approach) or decrease them (i.e., mobilization objective as in many SEAR applications). For example, distribution coefficients for phenanthrene and naphthalene to kaolinite are 0.002 and 0.0003 L/g, respectively (Table 4). Therefore, if enhanced mobilization of these HOCs in a similar type of aquifer system was desired, addition of a surfactant would have to bring the distribution coefficients below these values. However, as can be seen in Fig. 5, all distribution coefficients for the surfactant doses investigated here are larger than these values, even when the doses and subsequent aqueous surfactant concentrations are well above the CMC. This observation results from a combination of surfactant sorption followed by HOC partitioning to the sorbed surfactant.

A specific numerical example is enlightening. The K_D value of naphthalene for an initial Tween 80 dose of 7.63 mM is 0.0075 L/g (i.e., the last point in Fig. 5d corresponding to S_{surf} = 5.32 mM; this dose also corresponds to a sorbed Tween 80 concentration of ~29 μmol/g as can be seen in Fig. 1). This value greatly exceeds the K_D when no Tween 80 is present in the system. In other words, addition of Tween 80 to this model system or a similar type of aquifer system would lead to an increase in naphthalene retardation, not a decrease as would be desired for SEAR applications. Because the plateau for Tween 80 sorption occurs near the above dose (i.e., 770 × CMC, Fig. 1), it is expected that much higher doses would be needed to overcome the enhanced retardation effects caused by Tween 80 sorption to the aquifer matrix. These very high surfactant doses, however, would likely not be practical for real world applications. A more useful approach, therefore, would be to utilize the Tween 80 to increase retardation in this particular system; the optimum dose to use would correspond to a maximum K_D value.

In any evaluation of a remediation scheme utilizing surfactants, the effect of dose on HOC distribution coefficients must be quantified. Very often, only one coefficient value for HOC partitioning to sorbed surfactants has been reported in the literature, presumably because the experimental data covers only the sorption regions where the surfactant molecule interactions dominate at the surface (Nayyar et al., 1994; Park and Jaffé, 1993). However, all of the characteristic sorption regions will develop during an in-situ SEAR application as the surfactant front (i.e., mass transfer zone) advances through the porous medium. Therefore, the relative role of regional HOC partition coefficients to sorbed surfactant should be considered in any remediation process. Finally, the porosity or solid volume fraction for the particular subsurface system must be taken into account when surfactant sorption is quantified.

In addition to the equilibrium aspects discussed above, the rates of all forward and reverse processes, including HOC partitioning to surfactant micelles and sorbed surfactant, and HOC and surfactant sorption to aquifer materials, may need to be quantified (e.g., Deitsch and Smith, 1995; Sahoo and Smith, 1997; Yeom et al., 1995, 1996; Deitsch, following chapter). By taking into account all of these equilibrium and rate processes, numerical models may become effective tools for the screening/design (e.g., Ko and Schlautman, 1998) and/or application (e.g., Smith et al., 1997; Sahoo et al., 1998) of SEAR processes. To provide a truly realistic prediction/analysis of SEAR effectiveness, site-specific data including important HOC and surfactant rate processes must be acquired for input to the models. With the appropriate information and proper analysis, addition of surfactants can then be optimized to achieve increased performance, shorter remediation times, and/or lower overall remediation costs.

Selection of an appropriate surfactant is critical to the ultimate success of SEAR applications. Based on the results presented here as well as from other studies (e.g., Sahoo et al., 1998), it can be seen that surfactants having high HOC solubilization capabilities and low solid phase sorption potentials are desirable for SEAR processes. Unfortunately, these two desirable traits are often mutually exclusive. Therefore, other types of HOC solubilizing/facilitating agents may need to be considered in addition to conventional micelle-forming chemical surfactants for SEAR applications. For example, cyclodextrins have been shown to be effective for removing sorbed HOCs from contaminated aquifer systems because of their negligible sorption loss to the solid phase (Ko et al., 1999, 2000, and references therein); in some cases, the absence of cyclodextrin sorption will more than compensate the fact that they tend to have lower solubilization capabilities than do conventional surfactant micelles. Additionally, naturally-occurring organic materials (NOMs) such as humic and fulvic acids have been shown to be effective for removing sorbed HOCs from saturated aquifer materials (e.g., Johnson and Amy, 1995; Johnson, 2000). Although NOMs also adsorb to aquifer solids, their sorption tends to be less than that for conventional surfactants and, unlike conventional surfactants, the ability of HOCs to partition to sorbed NOMs is less than that for the dissolved NOM constituents (e.g., Schlautman and Morgan, 1993; Hur and Schlautman, 2000). Therefore, even though some fraction of NOMs may be lost to the immobile phase by sorption, overall HOC distribution coefficients will not show the large adverse impacts observed for conventional surfactants in SEAR processes as demonstrated in Fig. 5. Finally, the potential for aquifer plugging by surfactants or other facilitating agents must be considered during the selection process (e.g., Ko et al., 2000). For example, in some cases surfactant flushing of saturated soil or aquifer systems can disperse colloidal particles and thus lead to clogging of the pores, which significantly reduces flow through the contaminated subsurface material (Abdul et al., 1990).

One last consideration during the selection process of a suitable surfactant or other facilitating agent must be an examination of toxicity and biodegradability issues so that no adverse impact on the environment or on human health occurs. For example, upon completion of SEAR, any residual HOCs or surfactants remaining in the aquifer should be easily biodegradable or, at minimum, have a relatively low toxicity. Because NOMs and cyclodextrins are naturally-occurring materials, they may have less of an environmental impact than conventional chemical surfactants and may be more readily acceptable to the public for use in SEAR applications. Bioavailability of the HOCs partitioned to the facilitated phase may need to be considered if biological treatment of the effluent from a SEAR process is desired. For example, recent studies have shown that HOCs solubilized in surfactant micelles may or may not be readily available for biodegradation depending on the specific surfactant types and concentrations used (e.g., Laha and Luthy, 1991; Guha and Jaffé, 1996; Guha et al., 1998; Yeh et al., 1999); for cyclodextrins, a recent report suggests that HOC biodegradation will be enhanced when they are used as solubilizing agents (Wang et al., 1998). Finally, it has been reported that the biodegradation of some nonionic surfactants may actually stimulate HOC biodegradation in contaminated soils (Tiehm et al., 1997).

5. CONCLUSIONS

Sorption of an anionic surfactant, SDS, and a nonionic surfactant, Tween 80, on kaolinite resulted in highly nonlinear, S-shaped isotherms. SDS sorption on kaolinite plateaued slightly above its CMC value, with most of the sorption being attributed to hydrophobic interactions between SDS molecules. For Tween 80, sorption continued well beyond its CMC value; overall sorption for the nonionic surfactant can be attributed to hydrophobic interactions and/or hydrogen bonding between adjacent Tween 80 molecules or between the molecules and the kaolinite surface. SDS sorption on kaolinite showed strong pH and ionic strength dependency, with sorption increasing as pH decreased and ionic strength increased, respectively. For Tween 80, only pH changes affected its sorption on kaolinite for the conditions examined here.

Micellar partition coefficients (K_{mic}) increased with increasing ionic strength for SDS, but were not affected by changes in pH. For Tween 80, K_{mic} values showed no dependence on pH or ionic strength. Sorbed SDS and Tween 80 had very high affinities for phenanthrene and naphthalene; in most cases, the sorbed surfactant partition coefficients (K_{ss}) were higher than the corresponding K_{mic} values. The variations between K_{ss} and K_{mic} values are presumably due to geometric differences between the sorbed and micellar surfactant aggregates. In general, K_{ss} values showed a dependence on the amount and nature of sorbed surfactant that corresponded to the surfactant sorption isotherm. For sorbed SDS aggregates, organic carbon normalized K_{ss} values (i.e., K_{oc}) fell into two general regions depending on whether the pH was above or below the point of zero charge (PZC) of kaolinite. Plausible structural differences for sorbed SDS aggregates help to explain these regional K_{oc} values.

HOC distribution coefficients (K_D) between the immobile and mobile compartments varied as a function of solution chemistry and surfactant dose. K_D values increased with increasing surfactant dose at low surfactant concentrations, but decreased at higher doses. This trend indicates that competition for HOC partitioning occurs between sorbed- and micellar-phase surfactants. Overall, the results of this study demonstrate that retardation of HOCs by surfactants sorbed to the solid phase can be significant and must be considered for proper evaluation of treatment/remediation alternatives that use surfactants.

ACKNOWLEDGMENTS

We gratefully acknowledge the constructive comments of the three reviewers and their suggestions for improving the chapter. Funding of the work described here was provided by the Gulf Coast Hazardous Substance Research Center, which is supported under cooperative agreement R822721-01-4 with the United States Environmental Protection Agency. The contents of this chapter do not necessarily reflect the views and policies of the U.S. EPA nor does the mention of trade names or commercial products constitute endorsement or recommendation for use.

BIBLIOGRAPHY

Abdul, A.S., Gibson, T.L., and Rai D.N. (1990). "Selection of surfactants for the removal of petroleum products from shallow sandy aquifers." *Groundwater*, 28, 920-926.

Attwood, D., and Florence, A.T. (1983). *Surfactant Systems: Their Chemistry, Pharmarcy and Biology*. Chapman and Hall, New York.

Brownawell, B.J., Chen, H, Zhang, W. and Westall, J.C. (1997). "Sorption of nonionic surfactants on sediment materials." *Environ. Sci. Technol.*, 31, 1735-1741.

Chandar, P., Somasundaran, P., and Turro, N.J. (1987). "Fluorescence probe studies on the structure of the adsorbed layer of dodecyl sulfate at the alumina-water interface." *J. Colloid and Interface Sci.*, 117, 31-46.

Cummins, P.G., Penfold, J., and Staples, E. (1992). "Nature of the adsorption of the nonionic surfactant pentaethylene glycol monododecyl ether on a ludox silica sol." *J. Phys. Chem.*, 96, 8092-8094.

Deitsch, J.J., and Smith, J.A. (1995). "Effect of Triton X-100 on the rate of trichloroethene desorption from soil to water." *Environ. Sci. Technol.*, 29, 1069-1080.

Edwards, D.A., Adeel, Z. and Luthy, R.G. (1994). "Distribution of nonionic surfactant and phenanthrene in a sediment/aqueous system." *Environ. Sci. Technol.*, 28, 1550-1560.

Fuerstenau, D.W., and Wakamatsu T. (1975). "Effect of pH on the adsorption of sodium dodecane sulphonate at the alumina/water interface." *Faraday Discussions of the Chemical Society*, 59, 157-168.

Guha, S., and Jaffé, P.R. (1996). "Biodegradation kinetics of phenanthrene partitioned into the micellar phase of nonionic surfactants." *Environ. Sci. Technol.*, 30, 605-611.

Guha, S., Jaffé, P.R., and Peters, C.A. (1998). "Bioavailability of mixtures of PAHs partitioned into the micellar phase of a nonionic surfactant." *Environ. Sci. Technol*, 32, 2317-2324.

Gullick, R.W., Schlautman, M.A., and Weber, W.J. (1994). "Use of organo-clays and natural sorbents to retard organic contaminant transport through landfill barriers." In *Preprints of Papers Presented at the 207th ACS National Meeting, Division of Environmental Chemistry*, San Diego, CA. American Chemical Society, 34, 345-353.

Holsen, T.M., Taylor, E.R., Seo, Y.-C., and Anderson, P.R. (1991). "Removal of sparingly soluble organic chemicals from aqueous solutions with surfactant-coated ferrihydrite." *Environ. Sci. Technol.*, 25, 1585-1589.

Hur, J., and Schlautman, M.A. (2000). "Pyrene sorption to mineral-bound humic substances." Presented at the 220th ACS National Meeting, Division of Environmental Chemistry, Washington, DC, August 2000.

Israelachvili, J. N. (1992). *Intermolecular & Surface Forces*, 2nd edition, Academic Press, San Diego.

Jafvert, C.T. (1991). "Sediment-and saturated-soil- associated reactions involving an anionic surfactant (dodecylsulfate). 2. Partition of PAH compounds among phases." *Environ. Sci. Technol.*, 25, 1039-1043.

Jafvert, C.T., van Hoof, P.L., and Heath, J.K. (1994). "Solubilization of Nonpolar Compounds by Nonionic Surfactant Micelles." *Water Res.*, 28, 1009-1017.

Johnson, W.P. (2000). "Sediment control of facilitated transport and enhanced desorption." *J. Environ. Eng.*, 126, 47-57.

Johnson, W.P., and Amy, G.L. (1995). "Facilitated transport and enhanced desorption of polycyclic aromatic hydrocarbons by natural organic matter in aquifer sediments." *Environ. Sci. Technol.*, 29, 807-817.

Karcher, W.W. (1988). *Spectral Atlas of Polycyclic Aromatic Compounds Including Data on Physico-Chemical Properties, Occurrence and Biological Activity*, Vol. 1, Kluwer Academic Publishers, Boston.

Kibbey, T.C.G., and Hayes, K.F. (1993). "Partitioning and UV adsorption studies of phenanthrene on cationic surfactant-coated silica." *Environ. Sci. Technol.*, 27, 2168-2173.

Kile, D.E., and Chiou, C.T. (1989). "Water solubility enhancements of DDT and trichlorobenzene by some surfactants below and above the critical micelle concentration." *Environ. Sci. Technol.*, 23, 832-838.

Ko, S.-O. (1998). "Electrokinetic/surfactant-enhanced remediation of hydrophobic organic pollutants in low permeability subsurface environments." Ph.D. dissertation, Texas A&M University, College Station, TX.

Ko, S.-O., and Schlautman, M.A. (1998). "Partitioning of hydrophobic organic compounds to sorbed surfactants. 2. Model development/predictions for surfactant-enhanced remediation applications." *Environ. Sci. Technol.*, 32, 2776-2781.

Ko, S.-O., Schlautman, M.A., and Carraway, E.R. (1998a). "Effects of solution chemistry on the partitioning of phenanthrene to sorbed surfactants." *Environ. Sci. Technol.*, 32, 3542-3548.

Ko, S.-O., Schlautman, M.A., and Carraway, E.R. (1998b). "Partitioning of hydrophobic organic compounds to sorbed surfactants. 1. Experimental studies." *Environ. Sci. Technol.*, 32, 2769-2775.

Ko, S.-O., Schlautman, M.A., and Carraway, E.R. (1999). "Partitioning of hydrophobic organic compounds to hydroxypropyl-β-cyclodextrin: Experimental studies and model predictions for surfactant-enhanced remediation applications." *Environ. Sci. Technol.*, 33, 2765-2770.

Ko, S.-O., Schlautman, M.A., and Carraway, E.R. (2000). "Cyclodextrin-enhanced electrokinetic removal of phenanthrene from a model clay soil." *Environ. Sci. Technol.*, 34, 1535-1541.

Laha, S., and Luthy, R.G. (1991). "Inhibition of phenanthrene mineralization by nonionic surfactants in soil-water systems." *Environ. Sci. Technol.*, 25, 1920-1930.

Liu, Z., Edwards, D.A., and Luthy, R.G. (1992). "Sorption of nonionic surfactants onto soil." *Water Res.*, 26,1337-1345.

Miyagishi, S., Asakawa, T., and Nishida, M. (1987). "Influence of external environment on microviscosity in micelles." *J. Colloid and Interface Sci,*, 15, 1199-1205.

Miyagishi, S., Kurimoto, H., Ishihara, Y., and Asakawa, T. (1994). "Determination of the critical micelle concentrations and microviscosity with a fluorescence probe, auramine." *Bull. Chem. Soc. Jpn.*, 67, 2398-2402.

Mukerjee, P., Sharma, R., Pyter, R.A., and Gumkowski, M.J. (1995). "Adsorption of surfactants and solubilization in adsorbed layers." In *Surfactant Adsorption and Surface Solubilization*, Sharma R. Ed., ACS Symposium Series 615, American Chemical Society, Washington DC.

Nayyar, S.P., Sabatini, D.A., and Harwell, J.H. (1994). "Surfactant adsolubilization and modified admicellar sorption of nonpolar, polar, and ionizable organic contaminants." *Environ. Sci. Technol.*, 28, 1874-1881.

O'Haver, J.H., and Harwell, J.H. (1995). "Adsolubilization: some expected and unexpected results." In *Surfactant Adsorption and Surface Solubilization*, Sharma R. Ed., ACS Symposium Series 615, American Chemical Society, Washington DC.

Park, J.W., and Jaffé, P.R. (1993). "Partitioning of three organic compounds between adsorbed surfactants, micelles, and water." *Environ. Sci. Technol.*, 27, 2559-2565.

Pennell, K.D., Abriola, L.M., and Weber, W.J., Jr. (1993). "Surfactant-enhanced solubilization of residual of residual dodecane in soil columns: 1. Experimental investigation." *Environ. Sci. Technol.*, 27, 332-340.

Sahoo, D., and Smith, J.A. (1997). "Enhanced trichloroethene desorption from long-term contaminated soil using Triton X-100 and pH increases." *Environ. Sci. Technol.*, 31, 1910-1915.

Sahoo, D., Smith, J.A., Imbrigiotta, T.E., and McLellan, H.M. (1998). "Surfactant-enhanced remediation of a trichloroethene-contaminated aquifer: 2. Transport of TCE." *Environ. Sci. Technol.*, 32, 1686-1693.

Schlautman, M.A., and Morgan, J.J. (1993). "Binding of a Fluorescent Hydrophobic Organic Probe by Dissolved Humic Substances and Organically-Coated Aluminum Oxide Surfaces." *Environ. Sci. Technol.*, 27, 2523-2532.

Schwarzenbach, R.P., Gschwend, P.M., and Imboden, D.M. (1993). *Environmental Organic Chemistry*, John Wiley & Sons, Inc, New York.

Sharma, R. (1995). "Small-molecule surfactant adsorption, polymer surfactant adsorption, and surface solubilization: an overview." In *Surfactant Adsorption and Surface Solubilization*, Sharma R. Ed., ACS Symposium Series 615, American Chemical Society, Washington DC.

Smith, J.A., Sahoo, D., McLellan, H.M., and Imbrigiotta, T.E. (1997). "Surfactant-enhanced remediation of a trichloroethene-contaminated aquifer: 1. Transport of Triton X-100." *Environ. Sci. Technol.*, 31, 3565-3572.

Sun, S., Inskeep, W.P., and Boyd, S.A. (1995). "Sorption of nonionic organic compounds in soil-water systems containing a micelle-forming surfactant." *Environ. Sci. Technol.*, 29, 903-913.

Sun, S., and Jaffé, P.R. (1996). "Sorption of phenanthrene from water onto alumina coated with dianionic surfactants." *Environ. Sci. Technol.*, 30, 2906-2913.

Tiehm, A., Stiber, M., Werner, P., and Frimmel, F.H. (1997). "Surfactant-enhanced mobilization and biodegradation of polycyclic aromatic hydrocarbons in manufactured gas plant soil." *Environ. Sci. Technol*, 31, 2570-2576.

Wagner, J., Chen, H., Brownawell, B.J., and Westall, J.C. (1994). "Use of cationic surfactants to modify soil surfaces to promote sorption and retard migration of hydrophobic organic compounds." *Environ. Sci. Technol.*, 28, 231-237.

Wang, J.-M., Marlowe, E.M., Miller-Maier, R.M., and Brusseau, M.L. (1998). "Cyclodextrin-enhanced biodegradation of phenanthrene." *Environ. Sci. Technol*, 32, 1907-1912.

Xu, S., and Boyd, S.A. (1995). "Cationic surfactant sorption to a vermiculite subsoil via hydrophobic bonding." *Environ. Sci. Technol.*, 29, 312-320.

Yeh, D.H., Pennell, K.D., and Pavlostathis, S.G. (1999). "Effect of Tween surfactants on methanogenesis and microbial reductive dechlorination of hexachlorobenzene." *Environ. Toxicol. Chem.*, 18, 1408-1416.

Yeom, I.T., Ghosh, M.M., Cox, C.D., and Robinson, K.G. (1995). "Micellar solubilization of polynuclear aromatic hydrocarbons in coal tar-contaminated soils." *Environ. Sci. Technol.*, 29, 3015-3021.

Yeom, I.T., Ghosh, M.M., and Cox, C.D. (1996). "Kinetic aspects of surfactant solubilization of soil-bound polycyclic aromatic hydrocarbons." *Environ. Sci. Technol.*, 30, 1589-1595.

Chapter Ten

Surfactant-Enhanced Desorption of Organic Pollutants from Natural Soil

JAMES J. DEITSCH AND ELIZABETH J. ROCKAWAY
Department of Civil Engineering, University of Kentucky, Lexington, KY 40506-0281

Key words: ground water, remediation, surfactants, desorption

Abstract: The use of surface-active agents (i.e., surfactants) to increase the efficiency of pump-and-treat has been investigated for remediation sites where clean-up is limited by non-aqueous phase liquid dissolution and contaminant desorption. For both types of limitations, successful applications of the enhancement technologies have occurred at the laboratory scale and in a few small-scale field operations. For this chapter, emphasis is placed on assessing the use of surfactants to increase the rate of organic pollutant desorption from soil to water. The mechanisms of surfactant-enhanced desorption are introduced and factors affecting the efficiency and applicability of surfactant-enhanced remediation are discussed. To conclude the chapter, data are presented that show the effects of several different surfactants on the rate of trichloroethene (TCE) desorption from a peat soil. The surfactant Triton X-100 is shown to increase the rate of TCE desorption from a peat soil at several different concentrations. The increased rate of TCE desorption was caused by an increase in the desorption mass-transfer rate coefficient, as well as by increasing the concentration gradient driving desorption. This latter mechanism was only present at a surfactant concentration roughly twenty times the critical micelle concentration of Triton X-100. For the other surfactants tested – Tween 20, sodium dodecylsulfate, sodium dodecylbenzenesulfonate, and Triton X-405 – the rate of TCE desorption remained the same or was decreased. Results from this section emphasize the need for a better mechanistic understanding of the effects of surfactants on sorbed pollutants.

1. INTRODUCTION

In the past, pump-and-treat remediation was the predominant technology of choice for restoring aquifers contaminated with organic pollutants. Typically, in pump-and-treat systems, the contaminated water is removed from the subsurface using a system of extraction wells and the extracted water is treated by an appropriate unit operation. Unfortunately, pump-and-treat remediation of contaminated aquifers has not proven able to achieve EPA regulated health-based cleanup standards (Mackay and Cherry 1989). In fact, pump-and-treat systems are expensive to operate and generally are only able to remove modest amounts of the contaminants from the subsurface. The inability of pump-and-treat to meet stringent cleanup standards has fostered a tremendous amount of research to determine the reasons for the failure of pump-and-treat and to explore alternative treatment methodologies. The use of surface-active agents (i.e., surfactants) to increase the efficiency of pump-and-treat has been investigated for remediation limited by non-aqueous phase liquid dissolution and contaminant desorption (EPA 1995).

For organic contaminant plumes, there are several factors that can lead to the limited success of pump-and-treat operations. Two of the most commonly cited causes of poor pump-and-treat performance are 1) the slow dissolution of non-aqueous phase liquid pools into the ground water, and 2) the slow, kinetic desorption of organic pollutants from the subsurface soil to the ground water. Given the presence of non-aqueous phase liquids and sorbed compounds, it is not uncommon for only a small fraction of the subsurface contamination to be present in the ground-water phase. Thus, pump-and-treat remediation of a contaminated aquifer is limited by the slow rate of solute mass-transfer from non-aqueous phases to the aqueous phase. This chapter analyzes the use of surfactants to increase the rate of organic pollutant desorption from soil to water. Emphasis is placed on analyzing current theories regarding the rate-limited desorption of organic pollutants. Following the theory of kinetic desorption, the mechanisms of surfactant-enhanced desorption are introduced. As is shown, the implementation of surfactant-enhanced remediation technologies is a complicated task. Factors affecting the efficiency and applicability of surfactant-enhanced remediation are introduced. The final section of the chapter presents experimental data that show the effects of several different surfactants on the rate of trichloroethene desorption from a peat soil.

2. MECHANISMS OF RATE LIMITED DESORPTION

Because of the important role sorption and desorption have on the transport and remediation of organic compounds, extensive research has been conducted to elucidate the cause of rate-limited sorption and desorption. Most researchers believe that diffusional resistance encountered within the soil matrix is responsible for observed kinetic behavior ((Pignatello and Xing 1996), and references cited therein). Intraparticle pore diffusion (Ball and Roberts 1991; Farrell and Reinhard 1994; Harmon and Roberts 1994; Cunningham et al. 1997) and intra-organic matter diffusion (Brusseau et al. 1991a; Carroll et al. 1994; Huang and Weber 1997; Kan et al. 1997; Weber and Young 1997; Deitsch and Smith 1999) are often cited as rate-limited diffusion mechanisms. Intraparticle pore diffusion attributes the slow rate of desorption to restrictions encountered by a solute as it diffuses through a network of interconnected mesopores and micropores. Diffusive restrictions are a result of the tortuous path that the pore network creates, as well as the severe constriction of diffusive pathways within micropores. Intra-organic matter diffusions attributes the rate-limited desorption to diffusive restrictions encountered by the solute as it diffuses into and out the soil organic matter phase. Depending upon the soil characteristics, it is possible that either intra-particle pore diffusion or intra-organic matter diffusion may be the predominant rate-limiting mechanism.

In support of the intra-organic matter diffusion mechanism, the amorphous/glassy two-domain model is usually invoked to describe the structure of soil organic matter (Carroll et al. 1994; Huang and Weber 1997; LeBoeuf and Weber 1997; Weber and Young 1997; Xing and Pignatello 1997; Deitsch and Smith 1999). The soil organic matter is envisioned to be composed of two regions: an amorphous region that is capable of conformational changes in response to environmental changes (Carroll et al. 1994; LeBoeuf and Weber 1997), and a "more condensed, highly cross-linked" region that is "glassy" in naturc (Carroll ct al. 1994; LcBocuf and Weber 1997). Typically, the building blocks for the various components of soil organic matter (i.e., humic and fulvic acids, humins, etc.) are believed to be aromatic nuclei that are connected by aliphatic chains. Young and Weber (1995) state that "the diagenetic process results in the removal of oxygen-containing functional groups, increases in molecular weight, and increases in the fraction of carbon present in condensed aromatic nuclei compared to those present in aliphatic linkages." Thus, diagenesis transforms soil organic matter from an amorphous state to a condensed, crystalline state.

There is a growing body of experimental evidence that suggests that the rate-limited sorption and desorption of hydrophobic organic contaminants may be partially attributable to extremely slow diffusion of the solute through the condensed regions of the soil organic matter (Carroll et al. 1994; Young and Weber 1995; Huang and Weber 1997; LeBoeuf and Weber 1997; Xing and Pignatello 1997). In support of this concept, one study showed that solute diffusion coefficients in synthetic polymers were consistently two to three orders of magnitude lower in the polymer's glassy state than in its amorphous state (Berens 1989). In addition, it is believed that conformational changes in the soil organic matter after sorption has occurred may be a cause rate-limited desorption (Kan et al. 1994; Kan et al. 1997; Deitsch and Smith 1999). In addition, a fraction of the solute that is sorbed to soil appears to be irreversibly bound.

3. SURFACTANT-ENHANCED DESORPTION

To quantify the rate of solute desorption from soil, the following mathematical expression is often used:

$$\frac{dS}{dt} = -\alpha\left(S - K_D\, C\right) \tag{1}$$

where S is the concentration of the solute sorbed to soil (M/M), α is the mass-transfer rate coefficient (1/T), K_D is the sorption distribution coefficient (L^3/M), and C is the aqueous phase concentration of the solute (M/L^3) (Deitsch and Smith 1995). For nonionic organic contaminants at equilibrium,

$$S = K_D\, C \tag{2}$$

Therefore, equations (1) and (2) indicate that the rate of contaminant desorption from soil is a function of the mass-transfer coefficient, α, and the concentration gradient between the soil and aqueous phases. Surfactant-enhanced desorption focuses on these two factors. First, the interaction of surfactants with the soil organic matter is hypothesized to increase the magnitude of α (Deitsch and Smith 1995; Yeom et al. 1996; Sahoo and Smith 1997; Sahoo et al. 1998). Second, surfactant solutions may affect the apparent magnitude of K_D, thus altering the concentration gradient driving desorption from the solid phase to the aqueous phase (Deitsch and Smith 1995; Yeom et al. 1995; Yeom et al. 1996; Sahoo and Smith 1997; Sahoo et al. 1998). Equation (1) shows that the rate of desorption is directly proportional to the magnitude of the concentration gradient. These two mechanisms are now discussed in detail.

3.1 Surfactant Effects on α

Equation (1) quantifies the rate of desorption using a mass-transfer rate coefficient. An increase in the magnitude of α indicates a corresponding increase in the rate of solute diffusion within the sorbent matrix. The use of surfactants to increase desorption rates focuses on increasing the rate of solute diffusion through the soil organic matter. For a surfactant solution to increase the rate of solute diffusion through the condensed phase of the soil organic matter, the surfactants must cause conformational and structural changes within the soil organic matter.

Recent findings support the hypothesis that the glassy domain of soil organic matter is capable of undergoing conformational changes. First, Brusseau et al. (1991b) determined that the rate of organic pollutant desorption was increased by the ability of organic cosolvents to swell soil organic matter and reduce diffusive resistances. In another study (LeBoeuf and Weber 1997), conformational changes within the glassy domain of soil organic matter were assumed because the glass transition temperature, T_g, of the sorbent was decreased by solvent/sorbent interactions. As reported by LeBoeuf and Weber (1997) "T_g marks a second-order phase transition in which there is continuity of the free energy function and its first partial derivatives with respect to state variables such as temperature and pressure, but there is a discontinuity in the second partial derivatives of free energy." (The reader is referred to the preceding reference for a more detailed discussion of T_g.) The decrease in T_g represents a decrease in the degree of polymer cross-linking and as a result, there is a corresponding increase in the proportion of "soft" to "hard" carbon.

Such a mechanism could explain the effects of methanol on the rate of solute mass-transfer. LeBoeuf and Weber (1997) observed that a humic acid (solubility parameter, σ_p= 11.5 (cal/cm^3)$^{0.5}$) swelled to greater degree in the presence of water (σ_p=23.4 (cal/cm^3)$^{0.5}$) than did a poly(isobutyl methacrylate) polymer (σ_p=8.63 (cal/cm^3)$^{0.5}$). The magnitude of the solubility parameter is a measure of the phase's intermolecular bonding energy. Phases with similar solubility parameters will interact more favorably than phases with dissimilar solubility parameters. Thus, the increased swelling was attributed to the water being able to interact more favorably with the humic acid than with the poly(isobutyl methacrylate) polymer, based on the magnitude of the solubility parameters. Although σ_p

for a natural soil will not be as well defined as for a manufactured humic acid or polymer, it is reasonable to assume that σ_p for a natural soil could be within this range. Within this context, methanol with a σ_p of 14.5 $(cal/cm^3)^{0.5}$ would be expected to interact more favorably than water with the glassy domain, therefore causing a reduction in the average T_g of the soil organic matter. Thus, the methanol facilitates the transformation of the glassy region into an amorphous region where solute diffusion may be orders of magnitude higher.

Several recent studies have demonstrated that surfactant solutions can increase the rate of contaminant diffusion within the sorbent matrix during desorption compared to non-surfactant solutions (Deitsch and Smith 1995; Yeom et al. 1996; Sahoo and Smith 1997; Sahoo et al. 1998). Deitsch and Smith (1995) demonstrated that Triton X-100 solutions at different concentrations increased the magnitude of α for TCE desorption from a laboratory-contaminated peat soil by a factor of 2 to 3 compared to water. In another study (Yeom et al. 1995), researchers at the University of Tennessee showed that several different nonionic surfactants increased the magnitudes of polycyclic aromatic hydrocarbon diffusion coefficients for a weathered, coal tar-contaminated soil. Diffusion coefficients were up to two orders of magnitude higher in surfactant solutions compared to non-surfactant solutions. Sahoo and Smith (1997) quantified the rate of TCE desorption from a long-term field-contaminated soil. Experiments with Triton X-100 solutions showed that the mean value of α was increased 11.2% in batch experiments and 16.5% in column experiments compared to experiments sans Triton X-100. In a small-scale field test, Sahoo et al (1998) determined that flushing a section of a TCE-contaminated sandy aquifer with 300 mg/L of Triton X-100 increased the magnitude of α for TCE desorption by approximately 30%. Finally, Noordman et al (1998) reported that a rhamnolipid biosurfactant increased the rate constants associated with phenanthrene desorption in column experiments. In all of these studies, the increased rates of mass-transfer were attributed to the surfactants swelling the soil organic matter and increasing the rate of solute diffusion within the soil organic matter.

The hypothesized mechanism of enhanced desorption described in the previous paragraph can be understood within the framework of surfactant interactions with the soil organic matter. Depending upon the σ_p of a surfactant/water solution, surfactant solutions may transform portions of the "glassy" region of the soil organic matter into amorphous regions. Using an approach similar to LeBoeuf and Weber (1997), it may be possible to quantify the effect of different surfactant solutions on the T_g of commercial humic acid or other polymer matrixes. Such experimental data would provide compelling evidence for the hypothesized mechanism of surfactant-enhanced diffusion rates. Given the heterogeneous nature of natural sorbents, quantification of distinct values of T_g for natural soils is not possible.

3.2 Surfactant Gradient Effects

To understand the effects that surfactants have on the gradient driving the desorption of hydrophobic organic contaminants in soil/water systems, it is necessary to know how surfactant molecules distribute themselves among the different portions of the soil and the aqueous phase. Surfactant molecules may sorb to the different fractions of the soil. For example, surfactant molecules may sorb to and form aggregates at a soil's mineral surface. In addition, a fraction of the sorbed surfactant is solubilized within the organic matter phase of the soil. In the aqueous phase, surfactant molecules exist as monomers until a certain surfactant-specific aqueous concentration is reached. This concentration is known as the critical micelle concentration (CMC). Once CMC is reached, any additional surfactant molecules added to the system will join together to form micelles. Micelles may be envisioned as dynamic clusters of surfactant molecules arranged with their polar head groups oriented toward the aqueous solution and their non-polar tail groups oriented inward toward the interior of the cluster. With regard to hydrophobic organic contaminant sorption, the presence of sorbed surfactant may increase the apparent sorption distribution coefficient of the contaminant, whereas the presence of surfactant micelles in the aqueous phase may act to decrease the contaminant's apparent sorption distribution coefficient. These mechanisms are now discussed in greater detail.

The successful use of commercial surfactants above CMC to increase the apparent water solubility of otherwise slightly soluble nonionic organic compounds is well documented (Kile and Chiou 1989; Smith et al. 1991). Kile and Chiou (1989) determined that the apparent water solubilities of p,p'-DDT and 1,2,3-trichlorobenzene are increased by surfactants (Triton X-100, Triton X-114, Triton X-405, sodiumdodecylsulfate, and Brij 35) at concentrations above CMC. The apparent water solubility of p,p'-DDT was increased two orders of magnitude by Triton X-114. It is believed that the inner region of a micelle acts as a nonpolar micellar pseudophase into which slightly water-soluble organic compounds can be solubilized (Kile and Chiou 1989; Smith et al. 1991; DiCesare and Smith 1994).

The solubilization of nonionic compounds by surfactants can dramatically affect the sorption of these compounds to soil (Vigon and Rubin 1989; Smith et al. 1991). Vigon and Rubin (1989) determined that anthracene sorption from nonionic surfactant solutions above CMC was two orders of magnitude less than anthracene sorption from water. Smith and others (1991) found that Triton X-100 above CMC reduced the magnitude of tetrachloromethane's sorption distribution coefficient in water by approximately one-half. Given the relatively high water solubility of tetrachloromethane (800 mg/L), it is expected that more water-insoluble compounds will experience greater reductions in sorption (Smith et al. 1991).

A potential limitation of surfactant-enhanced desorption is the observation that sorbed surfactant molecules can increase the sorption of hydrophobic organic contaminants (Edwards et al. 1994; Sun et al. 1995; Ko et al. 1998). Sun et al. (1995) reported that the nonionic surfactant Triton X-100 increased the sorption of p,p'-DDT, 2,2',4,4',5,5'-PCB, and 1,2,4-trichlorobenzene to a soil (f_{oc}= 0.001) at concentrations below CMC. At concentrations above CMC, the distribution coefficients (K_D) of the DDT and PCB studied were reduced to levels below their respective values in pure water. However, at a surfactant concentration of five times CMC, the K_D of 1,2,4-trichlorobenzene was still a factor of three higher than K_D in pure water. Edwards et al. (1994) and Ko et al. (Ko et al. 1998) reported similar results for different groups of surfactants.

To account quantitatively for the preceding effects, Sun et al. (1995) developed a model to simulate the influence of a surfactant on the magnitude of the sorption of a hydrophobic organic contaminant. The governing equation is given by

$$K^* = K_D\left(\frac{1 + X_{S/OC}\,K_{S/OC}}{1 + X_{mn}\,K_{mn} + X_{mc}\,K_{mc}}\right) \tag{3}$$

where K^* is the surfactant altered distribution coefficient , K_D is the solute partition coefficient in the absence of surfactant, $K_{s/oc}$ is the solute partition coefficient between the sorbed surfactant and the native soil organic carbon, $X_{s/oc}$ is the fractional concentration of sorbed surfactant per unit mass of native soil organic carbon, K_{mn} is the solute partition coefficient between the surfactant monomers and the water, and K_{mc} is the solute partition coefficient between the surfactant micelles and the water. X_{mn} and X_{mc} are the fractional concentrations of surfactant monomers and micelles, respectively (Sun et al. 1995). K_{mn} and K_{mc} can be determined from solute solubility curves plotted as a function of aqueous surfactant concentration.

$K_{s/oc}$, as defined in the previous paragraph, represents the effectiveness of the sorbed surfactant to serve as a partition medium for a hydrophobic organic contaminant relative to the native soil organic matter. Mathematically, $K_{s/oc}$ is equivalent to

$$K_{S/OC} = \frac{K_S}{K_D} = \frac{K_S}{f_{OC}\,K_{OC}} \tag{4}$$

where K_s is the partition coefficient of a given hydrophobic organic contaminant between the sorbed surfactant and water, f_{oc} is the fractional organic carbon content of the soil, and K_{oc} is the organic-carbon normalized partition coefficient for a solute (Sun et al. 1995). Equation (4) can be substituted into equation (3). Thus, K_s is a fitting parameter that is used to simulate the functional dependence of K^* on surfactant concentration in the soil/water system. For modeling the rate of desorption, K^* from equation (3) can be substituted in for K_D in equation (1).

In previous studies, the solubilization of hydrophobic organic contaminants using surfactants has been shown to increase the rate of contaminant desorption from soil to water (Deitsch and Smith 1995; Yeom et al. 1995; Tiehm et al. 1997). A 3,000 mg/L solution of Triton X-100 (CMC ≈ 140 mg/L) increased the rate of desorption of laboratory-contaminated TCE from a peat soil (Deitsch and Smith 1995). However, the solubilization effect was secondary compared to the surfactant's effect on the desorption rate coefficient. Yeom et al (1995) developed a model that satisfactorily predicted the extent of polycyclic aromatic hydrocarbon solubilization from a coal tar-contaminated soil. Only at high surfactant dosages did the model fail to accurately predict the ability of different surfactants to solubilize polycyclic aromatic hydrocarbons. It was hypothesized that mass-transfer limitations encountered by the polycyclic aromatic hydrocarbons in the soil caused the observed differences between the data and the model simulations. In another study (Tiehm et al. 1997), two nonionic surfactants, Arkopal N-300 and Saogenat T-300, increased the rate of polycyclic aromatic hydrocarbon desorption from a field-contaminated soil. The primary mechanism for the enhanced desorption of polycyclic aromatic hydrocarbons was attributed to surfactant solubilization of the polycyclic aromatic hydrocarbons.

Presently, there are several important limitations to increasing the rate of hydrophobic organic contaminant desorption by surfactant solubilization. For example, to achieve a measurable reduction in K_D, surfactant concentrations must be significantly greater than CMC, particularly if the soil-water system is contaminated with chlorinated solvents or gasoline-range hydrocarbons that are characterized by aqueous solubilities greater than a few hundred milligrams per liter. Therefore, the material costs of using surfactants at concentrations above CMC may offset the benefits of increased desorption rates and shorter remediation times. In addition, during the operation of pump-and-treat systems, aqueous solute concentrations are typically low and the concentration gradient between the sorbed and aqueous phases may be close to a maximum. As a result, any reduction in K_D would have a minimal effect on the concentration gradient.

Regarding the ability of surfactants to increase the sorption capacity of a soil, this phenomena may not seriously affect the applicability of surfactant-enhanced desorption technologies. A recent study (Deitsch et al. 1998) indicated that the rates of 1,2-dichlorobenzene sorption and desorption to and from a clay modified with quaternary ammonium surfactants were exceedingly rapid. The quaternary ammonium surfactants used in the cited study are similar in molecular weight and structure to many of the surfactants mentioned in the previous paragraphs. Although to our knowledge no study has been conducted to determine the rate of solute desorption from sorbed nonionic surfactants, it is probably not a poor assumption to believe that the rate of solute desorption would be similar to the rates of desorption measured for the modified clays. If this is the case, the fast rate of solute desorption from the sorbed surfactants would counteract the effects of the increased sorption capacity resulting from the sorbed surfactants.

4. EXPERIMENTAL METHODOLOGIES AND APPROACH

In the previous sections, the use of surfactants to increase the rate of desorption of hydrophobic organic contaminants was discussed. For the current study, several different surfactants were tested to determine whether the rate of TCE desorption from a peat soil could be increased. The effects of the surfactants on the rate of TCE desorption was tested using a continuous-flow stirred-tank reactor (CFSTR) methodology. The observed data were simulated using a distributed-rate kinetic desorption model. The parameters determined from the model simulation were then use to discern the effects of the surfactants on the rate of TCE desorption from the peat soil. The experimental methodology and the modeling procedure are now described in detail.

4.1 Experimental Methodology

4.1.1 Soil

The soil used in this study was a peat collected at Picatinny Arsenal, New Jersey. The peat was air dried for 24 hr and then heated at 105°C for 24 hr.

It was then passed through a #10 sieve (2-mm openings) to insure that subsequent soil subsamples would be relatively homogeneous. The soil was then "conditioned" as described in the next section prior to kinetic experiments and quantification of soil organic-carbon contents. The organic-carbon content of the peat was quantified by Huffman Laboratories, Golden, CO and determined to be 24 percent.

4.1.2 CFSTR Methodology

A continuous-flow stirred tank reactor (CFSTR) apparatus was used to study the effect of several surfactants on the kinetic desorption of TCE from soil to water (Deitsch and Smith 1995). Individual CFSTRs consisted of a cylindrical glass reactor (1.0 cm i.d. X 15 cm) with Teflon end fittings and stainless-steel screens. The fittings were secured and formed airtight seals at the ends of the glass reactor. Omnifit two-way valves were connected to the inflow and outflow fittings of the CFSTR with stainless steel tubing. All the surfaces inside the CFSTRs were either glass or Teflon.

Each CFSTR contained 5 g of peat. Deionized, organic-free water was added to the reactor and mixed until the peat was completely wetted (i.e., all the air trapped in the peat was released) and the reactor was completely filled. To "condition" the peat, the CFSTRs were fastened to a rotary shaker and shaken continuously at 150 rpm at ambient room temperature. Distilled water containing 200 mg/L of sodium azide was pumped through each CFSTR at 12 mL/hr for 3 d. A Manostat cassette drive pump was used in all CFSTR experiments. The presence of the sodium azide solution eliminates microbiological activity in the CFSTRs under aerobic conditions, thus preventing biodegradation of TCE (Pavlostathis and Mathavan 1992).

The CFSTRs used in the TCE desorption experiments were contaminated with [^{14}C] TCE. 500 μCi of [^{14}C] TCE (specific activity equal to 6.2 mCi/mmol) was obtained from Sigma Chemical Co. and mixed with nonradioactive TCE to yield a net volume of 6 mL of neat liquid. The resultant chemical and radiochemical purity of the radioisotope was greater than 98 percent. A specific volume of [^{14}C] TCE was injected into each CFSTR.

Following the injection of TCE, the CFSTRs were continuously shaken at 150 rpm for a 1-week equilibration period and then sampled. The CFSTRs were separated into groups, which used either water or a specific concentration of surfactant as its inflow reservoir. Each reservoir also contained 200 mg/L of sodium azide. The different surfactants and the concentrations tested are listed in Table 1. The solutions were then pumped through the CFSTRs with a peristaltic pump at approximately constant flow rates (≈ 2 mL/hr) until the sampling was completed. The flow rates were measured throughout the course of the sampling period for individual CFSTRs. The variation of flow rates among CFSTRs was incorporated into the simulation models (see Section 4.2).

Table 1. Surfactant properties and the concentrations tested in the CFSTR experiments

Surfactants	MW[a]	CMC[b] (mg/L)	Concentrations Tested (mg/L)
Triton X-100	624	140	30 300 3,000
Triton X-405	1968	600	3,000
Tween 20	1227	ND	30 300 3,000
SDS[c]	288	2320	3,000
SDBS[d]	348	418	3,000

[a] Molecular Weight
[b] Critical Micelle Concentration
[c] sodium dodecylsulfate
[d] sodium dodecylbenzenesulfonate

Effluent from each CFSTR was collected periodically in 7-mL scintillation vials filled with 5 mL of Beckman Ready Safe scintillation cocktail. The mass of effluent captured (approximately 0.5 mL) was quantified gravimetrically. The radioactivity of the samples was measured with a Packard Tri-Carb 1900CA liquid-scintillation analyzer, and the corresponding TCE concentrations were calculated with a standard curve relating concentration to counts per minute. The sampling periods lasted between 7 and 10 d.

The assumption of a completely mixed reactor was tested by conducting a Br^- tracer experiment. The procedure used to test the assumption of complete mixing has been described in detail elsewhere (Deitsch and Smith 1995). Based on the results described in the preceding reference, it was determined that the CFSTRs were well approximated as completely mixed. Previously, the accuracy of the CFSTR desorption methodology was verified by conducting blank CFSTR experiments to determine whether losses of TCE occurred during the experimental procedure (Deitsch and Smith 1995).

4.2 Model Development

The desorption data collected during this investigation were analyzed with a distributed-rate model that has been described in detail elsewhere (Culver et al. 1997; Deitsch et al. 1998; Deitsch and Smith 1999). Briefly, to account for soil heterogeneity, the distributed-rate model replaces a single mass-transfer rate coefficient with a continuous distribution of rate coefficients. In this study, a Γ-probability density function (Γ-PDF) was used to generate the distribution of rate coefficients. The Γ-PDF is given by

$$p(\alpha) = \frac{\beta^{-\eta}\, \alpha^{\eta-1}}{\int_0^\infty x^{\eta-1} \exp(-x)\, dx} \exp\left(-\frac{\alpha}{\beta}\right) \tag{5}$$

where η (the shape parameter) and β (the scale parameter) are positive parameters and x is a dummy variable of integration. The shape and magnitude of the Γ-PDF are uniquely determined by altering the two parameters, η and β.

The sorbent was discretized into a finite number (NK) of sorbent compartments to numerically approximate the continuous distribution of mass-transfer rate coefficients. The mass of sorbent was divided equally among the NK sites. For a CFSTR, the distributed-rate model has the following governing equations (Culver et al. 1997):

$$\frac{dS_T}{dt} = \sum_{i=1}^{NK} \alpha_i (f\, K_D\, C - S_i) \tag{6}$$

$$\frac{dC}{dt} = \frac{Q}{V}(C_{in} - C) - \frac{M_s}{V}\left[\sum_{i=1}^{NK} \alpha_i (f\, K_D\, C - S_i)\right] \tag{7}$$

$$S_T = \sum_{i=1}^{NK} S_i \tag{8}$$

$$f = \frac{1}{NK} \tag{9}$$

S_T is the total sorbed concentration (M/M), α_i is the first-order mass-transfer rate coefficient for compartment i (1/T), f is the mass fraction of the solute sorbed in each site at equilibrium (assumed to be equal for all compartments), K_D is the distribution coefficient (L^3/M), C is the aqueous solute concentration (M/L^3), S_i is the mass sorbed in compartment i with respect to the total mass of the sorbent (M/M), Q is the volumetric flow rate through the reactor (L^3/T), C_{in} is the influent concentration of solute (M/L^3), M_s is the mass of sorbent in the reactor (M), and V is the aqueous reactor volume (L^3). Using the Γ-PDF, discrete values for the mass-transfer rate coefficients were generated for the NK compartments. The median value of the mass-transfer rate coefficient within each compartment was chosen as the representative value. The resulting system of ordinary differential equations was solved numerically using a 4th-order Runge-Kutta integration technique.

To numerically solve the distributed rate model when simulating the desorption data, the initial sorbed phase concentration of each of the NK sites must be estimated. For this study, it was assumed that the sorbed TCE was equally distributed among the NK sites. S_T was calculated based on the initial aqueous concentration of TCE and the other CFSTR parameters.

In Section 3.2 it was shown that surfactants can influence the magnitude of K_D. To use equation (3) to simulate changes in K_D, the following information is required: 1) the aqueous surfactant concentration in the CFSTR as a function of time; 2) the sorbed phase surfactant concentration in each of the NK sites; 3) the magnitudes of K_{mn} and K_{mc} for each surfactant; and, 4) the magnitude of K_S. Unfortunately, the required information to incorporate equation (3) into the distributed-rate model was not determined for this study. As a result, the influence of the surfactants on the distribution coefficient was not considered.

In a previous study (Deitsch and Smith 1995), the influence of Triton X-100 on K_D for the *same* Picatinny peat soil was limited to Triton X-100 concentrations above 300 mg/l. At 300 mg/L and below, Triton X-100 did not affect the magnitude of K_D. Since this study has not included the effects of the surfactants on K_D, it will not be possible to identify the mechanism of increased TCE desorption for surfactant solutions that influence the magnitude of K_D. However, for the 30 and 300 mg/L concentrations of Triton X-100, K_D can be assumed to be constant. As a result, the influence of the 30 and 300 mg/L Triton X-100 concentrations on the distribution of rate coefficients can be discerned.

In equation (7), the volumetric flow rate through the reactor is one of the parameters. Although the cassette pumps were set to flow at constant rates, there were slight fluctuations in flow rate during an experiment for a single CFSTR and among the different CFSTRs. To account for the flow variations during the course of an experiment, the flow rate was monitored as a function of time. The observed flow rates were then fit with a cubic spline interpolation polynomial to obtain a discretization of flow rates to be used as input for the distributed rate desorption model (see Figure 1). As seen in Figure 1, flow rates stabilized quickly for a given CFSTR. Also, the flow rates among CFSTRs typically varied by less then ± 5% from the average of all the CFSTRs.

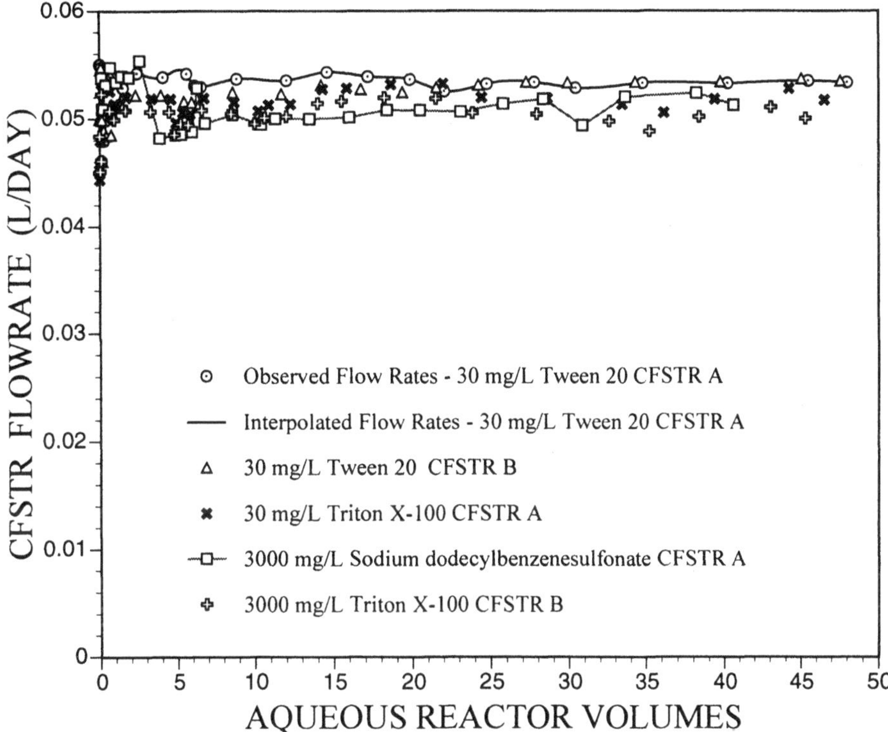

Figure 1. Observed CFSTR flow rates during the course of the desorption experiments for selected CFSTRs. For the 30 mg/L Tween 20 CFSTR, the cubic spline interpolation of the flow rate variations is shown.

The optimal model fit of the desorption data was determined by finding the values of η and β that minimized the percent error between the simulated and the observed TCE concentrations (Culver et al. 1997). The values of η and β that minimized the percent error were determined through an optimization routine used in previous studies (Deitsch et al. 1998; Deitsch and Smith 1999). The optimization program was routinely checked to insure that the global minimum for percent error was obtained for each CFSTR data set.

5. RESULTS & DISCUSSION

The raw data presented for the Triton X-100 experiments have been presented previously (Deitsch and Smith 1995) in a modified form. The distributed rate model was fit to the CFSTR desorption data (Figures 2 – 5). The optimal model parameters and percent errors are given for each experiment in Table 2. The distributed rate model was able to simulate the CFSTR data well. In several of the surfactant CFSTRs, the model underestimated the aqueous TCE concentrations at later times. This suggests that for some of the surfactant CFSTRs the amount of TCE flushed from the reactor may be greater than the amount predicted from the simulation.

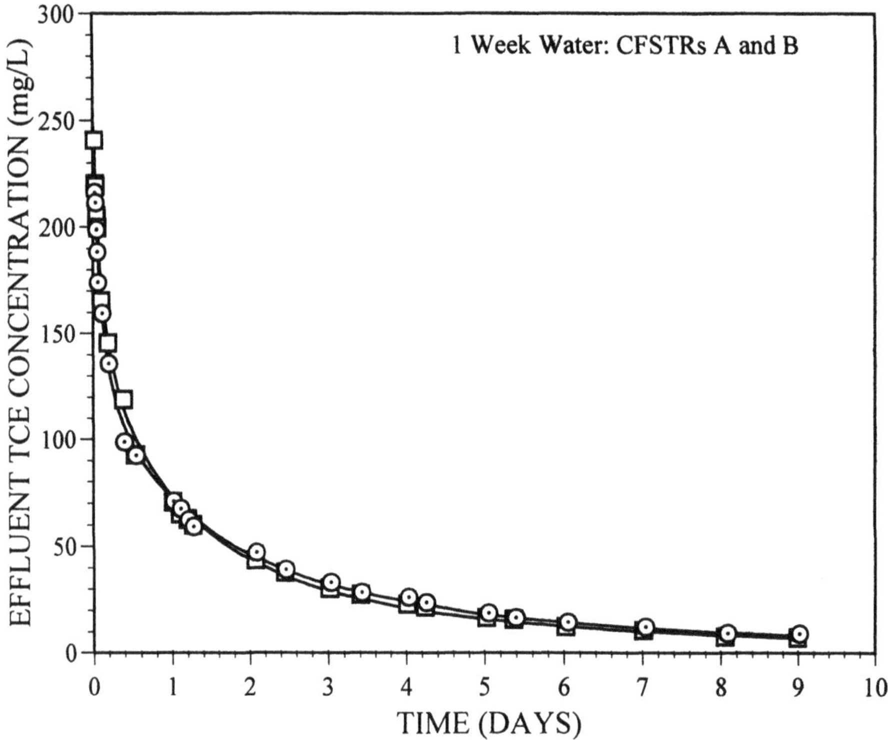

Figure 2. Distributed rate model simulations of the data obtained from the two CFSTRs flushed with the non-surfactant solutions.

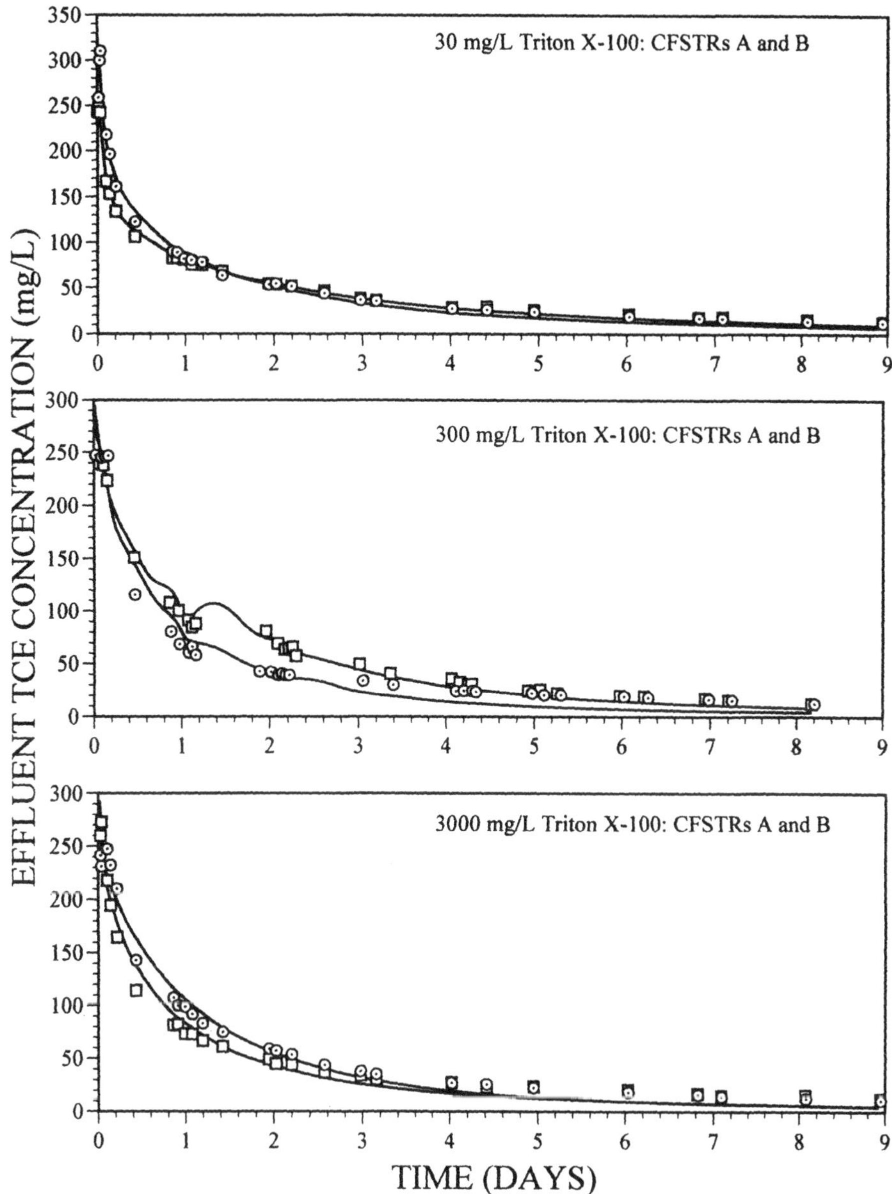

Figure 3. Distributed rate model simulations of the data obtained from the CFSTRs flushed with the 30 mg/L, 300 mg/L, and 3,000 mg/L Triton X-100 solutions.

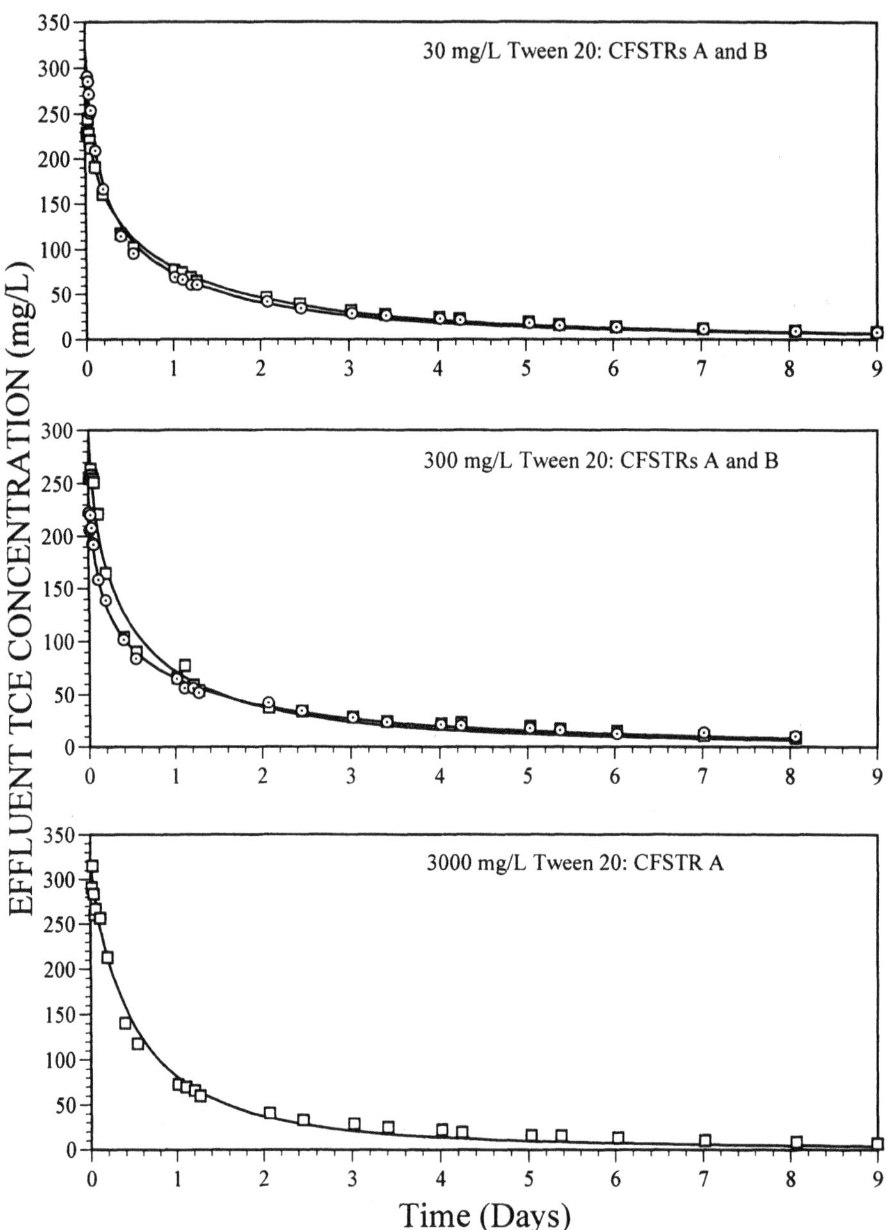

Figure 4. Distributed rate model simulations of the data obtained from the CFSTRs flushed with the 30 mg/L, 300 mg/L, and 3,000 mg/L Tween 20 solutions.

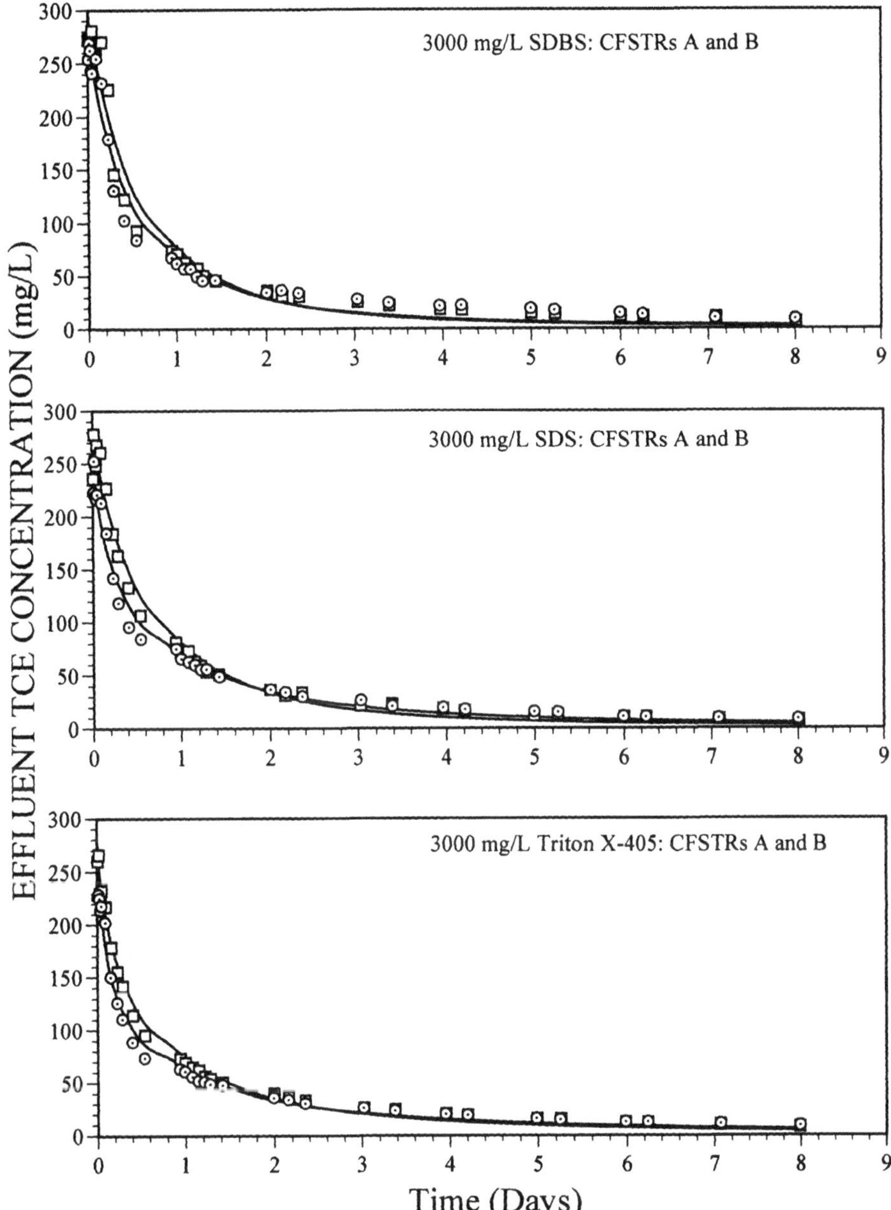

Figure 5. Distributed rate model simulations of the data obtained from the CFSTRs flushed with the 3,000 mg/L Sodiumdodecylbenzenesulfonate (SDBS), the 3,000 mg/L Sodiumdodecylsulfate (SDS), and the 3,000 mg/L Triton X-405 solutions.

Table 2. Minimum percent error and corresponding simulation parameters for the distributed-rate model

CFSTR	Γ-Model		
	η	β	% Error
Water A	0.114	173.9	3.223
Water B	0.156	37.13	2.864
30 mg/L Triton X-100 A	0.419	2.244	7.259
30 mg/L Triton X-100 B	0.500	1.820	9.669
300 mg/L Triton X-100 A	0.564	2.460	8.418
300 mg/L Triton X-100 B	0.472	5.917	14.07
3,000 mg/L Triton X-100 A	0.249	6.517	9.158
3,000 mg/L Triton X-100 B	0.238	38.061	9.558
30 mg/L Tween 20 A	0.354	3.103	4.309
30 mg/L Tween 20 B	0.289	2.686	4.410
300 mg/L Tween 20 A	0.267	4.138	8.410
300 mg/L Tween 20 B	0.316	2.097	4.032
3,000 mg/L Tween 20 A	0.172	16.413	6.637
3,000 mg/L Triton X-405 A	0.165	15.611	5.429
3,000 mg/L Triton X-405 B	0.190	4.705	6.672
3,000 mg/L SDS[a] A	0.0894	274.96	8.975
3,000 mg/L SDS B	0.151	10.889	7.978
3,000 mg/L SDBS[b] A	0.0765	323.30	12.35
3,000 mg/L SDBS B	0.0938	50.1113	11.21

[a] sodium dodecylsulfate
[b] sodium dodecylbenzenesulfonate

To determine whether the surfactant had enhanced the rate of TCE desorption, the amount of TCE flushed from each CFSTR as function of time was determined from the model simulations. The model simulations were averaged for each surfactant concentration. The model predictions for the Triton X-100 CFSTRs are shown in Figure 6. Based on Figure 6, it appears that the addition of Triton X-100 at all three of the concentrations increased the amount of TCE removed the CFSTRs when compared to the water CFSTRs. The results of this analysis are consistent with the results reported by Deitsch and Smith (1995). In the preceding reference, the same data were analyzed using a model that incorporated a time-varying mass-transfer rate coefficient. In that study, the rate coefficient was manually altered during the course of the simulation to obtain a "good" fit of the averaged experimental data. The modeling procedure employed by Deitsch and Smith (1995) enabled the CFSTR concentration profiles to be numerically integrated, and thus to develop mass-removed plots analogous to Figure 6. The present analysis using the distributed rate model and an optimization routine indicate that the conclusions of the original study were not biased by the fitting technique employed in the original study.

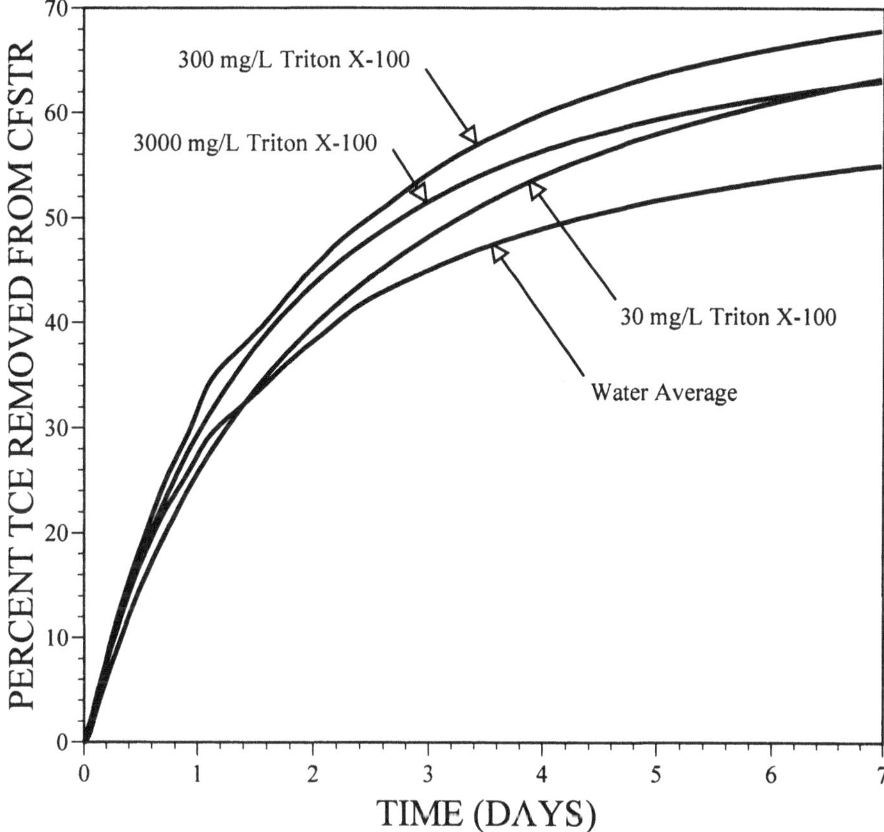

Figure 6. Calculated percentages (based on distributed rate model optimal simulations) of TCE removed from the CFSTRs flushed with the non-surfactant solution, the 30 mg/L Triton X-100 solution, the 300 mg/L Triton X-100 solution, and the 3,000 mg/L Triton X-100 solution. The removal profiles shown are averages of the replicate experiments.

The model predictions for the percent TCE removed from the Tween 20 CFSTRs are shown in Figure 7. In Figure 8, the model predictions for the percent TCE removed from the 3,000 mg/L sodiumdodecylsulfate CFSTRs, the 3,000 mg/L sodiumdodecylbenzenesulfonate CFSTRs, and the 3,000 mg/L Triton X-405 CFSTRs are shown. In contrast to the Triton X-100 CFSTRs, the surfactants tested appeared to have no effect on the rate of desorption (i.e., 30 mg/L Tween 20 CFSTR) or appeared to have inhibited the rate of TCE desorption (i.e., remaining surfactants). For example, the surfactants shown in Figure 8 reduced the amount of TCE flushed from the column by approximately 20%.

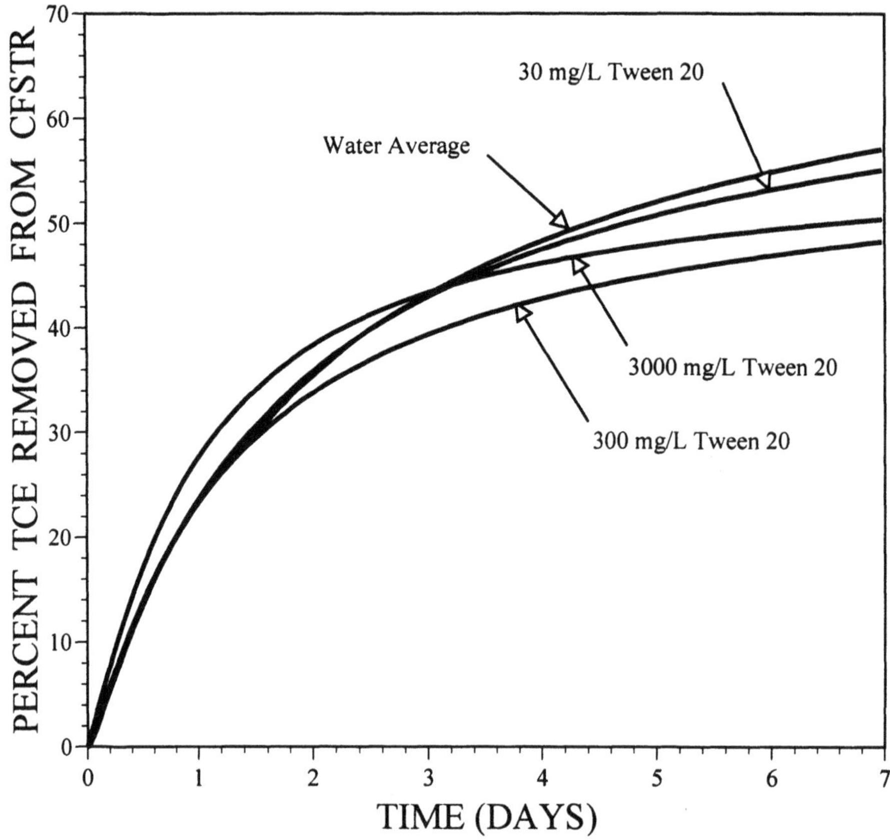

Figure 7. Calculated percentages (based on distributed rate model optimal simulations) of TCE removed from the CFSTRs flushed with the non-surfactant solution, the 30 mg/L Tween 20 solution, the 300 mg/L Tween 20 solution, and the 3,000 mg/L Tween 20 solution. The removal profiles shown are averages of the replicate experiments (except for the 3,000 mg/L Tween 20 profile that is from a single CFSTR).

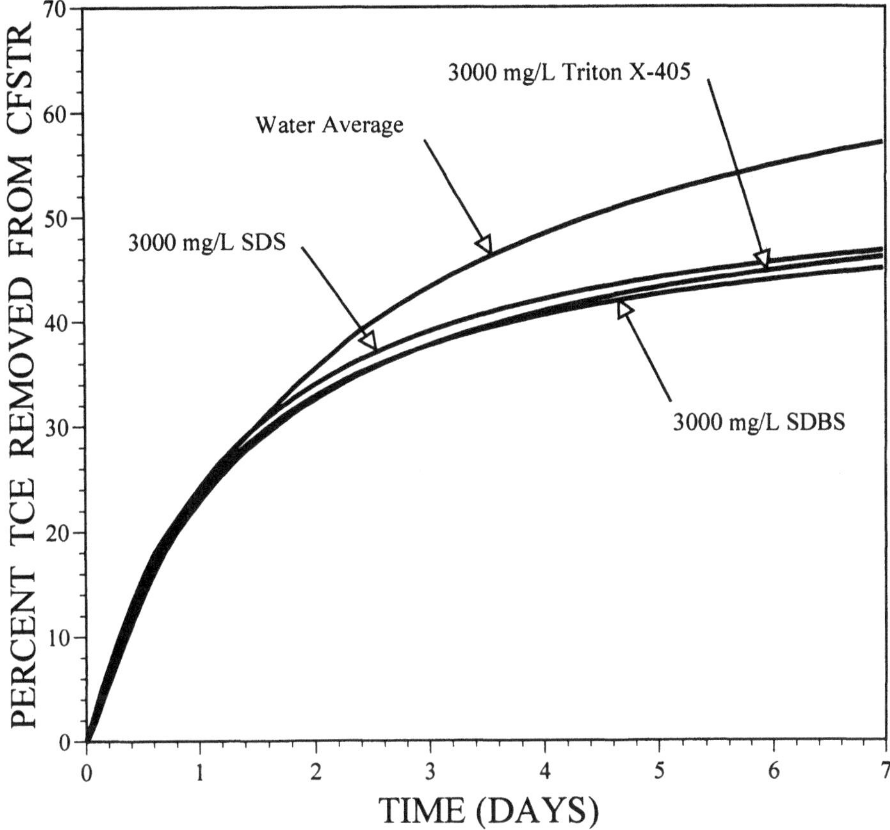

Figure 8. Calculated percentages (based on distributed rate model optimal simulations) of TCE removed from the CFSTRs flushed with the non-surfactant solution, the 3,000 mg/L Sodiumdodecylsulfate (SDS) solution, the 3,000 mg/L Sodiumdodecylbenzenesulfonate (SDBS) solution, and the 3,000 mg/L Triton X-405 solution. The removal profiles shown are averages of the replicate experiments.

Without knowing the effect of the surfactants on the distribution coefficient for TCE sorption to the peat soil, it is not possible to determine what mechanism caused the reduction in TCE removal. It is possible that the addition of sodiumdodecylsulfate, sodiumdodecylbenzenesulfonate, Tween 20, or Triton X-405 may have enhanced the sorption of TCE to the peat soil. If this were the case, the reduction in TCE removal would be attributable to a decrease in the magnitude of the concentration gradient driving the desorption process. Another possibility is that the addition of these surfactants may have inhibited the diffusion of the TCE from the peat soil to the aqueous phase. This last hypothesis could result if the surfactant molecules are not compatible with the glassy phase of the soil organic matter. If the surfactant molecules do not transform the glassy region of the soil organic matter to "rubbery" soil organic matter, the presence of the surfactant would not increase the rate of TCE diffusion from the glassy soil organic matter. In contrast, if the surfactants accumulate within the rubbery phase of the soil organic matter, the polymer density of the rubbery soil organic matter may increase and thus reduce the rate of solute diffusion through the soil organic matter.

As described previously, Triton X-100 at concentrations of 30 mg/L and 300 mg/L should not affect the magnitude of TCE sorption to the peat soil. Therefore, for these two surfactant concentrations, the optimal distributions of rate coefficients can be compared to the distribution of rate coefficients from the water CFSTRs. The average distribution of rate coefficients for these Triton X-100 CFSTRs and the water CFSTRs are shown in Figure 9.

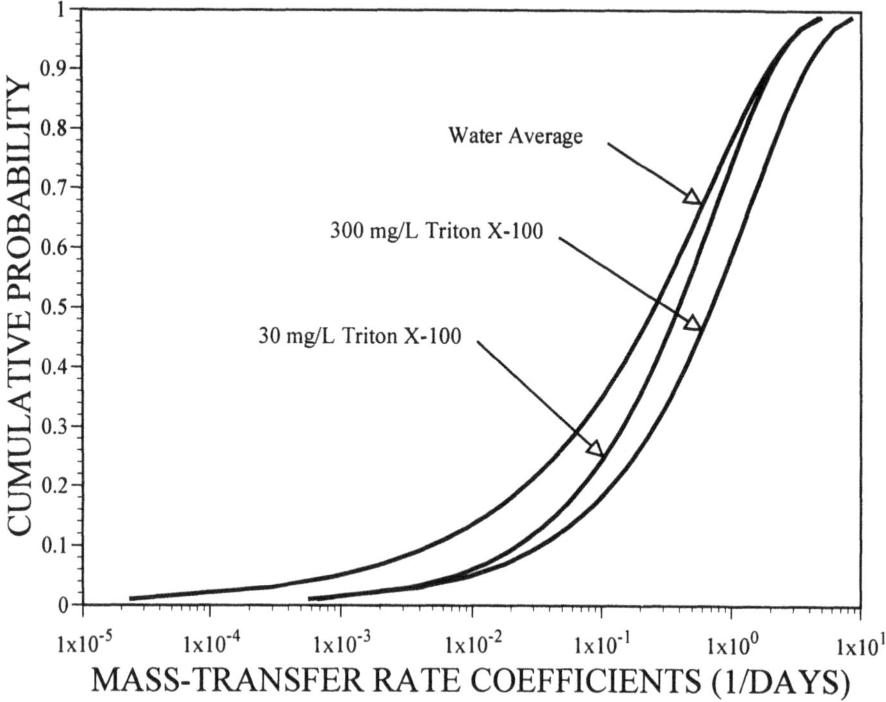

Figure 9. Optimal distributions of mass-transfer rate coefficients for the labelled CFSTR experiments. The distributions shown are the average distributions of the replicate experiments.

It is apparent from Figure 9 that the addition of Triton X-100 has shifted the distribution of rate coefficients to larger values when compared to the water CFSTRs. For the 300 mg/L Triton X-100 CFSTR, the spectrum of rate coefficients is shifted approximately one order of magnitude higher than the water CFSTR rate coefficient distribution. Therefore, the addition of Triton X-100 to the system has increased the rate of diffusion of the TCE from the peat to the aqueous phase. It is hypothesized that the addition of the Triton X-100 has transformed regions of glassy soil organic matter to rubbery soil organic matter, thus increasing the rate of solute diffusion through the soil organic matter. Evidence supporting this hypothesis has been given previously.

The results presented in this section indicate that all surfactants may not be suitable for surfactant-enhanced desorption. Multiple factors may influence the suitability of a surfactant for surfactant-enhanced desorption. These factors have been discussed previously. It is clear that additional research is needed to better understand how surfactant molecular structure and soil composition/chemistry affect the rate of solute desorption. However, the results presented in this chapter and in other studies indicate that surfactant-enhanced remediation of aquifers is a promising technology that needs to be explored.

BIBLIOGRAPHY

Ball, W. P., and Roberts, P. V. (1991). "Long-term sorption of halogenated organic chemicals by aquifer material. 2. Intraparticle diffusion." *Environ. Sci. Technol.*, 25, 1237-1249.

Berens, A. R. (1989). "Solubility and diffusion of small molecules in PVC." *Journal of Vinyl Technology*, 11(4), 171.

Brusseau, M. L., Jessup, R. E., and Rao, P. S. C. (1991a). "Nonequilibrium sorption of organic chemicals: Elucidation of rate-limiting processes." *Environ. Sci. Technol.*, 25, 134-142.

Brusseau, M. L., Wood, A. L., and Rao, P. S. C. (1991b). "Influence of organic cosolvents on the sorption kinetics of hydrophobic organic contaminants." *Environ. Sci. Technol.*, 25, 903-912.

Carroll, K. M., Harkness, M. R., Bracco, A. A., and Balcarcel, R. R. (1994). "Application of a permeant/polymer diffusion model to the desorption of polychlorinated biphenyls from Hudson River sediments." *Environ. Sci. Technol.*, 28(2), 253-258.

Culver, T. B., Hallisey, S. P., Sahoo, D., Deitsch, J. J., and Smith, J. A. (1997). "Modeling the desorption of organic contaminants from long-term contaminated soil using distributed mass transfer rates." *Environ. Sci. Technol.*, 31(6), 1581-1588.

Cunningham, J. A., Werth, C. J., Reinhard, M., and Roberts, P. V. (1997). "Effects of grain-scale mass transfer on the transport of volatile organics through sediments." *Water Resour. Res.*, 33(12), 2713-2726.

Deitsch, J. J., and Smith, J. A. (1995). "Effect of Triton X-100 on the rate of trichloroethene desorption from soil to water." *Environ. Sci. Technol.*, 29(4), 1069-1080.

Deitsch, J. J., and Smith, J. A. (1999). "Comparison of 1,2-DCB sorption and desorption rates for a peat soil." *Environ. Toxicol. Chem.*, 18(8), 1701-1709.

Deitsch, J. J., Smith, J. A., Arnold, M. B., and Bolus, J. (1998). "Sorption and desorption rates of carbon tetrachloride and 1,2-dichlorobenzene to three organobentonites and a natural peat soil." *Environ. Sci. Technol.*, 32, 3169-3177.

DiCesare, D., and Smith, J. A. (1994). "Effects of surfactants on the desorption rate of nonionic organic compounds from soil to water." *Rev. Environ. Contamin. Toxicol.*, 134, 1-29.

Edwards, D. A., Adeel, Z., and Luthy, R. G. (1994). "Distribution of nonionic surfactant and phenanthrene in a sediment/aqueous system." *Environ. Sci. Technol.*, 28(8), 1550-1560.

EPA. (1995). "In situ remediation technology status report: surfactant enhancements." *EPA542-K-94-003*, U.S. Environmental Protection Agency, Office of Solid Waste and Emergency Response, Washington, DC.

Farrell, J., and Reinhard, M. (1994). "Desorption of halogenated organics from model solids, sediments, and soil under unsaturated conditions. 2. Kinetics." *Environ. Sci. Technol.*, 28, 63-72.

Harmon, T. C., and Roberts, P. V. (1994). "Comparison of intraparticle sorption and desorption rates for a halogenated alkene in a sandy aquifer material." *Environ. Sci. Technol.*, 28, 1650-1660.

Huang, W., and Weber, W.J. Jr. (1997). "A distributed reactivity model for sorption by soils and sediments. 10. Relationships between desorption, hysteresis, and the chemical characteristics of organic domains." *Environ. Sci. Technol.*, 31(9), 2562-2569.

Kan, A. T., Fu, G., Hunter, M. A., and Tomson, M. B. (1997). "Irreversible adsorption of naphthalene and tetrachlorobiphenyl to Lula and surrogate sediments." *Environ. Sci. Technol.*, 31(8), 2176-2185.

Kan, A. T., Fu, G., and Tomson, M. B. (1994). "Adsorption/desorption hysteresis in organic pollutant and soil/sediment interaction." *Environ. Sci. Technol.*, 28, 859-867.

Kile, D. E., and Chiou, C. T. (1989). "Water solubility enhancements of DDT and trichlorobenzene by some surfactants below and above the critical micelle concentration." *Environ. Sci. Technol.*, 23(7), 832-838.

Ko, S.-o., Schlautman, M. A., and Carraway, E. R. (1998). "Partitioning of Hydrophobic Organic Compounds to Sorbed Surfactants. 1. Experimental Studies." *Environ. Sci. Technol.*, 32(18), 2769-2775.

LeBoeuf, E. J., and Weber, W.J. Jr. (1997). "A distributed reactivity model for sorption by soils and sediments. 8. Sorbent organic domains: Discovery of a humic acid glass transition and an argument for a polymer-based model." *Environ. Sci. Technol.*, 31(6), 1697-1702.

Mackay, D. M., and Cherry, J. A. (1989). "Groundwater contamination: Pump-and-treat remediation." *Environ. Sci. Technol.*, 23, 630-636.

Noordman, W. H., Ji, W., Brusseau, M. L., and Janssen, D. B. (1998). "Effects of rhamnolipid biosurfactants on removal of phenanthrene from soil." *Environ. Sci. Technol.*, 32(12), 1806-1812.

Pavlostathis, S. G., and Mathavan, G. N. (1992). "Desorption kinetics of selected volatile organic compounds from field contaminated soils." *Environ. Sci. Technol.*, 26, 532-538.

Pignatello, J. J., and Xing, B. (1996). "Mechanisms of slow sorption of organic chemicals to natural particles." *Environ. Sci. Technol.*, 30(1), 1-11.

Sahoo, D., and Smith, J. A. (1997). "Enhanced trichloroethene desorption from long-term contaminated soil using Triton X-100 and pH increases." *Environ. Sci. Technol.*, 31(7), 1910-1915.

Sahoo, D., Smith, J. A., Imbrigiotta, T. E., and McLellan, H. M. (1998). "Surfactant-enhanced remediation of a trichloroethene (TCE) contaminated aquifer: II. Transport of TCE." *Environ. Sci. Technol.*, 32(11), 1686-1693.

Smith, J. A., Tuck, D. M., Jaffé, P. R., and Mueller, R. T. (1991). "Effects of surfactants on the mobility of nonpolar organic contaminants in porous media." Organic Substances and Sediments in Water, R. Baker, ed., Lewis Publishers, Chelsea, Michigan, 201-230.

Sun, S., Inskeep, W. P., and Boyd, S. A. (1995). "Sorption of nonionic organic compounds in soil-water systems containing a micelle-forming surfactant." *Environ. Sci. Technol.*, 29(4), 903-913.

Tiehm, A., Stieber, M., Werner, P., and Frimmel, F. H. (1997). "Surfactant-enhanced mobilization and biodegradation of polycyclic aromatic hydrocarbons in manufactured gas plant soil." *Environ. Sci. Technol.*, 31, 2570-2576.

Vigon, B. W., and Rubin, A. J. (1989). "Practical considerations in the surfactant-aided mobilization of contaminants in aquifers." *J. Water Poll. Control Feder.*, 61(7), 1233-1240.

Weber, W.J. Jr., and Young, T. M. (1997). "A distributed reactivity model for sorption by soils and sediments. 6. Mechanistic implications of desorption under supercritical fluid conditions." *Environ. Sci. Technol.*, 31(6), 1686-1691.

Xing, B., and Pignatello, J. J. (1997). "Dual-mode sorption of low-polarity compounds in glassy poly(vinyl chloride) and soil organic matter." *Environ. Sci. Technol.*, 31, 792-799.

Yeom, I. T., Ghosh, M. M., and Cox, C. D. (1996). "Kinetic aspects of surfactant solubilization of soil-bound polycyclic aromatic hydrocarbons." *Environ. Sci. Technol.*, 30(5), 1589-1595.

Yeom, I. T., Ghosh, M. M., Cox, C. D., and Robinson, K. G. (1995). "Micellar solubilization of polynuclear aromatic hydrocarbons in coal tar-contaminated soils." *Environ. Sci. Technol.*, 29, 3015-3021.

Young, T. M., and Weber, W.J. Jr. (1995). "A distributed reactivity model for sorption by soils and sediments. 3. Effects of diagenetic processes on sorption energetics." *Environ. Sci. Technol.*, 29(1), 92-97.

Chapter Eleven

Surfactant-Enhanced Removal of Hydrophobic Oils from Source Zones

BIN WU[1], HEFA CHENG[1], JEFFREY D. CHILDS,[1,2] and DAVID A. SABATINI[1]

[1] *School of Civil Engineering and Environmental Science, University of Oklahoma, Norman, Ok. 73019*

[2] Corresponding Author

Key words: hydrophobic oils; equivalent alkane carbon number (EACN); surfactant-enhanced remediation

Abstract: This research has demonstrated that the more hydrophobic the oil (e.g., the higher its molecular weight), the more hydrophobic the surfactant system must be to achieve desirable phase behavior. The oil hydrophobicity can be characterized by its equivalent alkane carbon number (EACN). A new approach is presented for estimating the EACN of multi-component hydrophobic nonaqueous phase liquids (NAPLs), which can be used to help guide surfactant selection. Results demonstrate that achieving middle phase microemulsions is more complicated for high EACN oils (e.g. hexadecane with an EACN of 16). An Aerosol-OT (AOT)/ Tween 80 system was identified in batch studies and evaluated in column studies. In less than five injected pore volumes, this surfactant system removed greater than 99% of residual hexadecane from a vertical glass bead column by both mobilization and supersolubilization mechanisms. Counter-flow liquid-liquid extraction conducted in porous hollow fiber membranes was shown to effectively separate hydrophobic oils from surfactant solutions, thereby regenerating the surfactant for reuse. This research thus demonstrates that surfactant enhanced remediation of hydrophobic oils is a viable technology, worthy of further development.

1. INTRODUCTION

1.1 Phase Behavior of Surfactant/Water/Oil Systems

Microemulsions are transparent or translucent, thermodynamically stable "emulsion" systems (Griffin 1949). Forming a middle phase microemulsion (MPM) requires matching the surfactant system's hydrophobicity with that of the oil. The HLB (hydrophilic-lipophilic balance) number reflects the surfactant's partitioning between water and oil phases; higher HLB values indicate water soluble surfactants while lower values indicate oil soluble surfactants (Kunieda et. al. 1980, Abe et. al. 1986). While a balanced surfactant system produces middle phase microemulsions, an under-optimum surfactant system is too water soluble (high HLB) while an over-optimum system is too oil soluble (low HLB).

Figure 1 shows changes in the system phase behavior as its HLB value is systematically adjusted. The left side of the diagram represents a two-phase system with micellar-solubilized oil in equilibrium with an excess oil phase (Winsor Type I) (Winsor 1954). The right side of the diagram represents a different two-phase system with reversed micellar-solubilized water. In-between these two systems a third phase coemerges which contains enriched surfactant with solubilized water and oil. This new thermodynamically stable phase is known as a Winsor Type III middle phase microemulsion.

It is widely recognized that the system IFT reaches a minimum in the middle phase microemulsion region. At the same time, the solubilization parameter (σ), defined as mass of oil solubilized per unit mass of surfactant, is maximized in middle phase microemulsion systems (see Figure 1). This inverse relationship between the solubilization parameter (σ) and IFT (γ) has been defined by the Chun-Huh equation (Huh 1979, Sunwoo et. al. 1992, Abe et. al. 1987):

$$\sigma^2 \propto 1/\gamma \tag{1}$$

When equal volumes of water and oil are solubilized in the middle phase, the system is said to be at its optimum state (Heely et. al. 1974, Salager et. al. 1979) (which is denoted by superscript *), so defined because IFTs are minimized and thus the highest oil solubilization occurs.

The goal of surfactant enhanced subsurface remediation is to maximize the contaminant extraction efficiency while optimizing system economics. Since middle phase microemulsions maximize the solubilization while minimizing the oil-water interfacial tension, these systems are highly desirable, especially for NAPLs lighter than water, where downward migration of released oil is not a concern. Additional factors of concern are system capillary curves, viscosity, density, etc. For more details

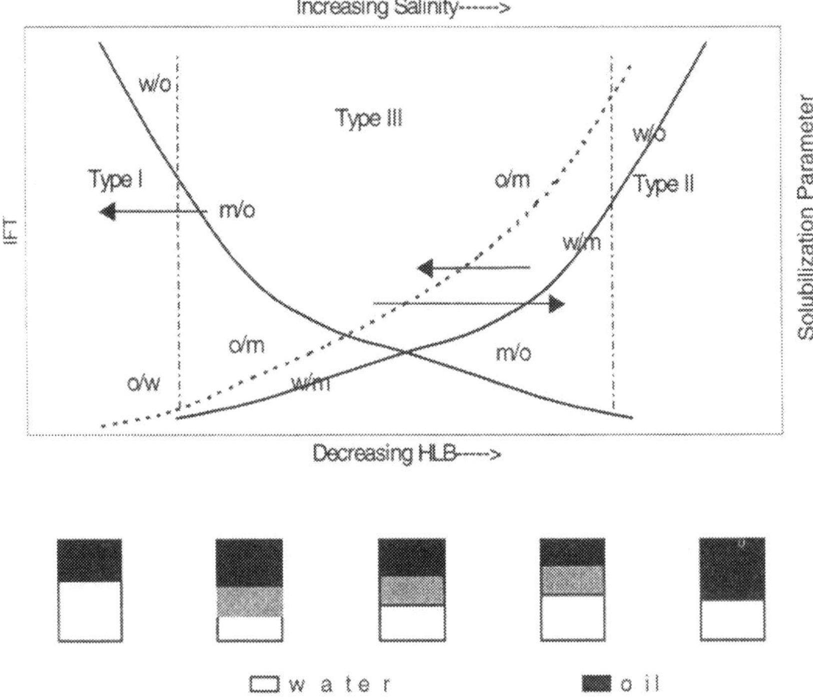

Figure 1. Illustration of Phase Behavior and Interfacial Tension (IFT) as the Scanning Variable is Altered for an Anionic Surfactant: Oil is o; w is Water; m is Middle Phase.

on properties affecting these systems, the interested reader is directed to Sabatini et. al. (1999) and Sabatini et. al. (2000), (e.g., viscosities of $<$ 3 cp and specific gravities of less than 1.03 are preferred).

1.2 Characterization of Oil Contaminants

Oil contaminants can range in both viscosity and molecular weight. The purpose of this work was to find optimal surfactant formulations to extract low viscosity ($<$ 100 cp), high molecular weight (hydrophobic) oils. In surfactant formulation it is common to define the oil molecular weight (hydrophobicity) by virtue of its equivalent alkane carbon number (EACN); aka, how many carbons would there be in an alkane oil of equivalent behavior. Thus, since some crude oils behave similarly to hexane, and since hexane has an alkane carbon number of 6, these crude oils also have an

EACN of six. We thus define high molecular weight (hydrophobic) oils as having EACN values >10 and low molecular weight oils (hydrophilic oils) as having an EACN of < 6.

1.3 High Viscosity Oil Contaminants

Creosotes and coal tars are common examples of high viscosity (100-1000 cp) (Pankow and Cherry, 1996) organic liquids with varying EACN values that are very challenging to remediate. Coal tar and creosotes, while slightly denser than water and ranging from low to high molecular weight, are much more viscous than chlorinated solvents. They are less soluble and mobile and are found in much larger volumes than chlorinated solvents, all making them ideal candidates for surfactant-based remediation systems. Below we summarize coal tar and creosote research to date, very little of which has focused on surfactant-enhanced extraction. We also briefly summarize SEAR research with chlorinated solvents, as we feel these results provide a precedent for our optimism and enthusiasm for extending the surfactant enhanced aquifer remediation (SEAR) technology to the more challenging hydrophobic oils.

Ortiz, et. al (1999) measured overall mass transfer coefficients for the aqueous dissolution of naphthalene, pyrene, and phenanthrene from oils ranging in viscosities from a light lubricating oil (86 cp) to a high viscosity oil (~1000cp). These measurements were performed in continuous-flow systems for time periods ranging from several months up to one year. The authors hypothesize that naphthalene, pyrene, and phenanthrene dissolution from a high viscosity oil (1000 cp) results in a depleted zone within the NAPL that increases with dissolution time.

Barranco et. al. (1999) hypothesized that film formation results from the adsorption of asphaltenes at the coal tar-silica interface. Barranco et. al. (1999) further postulate that these asphaltene components are responsible for pH-dependent interfacial properties observed in coal tar-water-quartz systems.

Peters and Luthy (1993, 1994) performed a detailed analysis of the equilibrium behavior of solvent coal tar water mixtures in work that was complementary to column studies performed by Roy, et. al. (1995). Peters and Luthy successfully modeled ternary phase diagrams of coal tar/n-butylamine/water systems. In addition, Peters and Luthy identified n-butylamine as the leading solvent for coal tar extraction. Pennell and Abriola (1993) report the solubilization of residual dodecane in Ottawa sand using a nonionic surfactant, polyoxyethylene sorbitan monooleate, which achieved a 5 order of magnitude increase over the aqueous solubility, but is still 7 times less than the equilibrium batch solubility with the same surfactant system.

Thermal technologies are also being considered for remediation of these wastes. In a recent summary, Jarsoch and Looney (1999) summarize the various forms of energy-based techniques for enhancing removal of organic contaminants. Udell (1998) discusses the application of in situ thermal remediation for DNAPL removal, with a focus on in situ steam generation. Cummings (1999) presented preliminary results of this technology at the Southern California Edison Visalia Pole Yard. The EPA document "Resource for Manufactured Gas Plant (MGP) Site Characterization and Remediation" identifies both thermal and soil washing techniques as viable for MGP sites (EPA, 2000).

1.4 Low Molecular Weight, Low Viscosity Oils

Surfactant-enhanced remediation of hydrophilic NAPLs (i.e., chlorinated hydrocarbons and gasoline fuels) is a more mature field. A brief summary is given below to demonstrate some important observations from surfactant-enhanced remediation of hydrophilic oil-contaminated soils, many of which will be equally germane to hydrophobic oils. Krebbs-Yuill et. al. (1995) highlighted the economic importance of minimizing surfactant losses, maximizing contaminant extraction, and regenerating the surfactant for reuse. Rouse et. al. (1993) showed that proper surfactant selection could greatly reduce surfactant losses (sorption, precipitation) and thus greatly improve system performance. Shiau et. al. (1994) demonstrated that middle phase microemulsions could be formed with chlorinated solvents using surfactants with direct food additive status, and that middle phase systems are much more efficient than simple micellar systems.. Lipe et. al. (1996) and Hasegawa et. al. (1997) showed that air stripping and liquid-liquid extraction could effectively regenerate surfactant systems laden with volatile and nonvolatile compounds, respectively, and that ultrafiltration could be used for surfactant reconcentration. These concepts have been reinforced and the technology proven through numerous field demonstrations, including the following: Traverse City, MI (Knox et. al. 1997, Sabatini et. al. 1997), three studies at Hill Air Force Base, UT (Knox et. al. 1997), and a study at Tinker Air Force Base, OK (Sabatini et. al. 1998).

Thus, surfactant enhanced subsurface remediation is a mature technology for remediating hydrophilic NAPL, as displayed at the field level. These successful field demonstrations provide encouragement for further evaluation of hydrophobic oils with a similar goal of field deployment. To this end, the current research evaluated laboratory batch and column studies for surfactant enhanced remediation of hydrophobic oil contamination, including phase behavior studies, column studies, and evaluating separation processes for remaining hydrophobic oils from surfactant solutions, all which are functions of the NAPL EACN.

2. DETERMINING EACN OF UNKNOWN NAPLS

Having stated that NAPL EACN (hydrophobicity) is critical to designing successful surfactant systems, a method is needed to determine the EACN of hydrophobic NAPLs. The EACN concept was introduced by Wade and Schecter and co-workers in the 1970s for understanding and correlating surfactants for enhanced oil recovery (Salager et. al. 1979, Bourrel and Schecter 1988) and later found to be of great value in understanding and evaluating surfactant enhanced subsurface remediation of chlorocarbons (Baran et. al. 1994). The EACN of a linear alkane is simply its carbon number. For example, the EACNs of n-octane and n-decane are 8 and 10, respectively. Given the EACN of an unknown oil, and using known parameters for a given surfactant, it is possible to estimate the optimal salinity for making a middle phase microemulsion with this system (Bourrel and Schecter 1988), as discussed below. Phase behavior experiments have been conducted for anionic surfactants and linear alkanes by varying the electrolyte concentration. The logarithm of the optimum electrolyte concentration, ($Ln(s^*)$), was found to be an increasing linear function of the alkane EACN. The optimum salinities for linear alkane mixtures were determined for the same anionic surfactant and electrolyte. Analysis of $Ln(s^*)$ versus Linear-Alkane-EACN showed the EACN of the mixture to be a mole fraction average of the individual linear-alkane-EACNs. For example, the optimum salinity for and EACN of an equal molar mixture of n-octane and n-decane is the same as the optimum salinity for and EACN of n-nonane. Later studies showed the EACNs of chlorocarbons and their mixtures to follow these same mixing rules (Baran et. al. 1994). Many chlorocarbon compounds have negative EACNs. For example the EACNs of TCE and DCB are –3.81 and –4.89, respectively (Dwarakanath et. al. 1998).

Thus, given the optimal salinity for an unknown oil, it is possible to back-calculate the oil EACN. However, this is not helpful in designing the surfactant system in the first place. It is desirable to characterize the oil EACN using an alternate approach. One such approach uses alcohol partitioning as an indicator of an unknown oil EACN. The partition coefficient (k_{ij}) for low concentrations of alcohol between water and NAPL is dependent upon the NAPL EACN where k_{ij} is defined as the concentration of the alcohol in the NAPL (mg/L) divided by the concentration of the alcohol in the water. The subscripts "i" and "j" refer to the NAPL and alcohol, respectively. For sufficiently low alcohol concentrations, the logarithm of the partition coefficient (Ln k_{ij}) is a linearly decreasing function of NAPL EACN. This relationship is analogous to the linearly increasing function of Ln(s^*) versus EACN described previously. By calibrating the alcohol's partitioning with oils of known EACN, one can use these parameters to estimate the EACN of unknown oils quickly, conveniently, and accurately.

Recent work evaluated the partition coefficients between 22 different alcohols and nine NAPLs ranging in EACN from chloroform (-15.13) to decane (10) (Dwarakanath 1998). Both plots of Ln (s*) vs. EACN and Ln(k_{ij}) vs. EACN provided good agreement for the EACN of each of the nine NAPLs. We have extended these results by measuring k_{ij} for four alcohols and 5 neat normal alkanes having EACN from 6 (hexane) to 16 (hexadecane). Next, we measured k_{ij} for each alcohol with diesel, and from calibration curves determined the EACN of diesel.

3. MATERIALS AND METHODS

3.1 Determining EACN of Unknown NAPLS

A total of 20 partition coefficients were measured using 1-pentanol, 2-methyl 2-hexanol, 2-methyl 3-hexanol, and 2,4-dimethyl 3-pentanol as the alcohols and C-6, C-8, C-10, C-12, and C-16 as the n-alkanes. For each alcohol, a 15-ml. aliquot of water with 1000 mg/L alcohol was placed in a 35 ml. vial with 15-ml. aliquot of NAPL and immediately sealed. The alcohol solutions were prepared quickly and immediately sealed with Teflon coated septa with no headspace to minimize volatilization. The samples were thoroughly shaken for 1 hour and allowed to separate for 24 hours. These mixing and separation times were sufficient for equilibration. Immediately following separation, the aqueous samples were analyzed for alcohol using a split injection Varian 3300 GC with FID detection. The partition coefficient was calculated by mass balance.

3.2 Phase Behavior Studies

The hydrophobic oils evaluated in phase behavior studies were diesel, dodecane, and hexadecane. A commercial Diesel was selected based on its occurrence as a subsurface contaminant and its hydrophobicity. Based on published chromatograms, the EACN of this diesel was expected to be between dodecane (EACN=12) and hexadecane (EACN=16). Phase behavior studies compared the properties of diesel with dodecane and hexadecane.

Surfactant systems used in hydrophobic oil phase behavior studies were chosen based on commercial availability and previous research in our laboratory. The HLB values of these surfactants are generally lower than surfactants employed in field studies described previously. Since surfactant enhanced solubilization will be too inefficient and thus not economically viable for hydrophobic oils, such as diesel and lubricating oils, surfactant-based mobilization systems are required to make this approach viable. Lower HLB surfactant systems must be designed that are capable of forming Winsor Type III systems with these hydrophobic oils while maintaining the water solubility of the surfactant system. Thus, while the high HLB Dowfax 8390 system has widely been used for remediating low EACN or hydrophilic NAPLs, it will not be effective for high EACN diesel and lubricating oils.

Table I summarizes properties and gives structures of the surfactants discussed in this chapter. Aerosol OT (AOT, 100% solid) was purchased from Fisher Scientific. TWEEN 80 (80.0% active) was purchased from Uniqema (Wilmington, DE). Tetrachloroethylene (PCE, 99% liquid), dodecane (99% liquid), and hexadecane (99% liquid) were purchased from Aldrich Co. (Milwaukee, WI). All chemicals were used as received.

Previous work has shown that binary surfactant systems containing Dowfax 8390 and the branched hydrophobic surfactant AOT can form Winsor III systems with both PCE and decane whereas DOWFAX 8390 by itself cannot (Wu et. al. 1999). This binary surfactant system was used in conjunction with hydrophobic octanoic acid to help with phase behavior and lessen the required concentration of $CaCl_2$. Since this formulation is rather complicated, questions about field robustness arise. Thus, for the phase behavior studies presented here, we used the simple binary system of the nonionic TWEEN 80 and the branched hydrophobic AOT, and we optimized the NaCl concentration to give the Winsor Type III system. The lesser electrolyte concentration requirement for the binary TWEEN 80/ AOT system helps to decrease the potential for undesirable phase behavior such as surfactant precipitation, thereby increasing surfactant system robustness.

Table I. Description of surfactants used in this research.

Surfactant	Classification	Average Molecular Weight	*Type*
Aerosol-MA80-I (AMA)	Sodium dihexyl sulfosuccinate	*388*	*Anionic*
Aerosol-OT (AOT)	Sodium dioctyl sulfosuccinate	445	*Anionic*
Tween 80	Polyoxyethylene sorbitan ester	*1300*	*Nonionic*
SDBS	Sodium dodecyl benzene sulfate	*348*	*Anionic*

All middle phase microemulsion and solubilization studies were carried out in 20 ml centrifuge tubes with Teflon screw caps. Experiments were conducted at controlled room temperature of 22°C. Equal volumes of aqueous surfactant solution and oil (10.00 ml each) were placed in the centrifuge tube. The tube was shaken gently on a wrist action shaker for 20 minutes and allowed to stand for at least 12 hours to ensure equilibrium. The volume of each phase was carefully recorded to within 0.01 mls. The surfactant and oil concentrations were quantified in each phase for a few select middle phase systems. The HPLC system used to quantify SDBS was a Shimadzu LC-10AD liquid chromatograph including SIL-10A autoinjector and SCL-10A controller. A Waters 486 UV detector was used for SDBS analysis. The wavelength selected for SDBS analysis was 225 nm and the mobile phase was composed of 80% methanol and 20% water. For quantification of dodecane dissolved in surfactant solutions, a Shimadzu 17A gas chromatograph was used with a Tekmar 7000 headspace autosampler and a Tekmar 7050 carrousel. The headspace equilibrium temperature was 65°C. The injector and detector temperatures were 250°C and 280°C respectively. The column temperature was ramped at the rate of 8°/min. from 40 to 230°C. FID detection was employed, and all samples and standards were normalized to a common surfactant concentration to account for surfactant-reduced sample volatility.

3.3 Column Studies

Glass-bead-packed one-dimensional column experiments were performed with vertically oriented 2.5 cm I.D., 15.0 cm long, 74.0 ml. total volume glass columns. The glass bead diameter ranged from 0.23 to 0.36 mm, and the packed-bed porosity was 0.36. Air was evacuated from the column by pumping twenty pore volumes of degassed, deionized water through the air-tight glass-bead-packed column prior to hexadecane contamination. A Cole-Parmer Masterflex digital peristaltic pump flushed degassed water through the uncontaminated column and flushed both water and surfactant solution through the hexadecane-contaminated column. Residual hexadecane saturations were established in the air-tight column by buretting hexadecane into the water-saturated glass-bead-packed column in a down-flow mode and then displacing the free hexadecane with water in an up-flow mode. Ten pore volumes of deionized water were injected into the contaminated column, and the volume of displaced free phase hexadecane was quantified before beginning the surfactant flush. Using this packing, contamination, and water flushing procedures, the residual hexadecane saturation in the contaminated zone was 22.5%. The volume of entrapped hexadecane was derived from a mass balance of the oil in the burette before and after the contamination procedure, and the free phase oil recovered during the water flush. Following the entrapment of hexadecane, the surfactant solution was pumped through the column in the up-flow mode at a flow rate of 0.3 ml./min. or pore water velocity of 2.4 m/day.

The column effluent was collected in 0.20 pore volume increments by a Spectra/Chrom fraction collector. The column effluent was categorized as one of the following: liberated free phase hexadecane, micellarly solubilized hexadecane, or microemulsified middle phase hexadecane. One column effluent sample contained all three phases, where the micellarly solubilized hexadecane was the most dense bottom phase, free phase hexadecane was the least dense top phase, with the microemulsified middle phase in between. Effluent samples of liberated free phase hexadecane were separated from both micellarly solubilized hexadecane and microemulsified middle phase hexadecane, and subsequently quantified with a graduated vial to within ±0.01 ml. Micellarly solubilized hexadecane was separated from microemulsified middle phase hexadecane, and both were diluted with methanol to a concentration below 500 mg/L which is in the linear region of the Varian 3300 GC using split injection and FID detection.

3.4 Separations

Liquid-liquid extraction of hydrophobic oil-laden surfactant solution was evaluated using counter-flow, porous hollow fiber membranes. Our liquid-liquid extraction experiments were conducted using Liqui-Cel Extra-Flow 2.5×8 Membrane Contactor purchased from Celgard LLC (Charlotte, NC). The dimensions of the column are 6.3 cm diameter and 20.3 cm. length with two ports at each end. Inside the column are approximately 7500 microporous polypropylene hollow fibers with 0.05 micron pore diameters, giving an effective surface area of 1.4 m^2 and 40% porosity. Both the aqueous surfactant solution and extraction solvent were pumped through the column using Cole Palmer peristaltic pumps. The extraction solvent chosen for this work was squalane (2, 6, 10, 15, 19, 23 hexamethyltetracosane, or $C_{30}H_{62}$). It has low toxicity, ultralow aqueous solubility, low viscosity for such a hydrophobic oil, and has been used in previous research. Furthermore, dodecane, diesel, and hexadecane are completely miscible with squalane. The upper limit of squalane EACN is 30 but due to branched methyl groups, it may be as low as 24.

Squalane was first pumped through the column in an upflow mode to saturate all extra-fiber pore space. The middle phase surfactant system (10% SDBS/17% IPA/12.4% NaCl/dodecane) was then pumped through the column in a downflow mode. The middle phase and squalane flow rates were 6.32 ml./min and 3.16 ml/min., respectively. The squalane and middle phase residence times in the column were 88.6 min. and 15.8 min., respectively.

SDBS concentrations were determined by HPLC as described previously. Dodecane concentrations in both the aqueous surfactant and squalane were determined by a GC/static Headspace technique.

4. RESULTS AND DISCUSSION

4.1 Determining EACN of Unknown NAPLS

Since the oil EACN is so critical to formulating surfactant middle phase microemulsions (MPMs), we will begin by discussing a method for characterizing the EACN of unknown NAPLs. The relation between the oil-water partitioning of alcohols (k_{ij}) and NAPL EACN is dependent upon the hydrophobicity of the alcohol, as given by:

$$\text{Ln } k_{ij} = C \times N_i + B \qquad (2)$$

Where the partitioning coefficient (k_{ij}) was previously defined and N_i is the NAPL EACN. “C” and “B” are the slope and intercept, respectively, for each alcohol obtained from regression of the experimental data. Figure 2 plots Ln k_{ij} against EACN for the four partitioning alcohols, and Table II

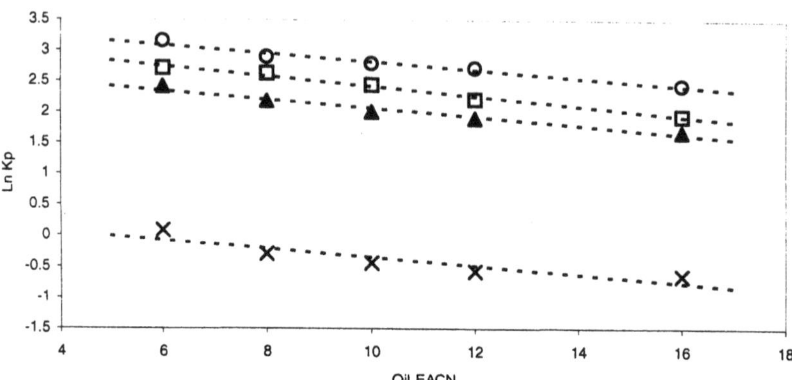

Figure 2. Plots of $Ln(k_{ij})$ versus EACN and best fit lines for each of the partitioning alcohols described in text.

summarizes the regression statistics. For a given oil, the oil-water partitioning (K_{ij}) for 1-pentanol is nominally 10-fold lower than for the other three partitioning alcohols. This is expected since 1-pentanol has two less carbons than each of the other three alcohols and is therefore the least hydrophobic and exhibits the lowest partitioning EACN. The lower partitioning coefficients and greater aqueous concentrations of 1-pentanol

Table II. Regression statistics for each partitioning alcohol in Figure 2 showing the correlation constants of Equation 2. Underneath the correlation constants are the upper and lower bounds of the 95% confidence interval.

1-pentanol R squared = 0.82		2-methyl 2-hexanol R squared = 0.97		2-methyl 3-hexanol R squared = 0.98		2,4-dimethyl 3-pentanol R squared = 0.96	
C	B	C	B	C	B	C	B
-0.07	0.3	-0.072	2.78	-0.082	3.25	-.066	3.49
(-0.13: -0.01)	(-0.3: 1.0)	(-0.096: -0.047)	2.51: 3.05	(-0.101: -0.063)	(3.04: 3.46)	(-0.043: -0.090)	(3.24: 3.75)

may explain the greater error in its GC quantification and thus its lower R squared value (0.82-Table II). Referring to Figure 2 and Table II, the 95% confidence intervals for "C" overlap. Thus, the coefficient "C" is the same for all four partitioning alcohols, indicating that the trend in alcohol partitioning is similar between the alcohols. Similar trends between partitioning alcohols (Figure 2) have been observed by others (Dwarakanath et. al. 1998).

The bilinear relation between k_{ij} and EACNs of both NAPL and alcohol can be shown as follows:

$$\text{Ln } k_{ij} = C \times N_i + M \times A_j + g \qquad (3)$$

"A_j" is the alcohol EACN, and g, M, and N are correlation constants obtained by regression of data from partitioning alcohols and NAPLs. Thus, "B" of equation (2) is equal to $(M \times A_j + g)$ in Equation (3), and "C" of equation (2) is equal to "C" of equation (3). Dwarkanath and Pope (1998) obtained values of M, g, and C for 22 different alcohols and nine NAPLs. Referring to "B" values in Table II, the alcohol EACN ranking is consistent with their alcohol EACN ranking. Furthermore, the EACNs of 2-methyl 2-hexanol, 2-methyl 3-hexanol, and 2,4-dimethyl 3-pentanol are similar since these alcohols contain the same number of carbons. The greatest EACN alcohol, 2,4-dimethyl 3-pentanol, exhibits greater branching than the other two alcohols. 1-pentanol is the lowest EACN alcohol from Table III, which is consistent with the fact that it has two less carbons than the other three alcohols listed above.

NAPL EACNs for commercial diesel oil, NAPL from Hill AFB OU1, and commercial motor oil were determined by measuring the alcohol partitioning into each NAPL and back-estimating the NAPL EACN via Figure 2 and Equation 2, with results summarized in Table III. The EACN of diesel is very close to the value estimated from the gas chromatographic analysis of commercial diesel fuels using the linear mixing rules described previously. (e.g., CHROM 2033, Alltech Catalog 450, 1999, P. 69). Both motor oil and Hill OU1 oil have EACNs greater than diesel, as expected. Thus, determining NAPL EACNs by partitioning alcohols is a very useful tool for hydrophobic oil characterization.

Table III. EACN determination of several multi-component hydrophobic oils.

	2,4dm 3 ptnl		2me 3hxnl		2me 2hxnl		Average NAPL EACN
	k_{ij}	EACN	k_{ij}	EACN	k_{ij}	EACN	
Diesel	13.86	12.9	9.54	13.7	6.30	13.1	13.2
Hill OU1	11.69	15.6	9.37	15.4	5.47	15.1	15.4
Motor oil	6.51	24.4	6.08	24.0	3.3	22.2	23.5

4.2 Phase Behavior

Hydrophilic PCE (EACN=2.9) was mixed with an equal volume of the hydrophilic surfactant system Aerosol-MA, isopropyl alcohol, and NaCl (AMA/IPA/NaCl) (Figure 3). With increasing salinity, a classical Winsor phase progression was observed from type I to type II, with type III in-between. When this same surfactant system was applied to dodecane, with an EACN of 12, the classical Winsor type I⇔III⇔II phase progression did not occur, and while a middle phase was achieved (data not shown), the optimum one-to-one ratio of oil volume to water/IPA volume was not realized, and Winsor II systems were not achieved. The phase behavior of a more hydrophobic surfactant system, AOT/TWEEN 80, with diesel (EACN≈13.3) as a function of salinity is shown in Figure 4. With increasing salinity, the Winsor phase progression (I⇔III⇔II) occurred (Figure 4). Thus, while the Winsor type I-III-II phase progression did not occur for AMA/IPA/diesel, it did indeed occur with the more hydrophobic surfactant system AOT/Tween 80/diesel, demonstrating the importance of matching the surfactant system hydrophobicity with that of the NAPL. The AOT/Tween 80 system was also able to form one-to-one oil-to-aqueous Winsor Type III systems with hexadecane (Figure 5), although it was not determined if Winsor Type II systems exist. Another example of a surfactant system that did not give the Winsor phase progression with either dodecane or diesel is SDBS/IPA (Figure 6). Although it is evident from Figure 6 that some dodecane was solubilized

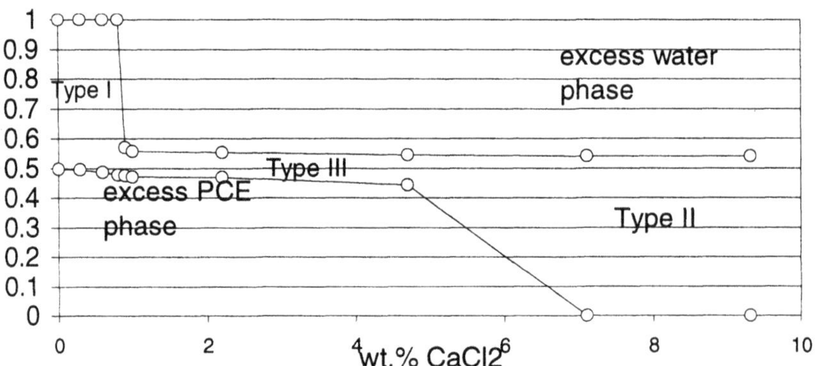

Figure 3. Phase diagram for PCE (EACN=2.9) at 22°C FOR 5.0% AMA, 2.5% IPA, and $CaCl_2$.

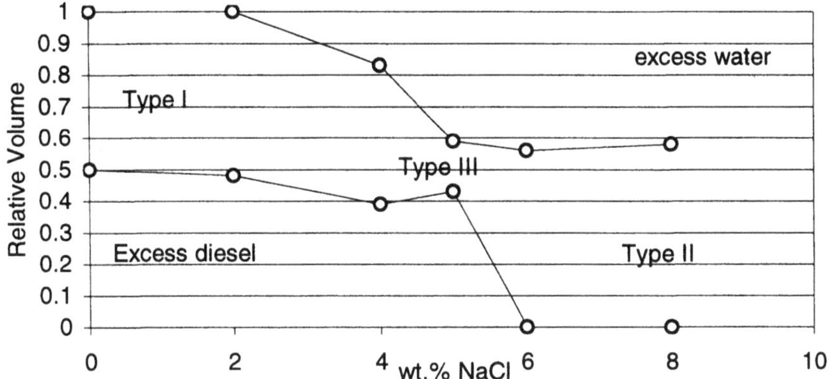

Figure 4. Phase Diagram for diesel (EACN=13.3) at 22°C for 2.0% AOT, 2.0% Tween 80, and NaCl.

into micelles in the type I region, this solubilization was well below equal volumes of dodecane and water/IPA anticipated for optimal Winsor type III systems. These results all indicate the sensitivity of the surfactant design to the contaminant hydrophobicity and demonstrate that proper surfactant system design promotes desirable phase behavior.

A more detailed qualititative observation for the system SDBS/IPA/dodecane/NaCl (Figure 6) is noteworthy. Three phases emerged

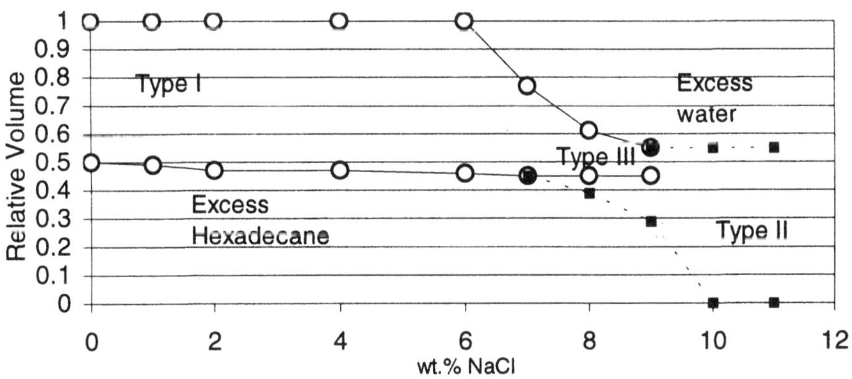

Figure 5. Phase Diagram for hexadecane (EACN=16) at 22°C for 2% AOT, 2% Tween 80, and NaCl. The open circles are data. The closed squares represent the phase diagram for a surfactant system which progresses through types I⇔III⇔II with neither phase separation nor precipitation.

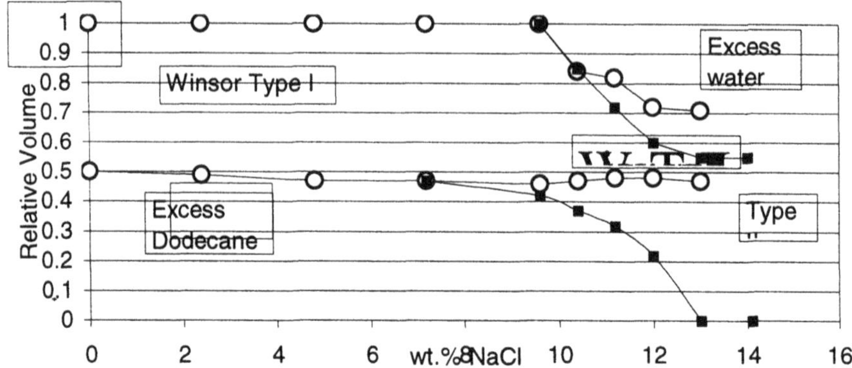

Figure 6. Phase Diagram for dodecane (EACN=12) at 22°C for 10% SDBS, 17% IPA, and NaCl. The open circle are data. Three phases emerge at ~10 wt.% NaCl and persist until precipitation which occurs at ~20 wt.% NaCl. The closed squares represent the phase diagram for a surfactant system which progresses through types I⇔III⇔II with neither phase separation nor precipitation.

at ~10 wt.% NaCl and persisted until precipitation at ~20 wt.% NaCl. Although a middle phase was observed at ~10 wt.% NaCl, it was not necessarily a strict Winsor Type III region. Furthermore, the volumetric ratio of oil to water and IPA in this "middle phase" never exceeded 0.2 for this system. The absence of a Winsor-type III phase for this system may be the result of the surfactant "salting out" at higher NaCl concentrations. For an aqueous solution of 10% SDBS and 17% IPA in the absence of dodecane, a phase separation occurs at ~15% NaCl, resulting in an IPA-deficient, surfactant-deficient bottom phase and an IPA rich, surfactant rich top phase.

Further quantitative information about the compositions of each phase in Figure 6 was obtained in the three-phase region for 12.4 wt.% NaCl. SDBS concentrations for each phase were determined. HPLC analysis of both middle and bottom aqueous phases revealed that 94.5% of the SDBS resided in the middle phase, while 4.9% resided in the bottom aqueous phase. The remaining 0.6% was presumed to be in the top dodecane phase, although HPLC concentrations were below detection limits in this phase. Headspace GC analysis revealed that most of the solubilized dodecane (95.8%) resided in the middle phase for a concentration of (10.3 mls dodecane)/(100 mls. middle phase) while the remaining 4.2% is solubilized in the bottom aqueous phase to give (0.50 mls. dodecane)/(100 mls. aqueous phase). Thus, the concentration of middle phase dodecane at 12.4 wt.% NaCl is a factor of 4 to 5 below the concentration obtained at optimum salinity for a system following the typical Winsor phase progression. Thus, quantitative results for the low middle phase dodecane concentration should correlate with the obvious deviation of experimental data in Figure 6 from a system that follows the classical Winsor phase progression.

Before further conclusions are made about deviation from Winsor phase progression due to surfactant "salting out" at high NaCl concentrations, the concentrations of IPA, NaCl, surfactant, and oil for each phase should be determined for a system that deviates from Winsor phase progression such as the system in Figure 6. For a system that follows a strict Winsor phase progression, all the surfactant, all the alcohol, all the salt, and all the solubilized oil reside in the middle phase for the optimum Winsor type III system. For three-phase regions, deviations from Winsor phase progressions coincide with substantial concentrations of IPA, salt, surfactant, and oil in bottom aqueous phase and/or substantial concentrations of IPA, salt, surfactant, and water in the top oil phase.

In summary, phase behavior was very sensitive to oil hydrophobicity. Systems that work for hydrophilic NAPLs such as chlorinated solvents will not necessarily work for more hydrophobic NAPLs, and surfactant systems that work for slightly hydrophobic NAPLs (e.g., dodecane) do not necessarily work for more hydrophobic NAPLs such as hexadecane. These results thus point to the need to characterize the NAPL hydrophobicity (EACN), as discussed in the previous section, and design the surfactant system appropriately.

4.3 Column Studies

Based on batch results, column studies were conducted with hexadecane-contaminated glass beads using the surfactant system 2% AOT/ 2% Tween 80/ NaCl; the results are summarized in Table IV and Figure 7. The percent of hexadecane removed from the column as micellarly solubilized, middle phase microemulsified, and free phase is plotted against the number of surfactant system pore volumes in Figure 7. Also plotted in Figure 7 is the concentration of micellarly solubilized oil. For the column study, a total of 3.0 mls of hexadecane contaminated approximately the top one-half of the glass-bead-packed column to give a residual saturation of 22.5% in the contaminated zone. As indicated in Table IV, the surfactant flushing was performed in two-stages. The first stage, from 0 to 2.6 injected pore volumes, employed 5.0% NaCl while the second stage, from 2.6 to 4.7 injected pore volumes, employed 7.0% NaCl. Referring to Figure 7 and batch results (Figure 5), the first stage 2% AOT/ 2% Tween 80/ 5.0% NaCl corresponds to Type I region approaching the middle phase region, and the second stage 2% AOT/ 2% Tween 80/ 7.0% NaCl corresponds to a middle phase region.

Since the surfactant system was pumped in the upflow mode, the mobilized free phase oil at the surfactant front was aided by buoyancy in its liberation from the column. Mobilization of hexadecane occurred from 0.9 to 1.3 injected pore volumes as evidenced by Figure 7. Since the glass-bead-packed column is saturated with water prior to surfactant flushing, the concentration of surfactant in the effluent is lower than in the injected solution during the first injected pore volume. From Figure 7, middle phase microemulsion is observed in the effluent after 1.3 injected pore volumes of this surfactant system. This observation must be reconciled with Figure 5 which shows that the system 2% AOT/ 2% Tween 80/ 5.0% NaCl/ hexadecane is type I. This surfactant system sufficiently lowers the aqueous/oil IFT to cause some mobilization of hexadecane that manifests itself as middle phase microemulsion. This must be either due to regions in the column where the concentrations of AOT, TWEEN 80, and NaCl give a different behavior than the injected solution or some other variation that causes this phase shift, pointing out the need to have a buffer zone if you are trying to avoid a certain type of phase.

The surfactant breakthrough and stage I maximum micellarly solubilized hexadecane (4200 mg/L) were observed at 1.5 injected pore volumes (Figure 7). The effluent micellarly solubilized hexadecane concentration maintained a plateau of 3000 to 4000 mg/L until nominally 1 pore volume of the second stage system (2% AOT/ 2% Tween 80/ 7.0% NaCl) was flushed through the column. Figure 7 shows a striking increase in the micellarly solubilized hexadecane accompanied by a lesser amount of free phase hexadecane and microemulsified middle phase hexadecane. The lower IFT of hexadecane with the second stage system corroborates the observed free phase oil, and Figure 5 corroborates the observed microemulsified middle phase oil. In addition the 10-fold increase in the micellarly solubilized oil is very encouraging and again illustrates solubility enhancement.

After the two-stage surfactant flush, a post-methanol extraction was performed to quantify the hexadecane remaining in the glass-bead-packed column. This extraction revealed that 99.4% of the hexadecane had been removed after both the 4.7 pore volume two-stage surfactant flush and 1.3 pore volume post-water flush. A mass balance of liberated free phase hexadecane, micellarly solubilized hexadecane, and microemulsified middle phase hexadecane from Figure 7 showed a 100.3% removal. It is very encouraging that the post-methanol extraction agrees well with the mass balance from Figure 7, and that both methods show that hexadecane removal was so high, and much higher than reported previously. (see background section)

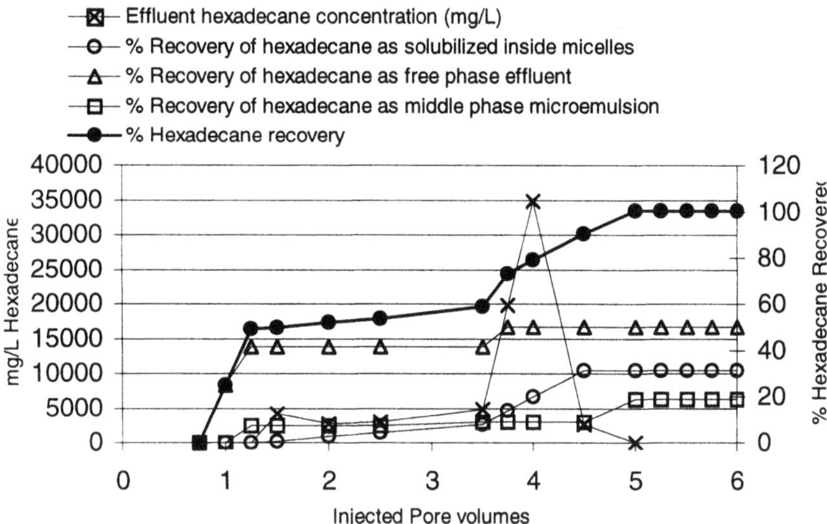

Figure 7. Plots of hexadecane removal percentage as micellarly solubilized, middle phase microemulsified, and free phase against the number of flushed surfactant system pore volumes. Also plotted is the effluent concentration of micellarly solubilized oil.

Table IV. The two-stage surfactant flushing results for Figure 7.

Surfactant System		IFT oil/surf (d/cm)	W. T.	% recovery by micellar solubilization	% recovery as free phase	% recovery as MPM	% recovery total	Pore vols. of flush
2% AOT/ 2% Tween 80/ 5.0% NaCl	Stage I	0.03	I	20.1	41.7	7.6	68.4	0 – 2.6
2% AOT/ 2% Tween 80/ 7.0% NaCl	Stage II	0.01	III	11.3	8.3	11.3	31.9	2.6 – 4.7
Combined stages	–	–	_	31.4	50	18.9	100.3	0 – 4.7
D.I. water	–	–	–	31.4	50	18.9	100.3	4.7 – 6.0

4.4 Separations

The objective of this portion of the research was to experimentally evaluate surfactant effects on the liquid-liquid separation of hydrophobic oils from a surfactant system. For pump-and-treat subsurface remediation in the absence of surfactant, contaminated ground water would be pumped from the subsurface and through a liquid-liquid extraction column where the contaminant partitions from the aqueous phase into an extraction solvent phase. In the absence of surfactant, the driving force for partitioning is a function of the contaminant hydrophobicity. In the presence of surfactants, the contaminant is subject to competitive partitioning (i.e., into the micelles and into the extracting oil).

By maximizing the contact area between extraction solvent and surfactant-solubilized hydrophobic oil contaminant through the use of state-of-the-art hollow fiber membrane columns, we hypothesize that hydrophobic oil contaminants can be separated from surfactant solutions without macroemulsification. For this research we were interested in the partitioning of the hydrophobic oil from the hydrophobic environment of the micelle via its aqueous concentration into a more preferred extracting solvent.

We began the batch squalane extraction studies by mixing 30 mls of the middle phase from the surfactant system 10% SDBS/17% IPA/12.4% NaCl/dodecane (Figure 6) with 30 mls. of squalane. The solutions were mixed for 10 minutes on a wrist action shaker and allowed to separate. Coalescence was rapid and complete in 5 minutes. The squalane rich phase was separated, and both phases were analyzed for dodecane and SDBS by the methods described previously. These results are summarized in Table V. The batch study was encouraging since 98% of the original solubilized dodecane from Figure 6 (12.4% NaCl) partitioned into squalane. All the surfactant remained in the aqueous phase, as expected.

Continuous flow column extraction was operated at 1:2 squalane volume to middle phase volume ratio which is lower than the 1:1 ratio for the batch studies. From Table VI it is encouraging that the separation of dodecane from surfactant is very good and not much different from the batch extraction study. This result may be attributed to the long contact time (15.8 min.) and the large contacting surface area (1.4 m^2/(100 ml solution)). These results demonstrate that counter-flow column extraction via state-of-the-art porous hollow fiber membranes effectively separate hydrophobic oils from surfactant solutions, thereby regenerating the surfactant for reuse. These results are thus very encouraging with respect to the overall system economics. (i.e., it is feasible to regenerate and reuse hydrophobic oil-laden surfactant systems)

Table V. Batch Squalane extraction results by mixing 30 mls of the middle phase from the system 10% SDBS/17% IPA/12.4% NaCl/dodecane (Figure 6) with 30 mls. of squalane.

Phase	Event	Volume (ml)	SDBS (g/100 ml.)	Dodecane (ml/100 ml.)
Middle Phase (Squalane deficient phase)	Before mixing	30	18.1	10.32
	After mixing and coalescence	27	20.2	0.19
Squalane rich phase	Before mixing	30	0	0
	After mixing and coalescence	33	Below detection limit	9.23

Table VI. The results for the liquid-liquid extraction via continuous counter-flow through porous hollow fiber membranes.

Phase	Event	Flow Rate (ml/min.)	SDBS (g/100 ml.)	Dodecane (ml/100 ml.)
Middle Phase (Squalane deficient phase)	Before Extraction	6.32 (during extraction)	18.1	10.3
	After Extraction	6.32 (during extraction)	19.9	0.28
Squalane rich phase	Before Extraction	3.16 (during extraction)	0	0
	After Extraction	3.16 (during extraction)	Below detection limit	20.1

5. CONCLUDING REMARKS

The classical Winsor phase type I ⇔III⇔II progression has been observed for SDBS/IPA and AMA/IPA with hydrophilic oils of EACN<10 (e.g., PCE with an EACN of 2.9). However, neither of these systems exhibited a classical Winsor phase progression with either dodecane or diesel (Figure 6 for example). Although a "middle phase" was observed, a "salting out" of the surfactant occurred before optimum type III was realized, and the type II system did not emerge. The partitioning alcohol EACN determination showed diesel oil to have an EACN of 12.9–13.7. Thus, based on the knowledge that neither SDBS/IPA nor AMA/IPA works for dodecane (EACN=12), and based on the partitioning alcohol EACN determination of diesel oil, we would predict that neither AMA/IPA nor

SDBS/IPA would have worked for diesel, as was observed experimentally. Neither SDBS/IPA nor AMA/IPA is predicted to exhibit the classical Winsor phase type I ⇔III⇔II progression for either the multi-component hydrophobic oils of Hill OU1 or commercial motor oil, both of which exhibited EACNs greater than diesel (Table IV).

The more hydrophobic surfactant system, AOT/TWEEN 80, exhibited the classical Winsor type I ⇔III⇔II phase progression for diesel (Figure 4). This surfactant system also formed a middle phase containing equal volumes of hexadecane and water. At present it has not been determined if the classical Winsor Type I-III-II progression occurs for commercial motor oil (EACN ≈ 22.3) with the AOT/Tween 80 system.

The surfactant system AOT/TWEEN 80 removed 99.4% of residual hexadecane from glass beads in 4.7 pore volumes with a total of 49.6% recovered as free phase, 31.1% recovered as micellarly solubilized , and 18.7% recovered as microemulsified in the middle phase. These results demonstrate the potential efficiency of supersolubilization (i.e., enhanced solubilization as the type I-III boundary is approached), and mobilization (even if just into the type III system but not yet optimal) in expediting extraction of hydrophobic NAPLs.

Research on separation processes demonstrated that hydrophobic oil-laden surfactant was efficiently regenerated through the use of state-of-the-art hollow fiber membrane columns. Specifically, the hydrophobic oil dodecane was effectively separated from the surfactant system SDBS/IPA/NaCl (Table VI). This separation was accomplished by counterflowing the oil/surfactant solution, comprised of 10.3 ml. dodecane per 100ml. of SDBS/IPA/NaCl solution, with the extraction solvent squalane. These results demonstrate the ability to effectively regenerate and reuse these surfactant systems.

In summary, research reported in this chapter illustrates not only the complexity of surfactant enhanced remediation of hydrophobic NAPLs, but also demonstrates the ability of properly designed surfactant systems to effectively remediate these hydrophobic oils (EACN of 10-20). Future research will further evaluate this area, including field studies. This research also explored surfactant systems for attacking even more hydrophobic, low viscosity NAPLs (EACNs〉20). Addressing the highly hydrophobic oils (EACNs〉〉20) and highly viscous oils may require combined approaches. (i.e., surfactants plus alcohols/solvents and/or temperature); such will also be the focus of future research.

BIBLIOGRAPHY

Abe, M., Schechter, D., Schechter, R. S.,Wade, W. H., Weerasooriya, U., and Yiv, S. (1986). *Journal of Colloid and Interface Science*, 114, 342.

Abe, M., Schechter, R. S., Sellia, R. D., Sheikh, B., and Wade, W. H. (1987). *Journal of Dispersion Science and Technology*, 8, 157.

Baran, J. R., Pope, G.A., Wade, W.H., Weerasooriya, and V. Yapa, A. (1994). *J. Colloid Interface Science*, 168, 67.

Barranco, F. T., and Dawson, H. E. (1999). "Influence of Aqueous pH on the Interfacial Properties of Coal Tar," *Environmental Science Technology*, 33, 1598-1603.

Bettahar, M., Schafer, G., and Baviere, M. (1999), "An Optimized Surfactant Formulation for the Remediation of Diesel Oil Polluted Sandy Aquifers," *Environmental Science and Technology*, 33, 1269-1273

Bourrel, M., and Schecter, R. S. (1988). *Microemulsions and Related Systems*, Marcel Dekker Inc.: New York,; Vol. 30, Surfactant Science Series.

Cummings, J. (1999). "Steam Enhanced Extraction – In Situ NAPL Recovery at the Southern California Edison Visalia Pole Yard," Presented at EPA Innovative Clean-up Approaches Conference, Indian Lakes Resort, Bloomingdale, IL November 2-4.

Dwarakanath, V., and Pope, G. A. (1998). "New Approach for Estimating Partition Coefficients between Nonaqueous Phase Liquids and Water". *Environ. Sci. Technol.*, 32, 1662-1666.

Edwards, D. A., Luthy, R. G., and Liu, Z. (1991). *Environmental Science and Technology*, 25, 127-133.

Ghoshal, S., Ramaswami, A., and Luthy, R. G. (1996) "Biodegradation of Napthalene from Coal Tar and Heptamethylnonane in Mixed Batch Systems," *Environmental Science Technology*, 30, 1282-1291.

Griffin, W. C. (1949). *Journal of the Society of Cosmetic Chemists*, 1, 311.

Hasegawa, M., Sabatini, D. A. and Harwell, J. H. "Liquid-Liquid Extraction for Surfactant-Contaminant Separation." *Journal of Environmental Engineering Division - ASCE*. 123(7), July 1997, 691-697.

Heely, R. N., and Reed, R. L. (1974) *Transactions of AIME* 257, 491.

Huh, C. (1979) *J. Colloid and Interface Sci.*, 71, 408.

Jarosch, T. R. and B. B. Looney, "Enhanced Recovery of Organics Using Direct Energy Techniques," *Innovative Subsurface Remediation: Field Testing of Physical, Chemical and Characterization Technologies*. ACS Symposium Series 725, American Chemical Society, Washington, DC, 24-35.

Knox, R. C., Shiau, B. J., Sabatini, D. A. and Harwell, J. H. (1999). "Field Demonstration of Surfactant Enhanced Solubilization and Mobilization at Hill Air Force Base, UT." *Innovative Subsurface Remediation: Field Testing of Physical, Chemical and Characterization Technologies*. ACS Symposium Series. In Press, Revised Paper Accepted October 15.

Knox, R. C., Sabatini, D. A., Harwell, J. H., Brown, R. E., West, C. C., Blaha, F., and Griffin, S. (November-December 1997) "Surfactant Remediation Field Demonstration Using a Vertical Circulation Well," Ground Water. 35(6), 948-953.

Krebbs-Yuill, B., Harwell, J. H., Sabatini, D. A., and Knox, R. C. (1995), "Economic Considerations in Surfactant-Enhanced Pump-and-Treat" in *Surfactant Enhanced Subsurface Remediation: Emerging Technologies*. ACS Symposium Series 594, American Chemical Society, Washington DC, 265-278.

Kunieda, H., and Shinoda, K. (1980) *Journal of Colloid and Interface Sci.*, 75, 601

Lipe, M., Sabatini, D. A., Hasegawa, M., and Harwell, J. H. (1996). "Micellar Enhanced Ultrafiltration and Air Stripping for Surfactant-Contaminant Separation and Surfactant Reuse." *Ground Water Monitoring and Remediation*. 16(1), , 85-92.

Ortiz, E., Kraatz, M., and Luthy, R. G. (1999). "Organic Phase Resistance to Dissolution of Polycyclic Aromatic Hydrocarbon Compounds," *Environmental Science and Technology*, 33, 235-242.

Pankow, J. F. and J. A. Cherry. (1996.). *Dense Chlorinated Solvents*, Waterloo Press, Portland, OR.

Pennell, K. D. Abriola, L. M., and Weber W. J. (1993), "Surfactant-Enhanced Solubilization of Residual Dodecane in Soil Columns 1. Experimental Investigation", *Environmental Science and Technology*, 27, 2332-2340.

Peters, C. A., and Luthy, R. G. (1993). *Environmental Science and Technology*, 27, 2831-2843.

Peters, C. A., and Luthy, R. G. (1994) "Semiempirical Thermodynamic Modeling of Liquid-Liquid Phase Equilibria: Coal Tar Dissolution in Water-Miscible Solvents," *Environmental Science and Technology*, 28, 1331-1340.

Rouse, J. D., Sabatini, D. A., and Harwell, J. H. (1993). "Minimizing Surfactant Losses Using Twin Head Anionic Surfactants in Subsurface Remediation." *Environmental Science and Technology*, 27, 2072-2078.

Roy, S. B., Dzombak, D. A., and Ali, M. A. (1995) "Assessment of in situ solvent extraction for remediation of coal tar sites: Column Studies," *Water Environment Research*, 67, 4-15.

Sabatini, D. A, Knox, R. C., Harwell, J. H. and Wu, B. "Integrated Design Of Surfactant Enhanced DNAPL Remediation: Effective Supersolubilization and Gradient Systems." Accepted, Journal of Contaminated Hydrology.

Sabatini, D. A., Harwell, J. H. and Knox, R. C. (1999) "Surfactant Selection Criteria for Enhanced Subsurface Remediation." In Innovative Subsurface Remediation: Field Testing of Physical, Chemical and Characterization Technologies. M. L. Brusseau, D. A. Sabatini, J. S. Gierke and M. D. Annable, eds. ACS Symposium Series 725, American Chemical Society, Washington, D.C., 8-23.

Sabatini, D. A., Knox, R. C., Harwell, J. H., Soerens, T. S., Chen, L., Brown, R. E. and West, C. (November-December 1997) "Design of a Surfactant Remediation Field Demonstration Based on Laboratory and Modeling Studies." Ground Water. 35(6), , 954-963.

Sabatini, D. A., Harwell, J. H., Hasegawa, M. and Knox, R. C. (1998) "Membrane Processes and Surfacant-Enhanced Subsurface Remediation: Results of a Field Demonstration." *Journal of Membrane Science*. 151(1) 89-100.

Salager, J. L., Vasquez, E., Morgan, J. C., Schechter, R. S., and Wade, W. H. (1979). *Society of Petroleum Engineering Journal*, 19, 107.

Salager, J. L., Morgan, J. C., Schecter, R. S. and Wade, W. H. (1979) "Optimum Formulation of Surfactant/Water/Oil Systems for Minimum Interfacial Tension or Phase Behavior". *Soc. Pet. Eng. J.*, 107.

Shiau, B. J., Sabatini, D. A., and Harwell, J. H. (1994). "Solubilization and Mobilization of DNAPLs using Direct Food Additive (Edible) Surfactants." *Ground Water*. 32(4), 561-569.

Sunwoo, C., and Wade, W. H. (1992). J. Dispersion Sci. Technol. 13, 49.

Udell, K. (1998). "Application of In Situ Thermal Remediation Technologies for DNAPL Removal," in Ground Water: Quality and Protection -- Proceedings of the Ground Water Quality '98 Conference, Universitaet Tuebingen, German, September 1998, IAHS Publication 250, 367-374.

Winsor, P. A. (1954). *Solvent Properties of Amphiphilic Compounds*, Butterworths, London.

Wu, B., Harwell, J. H., Sabatini, D. A., and Bailey, J. D. (1999). "Alcohol-Free Diphenyloxide Disulfonate (DPDS) Middle Phase Microemulsion Systems," Journal of Surfactants and Detergents.

Zhichu, B. Zhenshu, Z. Fei, X. Yueying, Q. and Jiayong, Y. (1999). "Wettability, Oil Recovery, and Interfacial Tension with an SDBS-Dodecane-Kaolin System," Journal of Colloid and Interface Science, 214, 368-372.

Chapter Twelve

In Situ Density Modification of Entrapped Dense Nonaqueous-Phase Liquids (DNAPLs) using Surfactant/Alcohol Solutions

TOHREN C. G. KIBBEY[1], C. ANDREW RAMSBURG[2], KURT D. PENNELL[2], and KIM F. HAYES[3]

[1]*School of Civil Engineering and Environmental Science, University of Oklahoma, Norman, OK*

[2]*School of Civil and Environmental Engineering, Georgia Institute of Technology, Atlanta, GA*

[3]*Department of Civil and Environmental Engineering, University of Michigan, Ann Arbor, MI*

Key words: DNAPLs; remediation; surfactants; mobilization; solubilization

Abstract: Proposed surfactant-based remediation techniques have the potential to be very effective at removing dense nonaqueous-phase liquids (DNAPLs) from contaminated sites. However, a risk associated with surfactant-based remediation of DNAPLs is the potential for unwanted downward mobilization of DNAPL contaminants, making them more difficult to remove from the subsurface. One approach that has recently been proposed to reduce this risk is to use hydrophobic alcohol solutions to reduce the densities of entrapped DNAPLs, converting them to light nonaqueous-phase liquids (LNAPLs). The work described here examines the use of surfactant/alcohol solutions to reduce the densities of DNAPLs prior to mobilization. Results indicate that surfactant selection can have a dramatic influence on rates of alcohol partitioning and associated density reduction, but that interfacial tension reduction is a potential source of difficulties for some systems.

1. INTRODUCTION

Dense nonaqueous-phase liquids (DNAPLs) are widespread at contaminated sites, where they are often distributed deep in aquifers due to their high densities [*US EPA*, 1993]. Surfactant-based remediation techniques have the potential to be very effective at removing DNAPLs from contaminated sites (e.g., [*Pennell, et al.*, 1994; *Baran, et al.*, 1994; *Pennell, et al.*, 1997; *Zimmerman, et al.*, 1999]). However, one risk associated with surfactant-based remediation of DNAPLs is the potential for unwanted downward mobilization of DNAPL to more difficult-to-reach sites in the aquifer. An approach that has recently been proposed to reduce this risk is the use of surfactant solutions containing hydrophobic alcohols or other agents to reduce the densities of entrapped DNAPLs, converting them to light nonaqueous-phase liquids (LNAPLs) [*Pennell*, 1998; *Kibbey, et al.*, 1998; *Lunn and Kueper*, 1999]. If a solution containing a sufficient quantity of an appropriate hydrophobic alcohol is pumped through a contaminated aquifer, the alcohol can partition into entrapped DNAPL, decreasing its density below that of water. For example, Figure 1 demonstrates this approach in a 2D aquifer cell. Approximately 40 mL of chlorobenzene is initially injected into the cell, and then a solution containing a mixture of Aerosol MA and Aerosol OT surfactants (excellent mobilizers, capable of producing ultra-low interfacial tensions) and 25% n-butanol is introduced from the left and allowed to flow through the contaminated zone. As the chlorobenzene is mobilized, alcohol partitions into it, converting it to an LNAPL and causing it to rise to the top of the aquifer cell where it is removed at the right of the cell. If this approach is used during a remediation operation, any mobilized NAPL will move upward in the aquifer, making it easier to remove. However, one potential concern with this approach is that its application is more difficult for more dense DNAPLs, like tetrachloroethylene (PCE). For the most dense DNAPLs, like PCE (ρ = 1.623 g/mL), a significant quantity of alcohol must partition into the DNAPL to reduce density below that of water; as a result, DNAPL mobilization may occur too rapidly for complete density conversion, and the DNAPL may drop in the aquifer. As such, the focus of the work described here is on developing a method for *in situ* conversion of DNAPLs to LNAPLs *prior* to mobilization. The work examines the use of surfactants that are not typically selected for mobilization applications, due to their limited interfacial tension lowering abilities.

This chapter describes recent work in our laboratories examining density modification of DNAPLs through a combination of batch non-equilibrium rate measurements and DNAPL displacement experiments in 2D aquifer cells. The objective of this work was to evaluate the applicability of nonionic surfactants as a delivery mechanism for introducing hydrophobic alcohols to convert the DNAPL to an LNAPL prior to mobilizing the NAPL. Three different nonionic surfactants were examined in combination with n-butanol and a range of DNAPLs. Overall, it was found that different surfactants can produce dramatically different rates of alcohol partitioning and density modification. However, for some systems interfacial tension reduction was found to be a problem, leading to unwanted downward migration of DNAPL prior to density conversion. Implications of observed results for remediation operations are discussed.

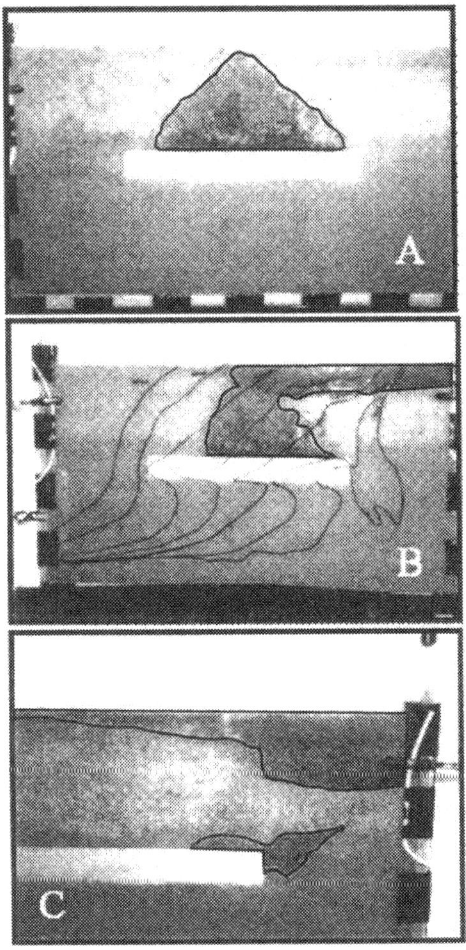

Figure 1. Density modified displacement of chlorobenzene with 4 % Aerosol MA/OT + 25 % n-butanol. (A) corresponds to initial injection of chlorobenzene, (B) and (C) correspond to increasing pore volumes of surfactant/alcohol flood ((C) is a close-up of the outlet of the cell). Dyed chlorobenzene is outlined in figures above. Note that lines are also visible in B indicating the movement of dyed surfactant/alcohol solution.

2. MATERIALS AND METHODS

2.1 Materials

The commercial nonionic surfactants used in this study included two alcohol ethoxylates, Witconol SN120 (Witco Corp., New York NY) and Brij 35 (ICI, Wilmington, DE), and an ethoxylated sorbitan fatty acid ester, Tween 80 (ICI). Brij 35 was purchased from Sigma (St. Louis, MO), while SN120 and Tween 80 were provided by their respective manufacturers. All surfactants were used as received.

DNAPLs examined include chlorobenzene (CB), 1,2-dichlorobenzene (1,2-DCB), trichloroethylene (TCE) and tetrachloroethylene (PCE). These DNAPLs were selected because of their relevance as environmental contaminants, and because they cover a range of polarities and densities. All DNAPLs were purchased from Sigma (St. Louis, MO) and were used as received.

HPLC Grade n-butanol (Mallinckrodt, Paris KY), and Milli-Q water (Millipore Corp., Bedford MA) were used for all experiments.

2.2 Visualization Procedures

An original objective of this work was to quantitatively determine rates of partitioning and density modification using an axisymmetric-drop shape analysis (ADSA) approach. However, vigorous motion of the DNAPL/water interface in the presence of alcohol/surfactant solutions precluded calculations. The batch measurements of DNAPL/LNAPL conversion rates described here were achieved by using a syringe to place a DNAPL drop of known volume on a glass slide in a newly-prepared surfactant/alcohol solution. A digital camera interfaced with a computer was set up to capture images at specified intervals. Time to conversion from DNAPL to LNAPL was determined as the time required for the bulk of the droplet to lift off from the glass slide. The aqueous solution was not mixed during experiments to avoid disturbing the NAPL droplet. Nevertheless, times determined with this method were found to be reproducible to within approx. 5%. In some cases, when low interfacial tensions or high initial densities were encountered (e.g., Fig. 5), a modified hanging drop slide was inverted and used in place of a glass slide to ensure the drop would remain in the field of view for the duration of the experiment.

Two-dimensional aquifer cell studies were conducted in glass cells 60.5 cm (L) x 35 cm (H) x 1.3 cm (W). The cells described here were packed with sand from Oscoda, Michigan, and had pore volumes of approx.

950 mL. DNAPL was added by injection approx. 4 mL below the water table using a syringe pump. DNAPL volumes were typically approx. 40 mL.

3. RESULTS AND DISCUSSION

The objective of *in situ* density modification is to use hydrophobic alcohol solutions to reduce the densities of entrapped DNAPLs. For this to work, significant quantities of alcohol must partition into the entrapped DNAPL. This is particularly true for high density chlorinated DNAPLs, such as tetrachloroethylene (PCE). For example, Table 1 shows the calculated percentage of n-butanol that must be present in a NAPL/n-butanol mixture for the density of the mixture to drop below 1 g/mL (calculated assuming constant molar volumes). For PCE, which has a density of 1.623 g/mL, n-butanol ($\rho = 0.81$ g/mL) must partition until the final NAPL is more than three quarters alcohol by volume. In order to deliver sufficient alcohol to the entrapped DNAPL, surfactants may be necessary. Surfactants can substantially increase the quantity of alcohol that can be dissolved, allowing more alcohol to be injected into the subsurface.

Table 1. DNAPL properties and time required for density conversion of a 15 mL drop by SN120/n-butanol.

DNAPL	density (g/mL)	n-BuOH needed (% of final volume)	$C_{12}E_{10}$ Partitioning Plateau (g/L NAPL)	Time to LNAPL Conversion (min)
CB	1.107	36 %	~230	2 min
1,2-DCB	1.305	62 %	~230	11 min
TCE	1.464	71 %	~360	15 min
PCE	1.623	77 %	~10	66 min

For the work described here, n-butanol was selected because of its balance of hydrophobicity and moderate water solubility. Water-miscible alcohols, such as methanol and ethanol, do not partition significantly from aqueous solution into NAPLs, because of their high affinity for water. Hydrophobic alcohols, such as hexanol and octanol, tend to partition substantially into NAPLs, but their aqueous solubilities can be relatively low, making it difficult to efficiently introduce sufficient alcohol to convert DNAPLs to LNAPLs. N-butanol was selected because its high water solubility (~6 % by weight) allows it to be injected in reasonable quantities in aqueous solution, yet its affinity for NAPLs is high enough for partitioning and the associated density reduction to occur. The nonionic surfactants used here were found to support concentrations of n-butanol up to approximately 14 % by weight, depending on surfactant and surfactant concentration. In addition to allowing greater quantities of alcohol to be added, the inclusion of surfactants was found to have a strong influence on the rate of density conversion, as will be discussed below.

When using an alcohol/surfactant solution to modify the density of DNAPLs, several processes occur simultaneously (Figure 2). Perhaps the most important process is alcohol partitioning, where alcohol which is initially solubilized in surfactant micelles partitions into the DNAPL,

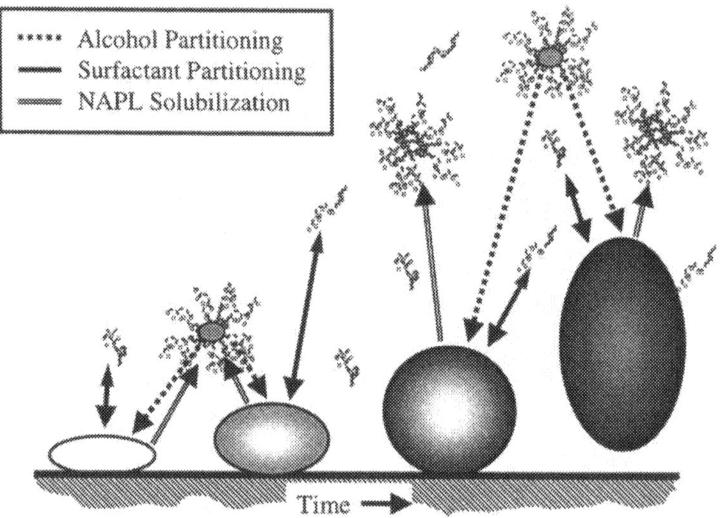

Figure 2. Processes involved in density modification of DNAPLs.

reducing the DNAPL's density and increasing its volume. However, other processes that also take place include surfactant partitioning and DNAPL solubilization. Previous work in our laboratories has found that partitioning of many nonionic surfactants into NAPLs can be significant for some DNAPLs, particularly for relatively polar DNAPLs, and can also lead to swelling of the DNAPL [*Butler and Hayes*, 1998; *Zimmerman, et al.*, 1999; *Cowell et al.*, 2000]. Solubilization of the DNAPL in surfactant micelles results in volume reduction of the DNAPL. All of these process occur simultaneously, and the extent to which they occur is influenced by the changing composition of the DNAPL which results from alcohol and surfactant partitioning and solubilization.

3.1 Influence of Surfactant Properties

The influence of surfactant properties on rates of density conversion is illustrated in Figure 3. Figure 3 shows density modification of a 15 μL volume of chlorobenzene using the three nonionic surfactant solutions, SN120, Brij 35 and Tween 80, all containing 11% n-butanol by weight. Note that the time scale of density modification for these three systems is very different, ranging from approximately two minutes for the SN120 (Figure 3A) to more than forty minutes for the Tween 80 (Figure 3C). Interestingly, the systems which produce the most rapid density modification are also those for which the surfactant partitions to the greatest extent into DNAPLs. Figure 4 shows time required for conversion of the 15 μL drop of

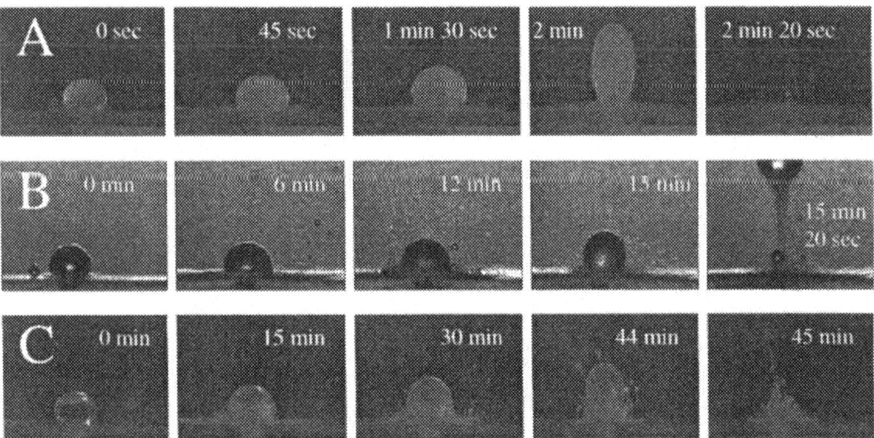

Figure 3. Density modification of chlorobenzene with surfactant/n-butanol solutions. Surfactant: (A) SN120, (B) Brij 35, (C) Tween 80.

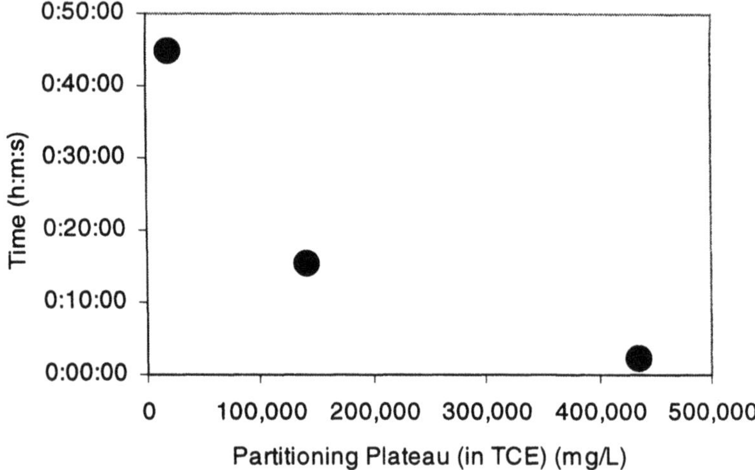

Figure 4. Time to LNAPL conversion of chlorobenzene vs. partitioning plateau (SN120, Brij 35, Tween 80 with n-butanol).

chlorobenzene from a DNAPL to an LNAPL as a function of the partitioning plateau for partitioning of the three surfactants between TCE and water. When surfactants partition between water and NAPLs, the amount of surfactant in the NAPL tends to reach a plateau at high surfactant concentrations, to some extent resulting from the formation of micelles in solution. For more polar NAPLs, the plateau tends to be higher (i.e., correspond to a greater quantity of surfactant in the NAPL) [*Zimmerman, et al.*, 1999; *Cowell et al.*, 2000]. Plateaus for surfactant partitioning into TCE are shown here because CB values were not available for the three surfactants examined; however, previous work has found that surfactant partitioning behavior is quantitatively similar for the two NAPLs. In addition, surfactant partitioning behavior tends to be qualitatively similar between surfactants for different NAPLs [*Cowell et al.*, 2000]. From Figure 4, it is apparent that the surfactant which partitions into NAPL to the greatest extent, SN120, corresponds to the most rapid rate of DNAPL to LNAPL conversion. The surfactant which has the lowest partitioning plateau, Tween 80, corresponds to the slowest rate of conversion. It is speculated that partitioning of the surfactant into the NAPL either (1) mixes the NAPL/water interface, facilitating mass transfer (the vigorous motion of DNAPL drops in the presence of SN120 and Brij 35 support this), and/or (2) creates a localized region around the DNAPL that is depleted in surfactant and supersaturated with alcohol, accelerating the alcohol partitioning process.

It is interesting to note that batch experiments suggest that many surfactants that are good solubilizers (such as SN120, Brij 35 and Tween 80) may produce significantly faster density modification than surfactants that are good mobilizers (such as the Aerosol series surfactants), which can take as long as several hours to reduce the density of DNAPLs (data not shown). From a practical standpoint this result suggests it might be useful to initially flood DNAPL-contaminated aquifers with solutions containing highly-partitioning solubilizers and alcohol to convert entrapped DNAPL to LNAPL *prior* to flooding with mobilizers to remove the NAPL. However, as will be discussed below, some difficulties result from the presence of alcohol, which can cause solubilizers to reduce interfacial tensions to a significantly greater extent than would be the case in the absence of alcohol.

3.2 Influence of NAPL properties

Table 1 shows the time required for a 15 μL drop of DNAPL to be converted to an LNAPL by a SN120/n-butanol solution (the system which produced the most rapid conversion). Note that while CB can be converted to an LNAPL in only 2 minutes, it takes about 11, 15 and 66 minutes for 1,2-DCB, TCE and PCE respectively. Moving down the table from CB to PCE, conversion to LNAPL becomes increasingly difficult because more dense DNAPLs require a greater total quantity of alcohol in order for their densities to be reduced (e.g., 77% for PCE vs. only 36 % for CB). In addition, because PCE is relatively nonpolar compared to the other three DNAPLs, it appears to be particularly difficult to convert to an LNAPL, because surfactants don't partition readily into the PCE. This is illustrated in Table 1 by the values of partitioning plateau for $C_{12}E_{10}$ (a linear alcohol ethoxylate surfactant similar to SN120). Note that surfactant partitioning is substantial for CB, 1,2-DCB and TCE (>200 g/L NAPL), but that is not the case for PCE (~12 g/L NAPL). This means that neither of the two proposed mechanisms of accelerating alcohol partitioning through surfactant partitioning (mixing of the NAPL/water interface, and local supersaturation caused by alcohol depletion) will be particularly effective with PCE.

Figure 5 shows conversion of PCE from a DNAPL to an LNAPL using the SN120/n-butanol solution. Unlike the CB, which only needs to contain 36 % n-butanol to be converted to an LNAPL, PCE requires that the final composition be approximately 77 % n-butanol by volume. This leads to a situation where the resulting LNAPL has a composition that is more than three quarters n-butanol. Because the interfacial tension of n-butanol with water (1.8 dynes/cm [*Demond and Lindner*, 1993]) is very low compared to that of PCE (47.5 dynes/cm [*Demond and Lindner*, 1993]) (and note that both interfacial tensions will be lower in the presence of the SN120 surfactant), the droplet is unable to hold together as it is converted to an LNAPL, and it streams upward rather than being converted as a single drop as was the case with CB.

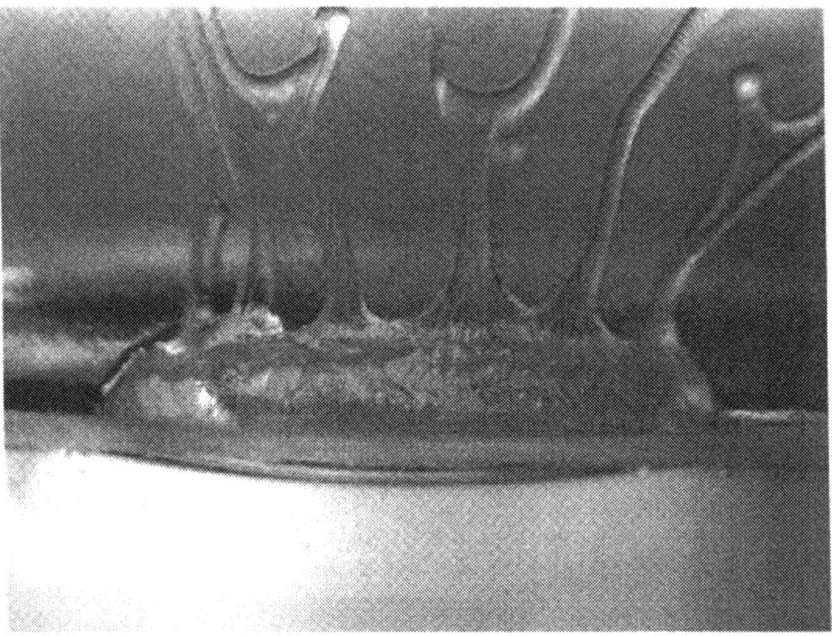

Figure 5. Density modification of PCE with SN120/n-butanol.

NAPL volume is an important consideration in density modification, because diffusion of alcohol into the bulk DNAPL can potentially limit the rate at which density can be converted. Figure 6 shows the time required for an SN120/n-butanol solution (86 g/L SN120/11 % n-butanol) to convert a drop of chlorobenzene to an LNAPL, as a function of initial drop volume. As would be expected, larger drops take longer to convert, due to their greater volume which requires alcohol to diffuse farther to reach the center of the drop. In actual remediation scenarios, large pools of DNAPL will be more difficult to convert to LNAPL than will smaller volumes of DNAPL. As a result, application of *in situ* density modification is most likely to be most directly applicable to moderately contaminated sites than for sites with significant DNAPL pools.

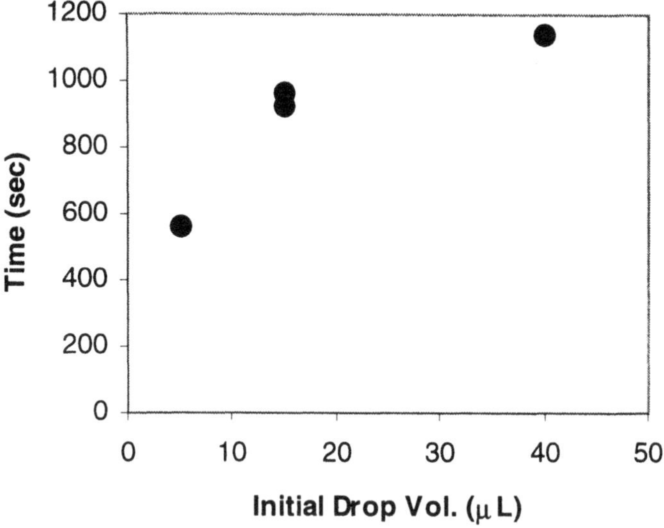

Figure 6. Time required for conversion of chlorobenzene by SN120/n-butanol, as a function of initial drop volume.

3.3 Interfacial Tension Reduction

Testing the surfactant formulations described above in two-dimensional aquifer cell studies produced an unexpected result: in the presence of alcohol, all three surfactants caused mobilization of DNAPL. Although DNAPL/water interfacial tensions for surfactant in the absence of alcohol are too high for mobilization to occur, addition of alcohol to the surfactant solution leads to a dramatic interfacial tension reduction. A likely reason for the reduction is that as alcohol partitions, the DNAPL/water interface becomes more like an n-butanol water interface. In the absence of surfactant the n-butanol/water interfacial tension is 1.8 dynes/cm – significantly lower than most DNAPL/water interfacial tensions (e.g., 47.5 dynes/cm for PCE). As the interface becomes more alcohol-like, the impact of surfactants on interfacial tension reduction is more dramatic, and surfactants that are typically not good mobilizers can cause mobilization. Note that although the NAPL interface may be alcohol-like, the bulk of the NAPL may still be predominantly a DNAPL, depending on the rates of partitioning. A two-dimensional aquifer cell study (not shown) using SN120/n-butanol to convert PCE to an LNAPL led to significant unwanted downward mobilization of PCE. Small scale experiments with the other two surfactants produced the same result. Only aquifer cell experiments with 6% alcohol in the absence of surfactant (data not shown) were able to successfully reduce DNAPL densities without producing mobilization, although multiple pore volumes of alcohol solution were required. Other approaches, such as flooding with surfactant/alcohol macroemulsions, have shown promise and will be described elsewhere. However, interfacial tension reduction remains a concern for *in situ* density reduction with surfactant/alcohol solutions.

CONCLUSIONS

The work presented here illustrates that surfactant selection can have a substantial impact on the rates of alcohol partitioning and associated reduction of DNAPL density. More work is needed to develop surfactant/alcohol systems which can minimize interfacial tension reduction while still providing acceptable density conversion rates. Because the potential benefits of *in situ* density modification are substantial, future work will be directed at system design to minimize interfacial tension reduction.

ACKNOWLEDGEMENTS

Funding for the research was provided by the Great Lakes and Mid-Atlantic Hazardous Substance Research Center under grant R819605-01 from the Office of Research and Development, U.S. Environmental Protection Agency. Partial funding of the research activities of the Center was provided by the State of Michigan Department of Environmental Quality. The content of this publication does not necessarily represent the views of either agency.

REFERENCES

Baran, J. R.; Pope, G. A.; Wade, W. H.; Weerasooriya, V.; Yapa, A. "Microemulsion Formation with Chlorinated Hydrocarbons of Differing Polarity." *Environ. Sci. Technol.* **1994**, *28*, 1361-1366.

Butler, E. C.; Hayes, K. F. "Micellar Solubilization of Non-Aqueous Phase Liquid Contaminants by Nonionic Surfactant Mixtures: Effects of Sorption, Partitioning and Mixing," *Water Research*, **1998**, *32*, 1345-1354.

Cowell, M. A.; Kibbey, T. C. G.; Zimmerman, J. B.; Hayes, K. F. "Partitioning of Ethoxylated Nonionic Surfactants in Water/Non-Aqueous Phase Liquid (NAPL) Systems: Effects of Surfactant and NAPL Properties," *Environ. Sci. Technol.*, **2000**, *34*, 1583-1588.

Demond, A. H; Lindner, A. S. "Estimation of Interfacial Tension Between Organic Liquids and Water," *Environ. Sci. Technol.*, **1993**, *27*, 2318-2331.

Kibbey, T. C. G.; Ramsburg, C. A.; Pennell, K. D.; Hayes, K. F. "Effects of Surfactant Properties on Equilibrium and Non-equilibrium Alcohol Partitioning into Dense Nonaqueous-Phase Liquids (DNAPLs) for *In Situ* Density Modification Applications," presented at the Fall American Geophysical Union national meeting, San Francisco, California, December, **1998**.

Lunn, S. R. D.; Kueper, B. H. "Risk Reduction during Chemical Flooding: Preconditioning DNAPL Density In Situ Prior to Recovery by Miscible Displacement," *Environ. Sci. Technol.*, **1999**, *33*, 1703-1708.

Pennell, K. D.; Jin. M.; Abriola, L. M.; Pope, G.A. "Surfactant Enhanced Remediation of Soil Columns Contaminated by Residual Tetrachloroethylene." *J. Contam. Hydrol.* **1994**, *16*, 35-53.

Pennell, K. D.; Adinolfi, A. M.; Abriola, L. M.; Diallo, M. S. "Solubilization of Dodecane, Tetrachloroethylene, and 1,2-Dichlorobenzene in Micellar Solutions of Ethoxylated Nonionic Surfactants." *Environ. Sci. Technol.* **1997**, *31*, 1382-1389.

Pennell, K. D., U.S. Conventional Patent Application (Patent Pending) "Density Modification Displacement to Remediate Contaminated Aquifers" Serial # 09/056,487, Filed April 7, **1998**.

Evaluation of the Likelihood of DNAPL Presence and NPL Sites; U.S. Environmental Protection Agency; Washington DC, **1993**; 9355.4-13, EPA 15401R-93-073, PB93-963343.

Zimmerman, J. B.; Kibbey, T. C. G.; Cowell, M. A.; Hayes, K. F. "Partitioning of Ethoxylated Nonionic Surfactants into Nonaqueous-Phase Organic Liquids: Influence on Solubilization Behavior," *Environ. Sci. Technol.*, **1999**, *33*, 169-176.

Chapter Thirteen

Effects of Cosolvent Addition on Surfactant Enhanced Recovery of Tetrachloroethene (PCE) from a Heterogeneous Porous Medium

TAMMY P. TAYLOR[1, 2] and KURT D. PENNELL[1]

[1] *School of Civil and Environmental Engineering, Georgia Institute of Technology, Atlanta, GA 30332-0355*
[2] *current address: Chemical Science and Technology Division, Los Alamos National Laboratory, Los Alamos, NM 87545*

Key words: surfactant, remediation, ethanol, DNAPL, tetrachloroethene

Abstract: The ability of surfactant formulations containing ethanol (EtOH) to enhance the recovery of a representative dense nonaqueous phase liquid (DNAPL), tetrachloroethene (PCE), was evaluated in batch, column, and heterogeneous, two-dimensional (2-D) systems. The experimental studies were designed to investigate the influence of EtOH addition, rate-limited mass transfer and subsurface layering on the micellar solubilization of PCE by aqueous solutions (4% wt.) of polyoxyethylene (20) sorbitan monooleate (Tween 80). In completely mixed batch reactors, the solubility of PCE in Tween 80 solutions (0.5 to 4% wt.) increased incrementally as the EtOH concentration was raised from 0% to 5% and 10%. Results of one-dimensional column studies demonstrated that solubilization of residual PCE was rate-limited regardless of the EtOH concentration. Effluent data were used to develop effective PCE mass transfer coefficients (K_e) as a function of EtOH concentration, Darcy velocity, and duration of flow interruption. For the heterogeneous 2-D system, solutions containing 4% Tween 80 or 4% Tween 80 + 5% EtOH were injected into rectangular boxes packed with 20-30 mesh Ottawa sand and three low permeability layers. Visual observation of surfactant fronts and effluent concentration data demonstrated that the addition of 5% EtOH resulted in density over ride of the injected solution, and failed to enhance PCE recovery compared to the 4% Tween 80 solution alone. The results of this work indicate that although relatively small additions of EtOH can improve the solubilization capacity of surfactant formulations, differences between flushing and resident solution density must be carefully accounted for when utilizing cosolvent-amended surfactant formulations for DNAPL remediation.

1. INTRODUCTION

The failure of conventional of pump-and-treat systems to remediate DNAPL-contaminated aquifers has prompted the development of alternative treatment technologies to expedite subsurface restoration. Pump-and-treat enhancements incorporating surfactant or cosolvent flushing have shown considerable promise in laboratory-scale tests (Pennell *et al.*, 1993; Shiau *et al.*, 1994; Edwards *et al.*, 1994; Rouse *et al.*, 1995; Waddill and Parker, 1997). This remediation strategy is based on the ability of surfactants or cosolvents to (a) increase the aqueous solubility of NAPLs as a result of micellar solubilization, and/or (b) displace entrapped NAPLs due to interfacial tension (IFT) reductions. In practice, the use of surfactant and cosolvent formulations exhibiting low IFTs (e.g., < 1 dyne/cm) are typically avoided because of the risk for downward migration of the displaced DNAPL free product. Pilot-scale surfactant and cosolvent flushing trials involving DNAPL have yielded mixed results, and in a number of cases failed to reach desired clean up levels (Fountain *et al.*, 1995; Fountain *et al.*, 1996; Fountain, 1997; AATDF, 1997; Smith *et al.*, 1997). One of the most important factors contributing to low recovery rates is subsurface heterogeneity, typically in the form of low-permeability layers or lenses. Due to the relatively large entry pressure of low permeability media, DNAPLs often accumulate to form "pools" above such lenses. The resulting DNAPL pools possess low specific interfacial area, thereby limiting mass transfer into the aqueous phase and reducing the effective permeability of the source zone. Thus, any evaluation of flushing strategies for DNAPL source zone treatment must account for the potential effects of subsurface heterogeneity on flushing solution delivery and recovery, and on DNAPL mass transfer rate limitations.

A limited number of studies have considered the use of surfactant and cosolvent mixtures to enhance the recovery of NAPLs (Martel *et al.*, 1993; Martel and Gelinas, 1996). Martel *et al.* (1993) and Martel and Gelinas (1996) employed ternary phase diagrams to select surfactant+cosolvent formulatons for treatment of NAPL-contaminated aquifers. The surfactant+cosolvent formulations used in their work, which included lauryl alcohol ethersulfate/n-amyl alcohol, secondary alkane sulfonate/n-butanol, and alkyl benzene sulfonate/n-butanol, were shown to be effective solubilizers of residual trichloroethene (TCE) and PCE in soil columns (Martel *et al.*, 1993). However, very little information is available regarding the effect of cosolvents on the solubilization capacity and phase behavior of ethoxylated nonionic surfactants.

The use of cosolvents to achieve neutral-buoyancy surfactant floods has also been investigated (Shook *et al.*, 1997, 1998; Kostarelos *et al.*, 1998).

This approach is specifically designed to minimize downward migration of surfactant plumes following DNAPL solubilization. Shook *et al.* (1997) performed 2-D sand tank experiments (1.03 m x 0.41 m x 0.051 m) to assess the migration of a TCE-surfactant plume with and without cosolvent (isopropanol) addition. A surfactant formulation consisting of 4% sodium dihexyl sulfosuccinate (Aerosol MA) and 8% isopropanol resulted in stable miscible displacement of a TCE-rich surfactant plume and full capture of TCE at extraction ports. When isopropanol was not added to the surfactant solution, the solubilized TCE plume contacted the lower boundary of the tank approximately 0.15 m prior to reaching the extraction ports (Shook *et al.*, 1997). One drawback of alcohol addition is the potential for density over ride of the flushing solution, which will usually have a lower density than the resident solution (i.e., groundwater). This issue is particularly relevant for DNAPLs, which may accumulate above the lower confining layer of an aquifer formation. Therefore, careful consideration of contrasting solution density will be necessary in order to design effective surfactant flushing schemes when alcohol is added as a cosolvent.

The overall objective of this research was to evaluate the effects of cosolvent addition on the ability of an ethoxylated nonionic surfactant to recover PCE from a heterogeneous, 2-D system. The specific tasks of this work were to (a) quantify the PCE solubilization rate and capacity in the presence and absence of a representative cosolvent (EtOH) and (b) investigate the effects of EtOH addition on surfactant delivery, plume migration and PCE recovery in a 2-D, layered sand tank. A representative nonionic surfactant, polyoxyethylated (POE) (20) sorbitan monooleate (Tween 80), was selected for study because of its capacity to solubilize PCE (~ 0.7 g PCE/g surf at 20°C) and relatively high interfacial tension with PCE (5 dynes/cm). EtOH was chosen as the representative cosolvent because of its relatively low density ($\rho = 0.79$ g/cm^3) and regulatory acceptance.

2. EXPERIMENTAL METHODS

2.1 Materials

Polyoxyethylene (POE) (20) sorbitan monooleate (Tween 80, Lot 36218, ICI Surfactants, Inc.) was used as received with no further purification. The average molecular weight of Tween 80 is 1310 g/mole, the density is 1.07 g/cm^3, the hydrophile-lipophile balance (HLB) is 15, the critical micelle concentration (CMC) is 13 mg/L, and the micelle aggregation number is 110 (Pennell *et al.*, 1997). HPLC-grade PCE was obtained from Aldrich Chemical Co., and was used as received. PCE has a density of 1.623 g/cm^3, a molecular weight of 165.8 g/mole, and an aqueous solubility of approximately 150 mg/L at 25°C (Verschueren, 1983). Denatured EtOH was obtained from the Chemistry Department at the Georgia Institute of Technology. The density of EtOH was determined to be 0.79 g/cm^3 at 20°C using a calibrated 25 mL glass pycnometer. Oil-red-O, an organic soluble dye (Fisher Scientific), was used at a concentration of 10^{-4} M to facilitate visual observation of PCE during 1-D column and 2-D box experiments. The addition of 10^{-4} M Oil-Red-O had minimal effects (+/- 5%) on measured values of IFT and PCE solubility. In the 2-D box studies, surfactant solutions were dyed with Brilliant Blue (Fisher Scientific) to observe the location and shape of the surfactant front. Independent experiments showed that the Brilliant Blue dye did not alter the viscosity, IFT, and equilibrium solublization capacity of the surfactant solution.

Aqueous solutions were prepared with purified water processed using a Barnstead Nanopure® Analytical Deionization system. Sodium azide (NaN_3) was used at a concentration of 1 g/L in all solutions to inhibit microbial activity. Calcium chloride ($CaCl_2$), at a concentration of 0.5 g/L, was employed as background electrolyte. The total ionic strength of the background solution ($CaCl_2$ + NaN_3) was 0.0289 mole/L. The density of Nanopure® water containing 0.5 g/L $CaCl_2$ + 1 g/L NaN_3 was found to be 0.998 g/cm^3 at 20°C. Potassium iodide (KI) (Fisher Scientific), served as a non-reactive, conservative tracer for determination of hydrodynamic dispersion coefficients in 1-D column and 2-D sand tank experiments.

Ottawa 20-30 mesh sand (0.60-0.85 mm; ASTM designation C778), obtained from U.S. Silica Co., Ottawa, IL, was used as the porous medium for column experiments and as the background medium for box studies. Low-permeability layers were packed within Ottawa 20-30 mesh sand in 2-D box studies. The rectangular layers consisted of a natural aquifer material obtained from Wurtsmith AFB in Oscoda, Michigan and F-70 sand (Ottawa 40-270 mesh) obtained from U.S. Silica Co. Table 1 summarizes relevant properties of the porous media used in the column and 2-D box experiments.

2.2 Batch Reactor Procedures

The solubility of PCE in aqueous solutions of Tween 80 and Tween 80 containing 5% and 10% EtOH was measured in batch reactors maintained at room temperature (20°C +/- 1°C), following procedures described by Pennell *et al.* (1997). Thirty-five mL borosilicate glass centrifuge tubes, capped with closed top, Teflon®-backed caps, were filled with approximately 2 mL of neat liquid PCE and 30 mL of surfactant or surfactant/cosolvent solution. The concentration of Tween 80 in the aqueous phase ranged from 5,000 to 40,000 mg/L. All batch reactors were prepared in triplicate. The solutions were mixed for 7 days on an oscillating shaker (LabQuake®). Macroemulsions that formed during mixing were separated by centrifugation at 2,500 rpm for 30 minutes at 20°C, using a Beckman® Avanti J-25 temperature-controlled centrifuge.

Table 1. Selected properties and Tween 80 Langmuir sorption parameters for Ottawa 20-30 mesh sand, Ottawa F-70 sand and Wurtsmith aquifer material.

Porous Medium	OC Content (%)	Specific Surface Area (m^2/g)	Permeability (m^2)	$C^s_{s,max}$ (mg/g)	b (L/mg)
20-30 mesh Ottawa sand	ND	0.10	3.9(-10) [a]	0.156 [b]	0.010 [b]
F-70 Ottawa sand	ND	0.16	8.2(-12) [a]	0.160 [b]	0.029 [b]
Wurtsmith Aquifer material	0.02	0.51	4.2(-11) [b]	0.194 [b]	0.008 [b]

[a]Kueper and Frind (1991)
[b]Taylor (1999)
ND = non-detectable

2.3 Column Procedures

The column apparatus consisted of a borosilicate glass preparative chromatography column (4.8 cm i.d.) equipped with an adjustable endplate (Figure 1). The bottom endplate was fitted with two 40-mesh screens while the top endplate was fitted with two Whatman® No. 42 filters. The columns were packed with 20-30 mesh Ottawa sand under vibration in 2 cm increments. After packing was complete, the columns were purged with carbon dioxide for 20 minutes to facilitate dissolution of entrapped gas during water saturation. Complete water saturation was achieved after flushing the column with approximately 15 pore volumes of de-aired Nanopure® water containing 0.5 g/L $CaCl_2$ and 1 g/L NaN_3.

For the non-reactive tracer tests, a solution containing KI was introduced to the bottom of the column (up-flow mode) using a Rainin® Model SD-200 solvent delivery pump. The first non-reactive tracer test was performed after complete water saturation of a column. PCE was then introduced into the bottom of the column using a Harvard Apparatus® Model 22 syringe pump at a flow rate of 0.33 mL/min. When approximately 70% of the pore volume was occupied by PCE, the flow direction was reversed and water was pumped downward from the top of the column. PCE existing as mobile free product was displaced from the column by pumping approximately two pore volumes of water at 1 mL/min, followed by ten pore volumes of water at 5 mL/min. A second non-reactive tracer test was performed following the establishment of residual PCE. Final PCE residual saturations were determined gravimetrically and verified based on results of the second tracer test. Measured tracer breakthrough curves were fit to an analytical solution of the one-dimensional advective-dispersive reaction (ADR) transport equation using CXTFIT (van Genuchten, 1981). An aqueous solution of 4 % (wt.) Tween 80, prepared with 0%, 5% or 10% EtOH, was then flushed through columns in a down-flow mode. Effluent samples were collected in 20 mL scintillation vials using an Isco® fraction collector. The scintillation vials were immediately capped and samples were prepared for analysis by gas chromatography (GC). At the end of each column study, the contents of the soil column were removed and washed with isopropanol to extract the remaining PCE. The drained solution and isopropanol wash solution were then analyzed for PCE by GC. Total PCE mass balances of greater than 94% were achieved for all column experiments.

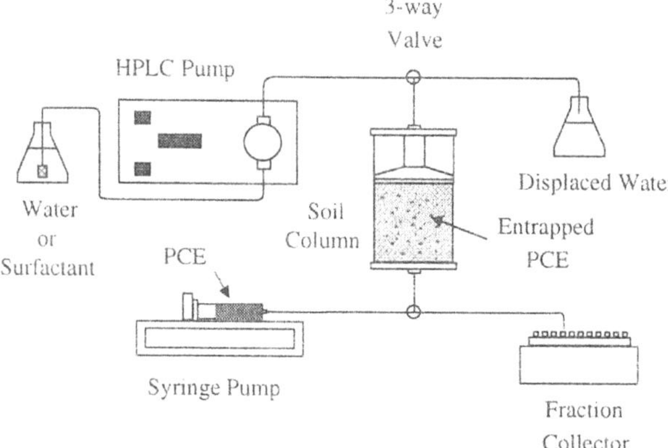

Figure 1. Schematic diagram of the 1-D column apparatus.

2.4 2-D Box Procedures

The effect of rate-limited micellar solubilization, low-permeability layers, and solution phase density on the recovery of PCE by 4% Tween 80 and 4% Tween 80 with 5% EtOH was evaluated in two separate 2-D box studies. The inside dimensions of the rectangular box were 32 cm (height) x 61.7 cm (length) x 1.4 cm (width). Rectangular end chambers, constructed of aluminum alloy, were screened over the entire vertical height of the box. An unscreened section of aluminum served as the bottom of the box and parallel glass plates served as the front and back of the box (Figure 2). Ottawa sand (20-30 mesh) was packed under water-saturated conditions in 3 cm increments. Three horizontal, rectangular low-permeability layers (~5 cm height X ~18 cm length) were packed within the Ottawa 20-30 mesh sand. After each box was packed, water flow was established by connecting a 2 L constant head bottle to the influent end chamber and a collection bottle to the effluent end chamber. The height of the constant head bottle and the exit tubing leading to the effluent collection bottle were adjusted to control flow through the box. Hydraulic conductivity was determined by measuring flow rate versus head drop across the box. A non-reactive tracer (KI) test was then performed to measure the overall hydrodynamic dispersion coefficient.

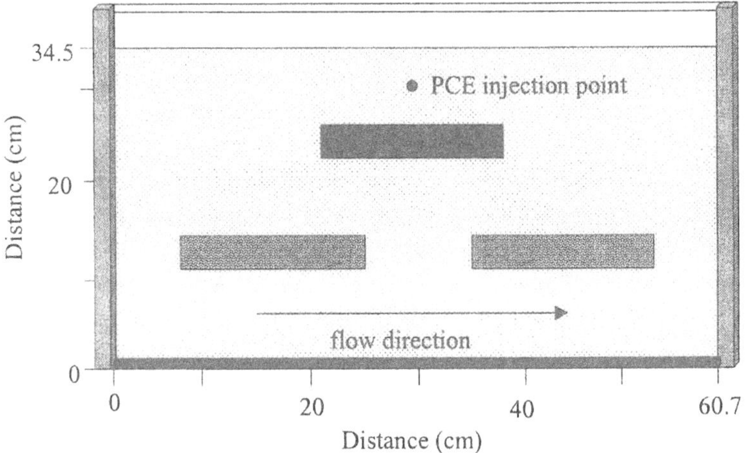

Figure 2. Schematic diagram of the 2-D sand tank apparatus.

Dyed PCE was injected into each box near the top-center using a Harvard Apparatus® Model 22 syringe pump. PCE was injected at a flow rate of 0.33 mL/min. When PCE began to spill over the bottom Wurtsmith layers and flow toward the end chambers, PCE injection was terminated. The PCE was then allowed to redistribute within the box for approximately 15 hours. A second non-reactive tracer (KI) test was conducted following PCE injection and redistribution. Aqueous solutions of 4 % Tween 80 and 4% Tween 80 + 5% EtOH dyed with Brilliant Blue were flushed through Box A and Box B at a flow rate of approximately 5 mL/min. Aqueous effluent samples were collected over time and analyzed for PCE and Tween 80 using GC and high pressure liquid chromatography (HPLC), respectively.

Approximately 7.9 and 7.6 pore volumes (8.7 L and 8.3 L) of surfactant solution were flushed through Box A and Box B, respectively. The effects of non-equilibrium mass transfer on PCE recovery were assessed through a series of flow interruptions, lasting from 12 to 17 hours. PCE remaining in each box after the surfactant flushing procedure was extracted with isopropanol and analyzed by GC. Total PCE mass balances of greater than 94% PCE was achieved in both box studies.

2.5 Analytical Methods

Aqueous-phase concentrations of PCE were determined by GC using a Hewlett-Packard (HP) Model 6890 gas chromatograph equipped with an autosampler, a precolumn surfactant trap, an HP-5 crosslinked 5% PH Siloxane® column, and a flame ionization detector (FID). The precolumn trap was installed to prevent surfactant fouling of GC inlets, column and detector. Samples were prepared by adding approximately 0.4 mL of aqueous sample to 1.2 mL of isopropanol in glass autosampler vials. The GC vials were sealed with Teflon® backed aluminum caps to minimize volatilization. Triplicate injections of each sample were performed.

Analysis of Tween 80 was performed using a Hewlett Packard 1100 series HPLC equipped with a Sedex 55® Evaporative Light Scattering Detector (ELSD). The mobile phase consisted of 80% acetonitrile and 20% water. Duplicate injections (5 μL) of each sample were evaluated by HPLC. Potassium iodide, used for the 1-D column and 2-D box tracer studies, was analyzed with a continuous flow Isco $V^{4®}$ variable UV wavelength absorbance detector equipped with an EZChrom® Chromatography data acquisition system.

3. EXPERIMENTAL RESULTS AND DISCUSSION

3.1 Batch Experiments

Equilibrium solubility relationships for PCE in micellar solutions of Tween 80, Tween 80 + 5% EtOH and Tween 80 + 10% EtOH at 20°C are shown in Figure 3. Over the surfactant concentration range investigated (0.5 to 4% by wt.), all systems exhibited Winsor Type I behavior characteristic of stable oil-in-water microemulsions. Regardless of the EtOH concentration, the equilibrium solubility of PCE increased linearly over the concentration range investigated. A weight solubilization ratio (WSR), defined as the weight of PCE solubilized per weight of surfactant in micellar form, was determined for each surfactant formulation using a least-squares linear regression procedure (SYSTAT®) applied to each equilibrium solubility curve above the CMC. Molar solubilization ratios (MSRs) were then calculated from the corresponding WSR valves using the average molecular weight of the surfactant and the molecular weight of the PCE.

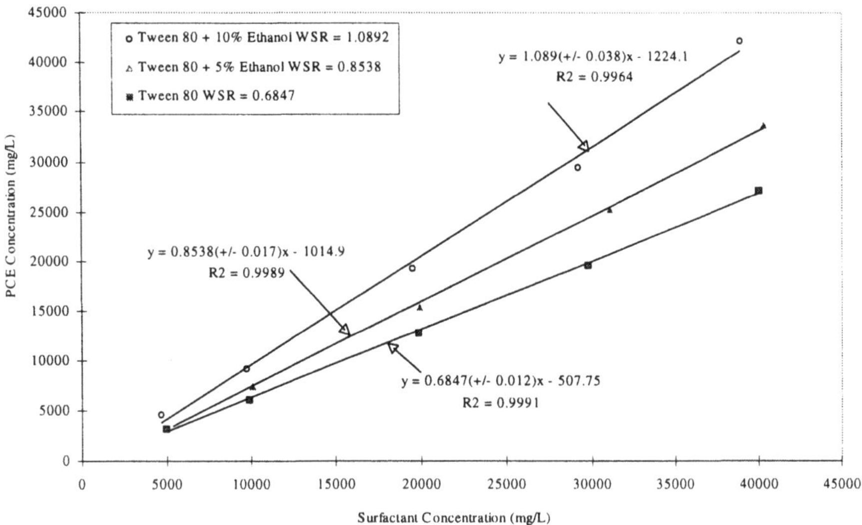

Figure 3. Equilibrium solubility of PCE in solutions of 4% Tween 80, 4% Tween 80 + 5% EtOH, and 4% Tween 80 + 10% EtOH.

The resulting MSRs for Tween 80, Tween 80 + 5% EtOH and Tween 80 + 10% EtOH were 5.41, 6.67, and 8.52 mole PCE/mole of surfactant, respectively. The equilibrium solubilities of PCE in the three surfactant formulations consisting of 4% Tween 80, 4% Tween 80 + 5% EtOH and 4% Tween 80 + 10% EtOH were 26,900, 33,100, and 42,300 mg/L, respectively.

3.2 Column Experiments

Three 1-D column studies were conducted to investigate the effects of EtOH addition (0, 5, and 10% wt) on the micellar solubilization of residual PCE by 4% Tween 80 in a uniform porous medium. The applied Darcy velocities ranged from 3.3 to 4.9 cm/hr. Periods of flow interruption (8 and 15 hours) were utilized to evaluate PCE mass transfer under no flow conditions. A summary of the experimental conditions and physical properties of three representative soil columns (1, 2 and 3) is presented in Table 3. Concentrations of PCE in column effluent versus the cumulative effluent volume collected are presented in Figure 4. At early time (less than ~500 mL), effluent PCE concentrations remained constant for all systems investigated, indicative of steady-state conditions (i.e., $\partial C/\partial t \cong 0$). Regardless of the amount of EtOH added to the surfactant solution, measured effluent PCE concentrations were considerably less than the equilibrium value determined in batch experiments (shown as a dashed horizontal line in Figures 4a, 4b and 4c). For example, the equilibrium solubility of PCE in 4% Tween 80 was 26,900 mg/L, while the steady-state effluent concentration of PCE observed in column 1 was approximately 8,000 mg/L (Figure 4a). Similar trends were observed for 4% Tween 80 solutions containing 5 and 10% EtOH, although the actual concentrations were slightly higher due to the greater solubility of PCE in the presence of EtOH (Figures 4b and 4c, respectively).

Table 2. Summary of experimental conditions and physical properties of Columns 1, 2 and 3.

Experimental Parameter	Column 1 4% Tween 80	Column 2 4% Tween 80 + 5% EtOH	Column 3 4% Tween 80 + 10% EtOH
q (cm/hr)	4.887	4.588	3.3314
v (cm/hr)	15.893	15.002	10.926
Duration of flow interrupt (hr)	8	8	15
L (cm)	10.0	9.7	9.9
ρ_b(g/cm^3)	1.735	1.745	1.736
n (-)	0.345	0.342	0.345
Pore volume (mL)	69.05	63.10	65.30
α_i (cm)	0.036	0.035	0.039
S_n (%)	10.87	10.58	12.57
α_o (cm)	0.065	0.064	0.059
Total trapping number (N_T)	2.59e^{-6}	2.69e^{-6}	2.15e^{-6}
PCE Recovery at 500 mL (%)	28.0	33.7	40.0

$N_T = \left|N_{Ca} + N_B\right|$ (Pennell et al., 1996)

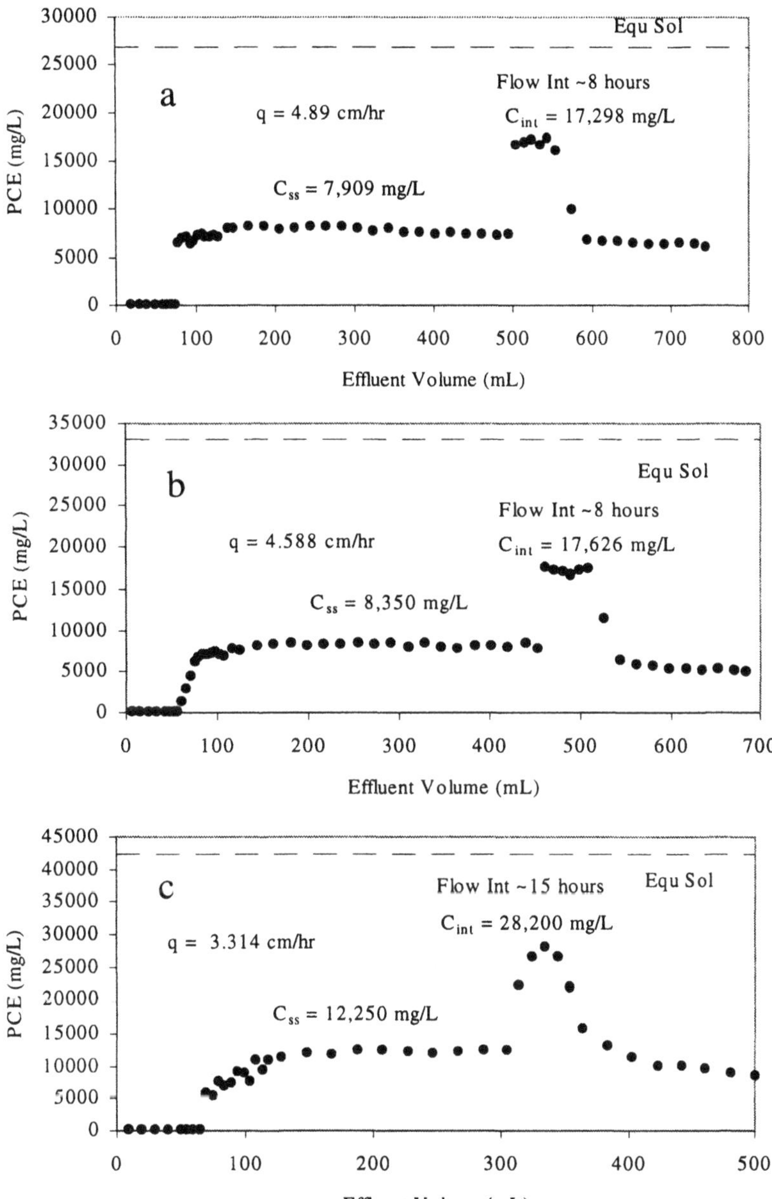

Figure 4. Measured effluent PCE concentrations versus effluent volume for a) Column 1 (4% Tween 80), b) Column 2 (4% Tween 80 + 5% EtOH), and c) Column 3 (4% Tween 80 + 10% EtOH).

The solubilization and aqueous-phase transport of a NAPL in a homogeneous soil column can be represented using a one-dimensional form of the advective-dispersive reactive (ADR) transport equation. Employing a linear driving force expression to describe NAPL dissolution into the aqueous phase, the 1-D ADR equation may be written as:

$$ns_a \frac{\partial C_a^o}{\partial t} + \rho_m \frac{\partial C_s^o}{\partial t} = \frac{\partial}{\partial x}\left(ns_a D_H^o \frac{\partial C_a^o}{\partial x} \right) - q\frac{\partial C_a^o}{\partial x} + K_e(C_{a,sat}^o - C_a^o) \qquad (1)$$

where n is porosity [L^3/L^3], s_a is the aqueous-phase saturation [L^3/L^3], C_a^o is the aqueous-phase concentration of the organic solute [M/L^3], t is time [t], ρ_m is the bulk density of the soil matrix [M/L^3], C_s^o is sorbed-phase concentration of the organic solute [M/M], x is distance [L], D_H^o is the hydrodynamic dispersion coefficient [L^2/t], q is the Darcy velocity [L/t], K_e is the effective mass transfer coefficient [1/t], and $C_{a,sat}^o$ is the equilibrium or saturated solubility of the organic solute in the aqueous phase [M/L^3]. The effective mass transfer coefficient, K_e, represents the product of NAPL interfacial area (a_n) [L^2/L^3] and a NAPL-water mass transfer coefficient (k_l) [L/t]. Assuming that changes in the sorbed-phase concentration of the organic species are minimal ($\partial C_s^o / \partial t = 0$) and that the effluent concentration of the organic species is constant over time (i.e., steady-state conditions apply where $\partial C_a^o / \partial t = 0$), Equation 1 may be expressed as:

$$0 = ns_a D_H^o \frac{\partial^2 C_a^o}{\partial x^2} - q\frac{\partial C_a^o}{\partial x} + K_e(C_{a,sat}^o - C_a^o) \qquad (2)$$

The analytical solution to Equation 2 for a semi-infinite medium with first-type boundary conditions may be written in dimensionless form as (Powers *et al.*, 1991):

$$C^* = 1 - \exp\left(\frac{1 - \sqrt{1 + 4D_a / Pe}}{2 / Pe} X \right) \qquad (3)$$

where C^* is the normalized concentration($C_a^o / C_{a,sat}^o$), D_a is the Damköhler number ($D_a = K_e L/q$), Pe is the Peclet number ($Pe = vL/D_H^o$), v is pore-water velocity [L/t], and X is x/L, where L is column length. The resulting steady-state effective mass transfer coefficients ($K_{e,ss}$), obtained for each column experiment using early-time effluent concentration data, are given in Table 3.

Table 3. Effective mass transfer coefficients obtained for Columns 1, 2 and 3 under flowing and flow interrupt conditions.

Column Experiment	Darcy Velocity (cm/hr)	$K_{e,ss}$ (1/hr)	Flow Interrupt (hrs)	$K_{e,int}$ (1/hr)
Column 1 4 % Tween 80	4.88	0.17	8.0	0.026
Column 2 4% Tween 80 + 5% EtOH	4.59	0.14	8.0	0.018
Column 3 4% Tween 80 + 10% EtOH	3.31	0.11	15.0	0.015

As the concentration of EtOH increased from 0 to 10%, the effective steady-state mass transfer coefficient declined from 0.17 1/hr to 0.11 1/hr, which was due in part to the change in Darcy velocity. Using correlations developed for $K_{e,ss}$ as a function of Darcy velocity and alcohol concentration (Taylor, 1999), the effect of EtOH concentration can be evaluated at a single, representative Darcy velocity. For example, using a Darcy velocity of 4.0 cm/hr, the value of $K_{e,ss}$ would be 0.14, 0.13, and 0.13 for 4% Tween 80, 4% Tween 80 + 5% EtOH and 4% Tween 80 + 10% EtOH, respectively. Thus, the addition of EtOH to 4% Tween 80 had no discernable influence on the effective steady-state mass transfer coefficient. It should be recognized, however, that although the mass transfer coefficient remained essentially unchanged, the steady-state concentration of PCE in the column effluent (C_a^o) and the cumulative PCE mass recovery increased substantially as a result of EtOH addition (Table 2). This behavior can be explained by the fact that the equilibrium solubility of PCE ($C_{a,sat}^o$) increased by more than 50%, from 26,900 mg/L to 42,300 mg/L, with the addition of 10% EtOH.

To obtain values for the effective mass transfer coefficient during flow interruption ($K_{e,int}$), Equation 1 was written assuming that flow was zero (q = 0), the change in PCE sorption over time was negligible ($\partial C_s^o / \partial t = 0$); and the change in PCE concentration along the column was negligible ($\partial C_a^o / \partial x = 0$):

$$n s_a \frac{\partial C_a^o}{\partial t} = K_{e,int}(C_{a,sat}^o - C_a^o) \qquad (4)$$

Integrating equation 4 with respect to time yields the following relationship:

$$ns_a \ln\left[\frac{(C^o_{a,sat} - C^o_{a(i)})}{(C^o_{a,sat} - C^o_{a(t)})}\right] = K_{e,int}\, t \qquad (5)$$

where $C^o_{a(i)}$ is the initial effluent concentration at the start of flow interruption and $C^o_{a(t)}$ is the average peak concentration observed after flow resumes. Values of $K_{e,int}$ were obtained by applying equation 5 to the flow interruption data shown in Figure 4. For 4% Tween 80, 4% Tween 80 + 5% EtOH and 4% Tween 80 + 10% EtOH the values of $K_{e,int}$ were 0.026, 0.018, and 0.015 1/hr, respectively (Table 3). Although these data indicate that the effective mass transfer coefficient obtained for no flow conditions ($K_{e,int}$) decreased in the presence of EtOH, the change in aqueous phase PCE concentration with time (i.e., dC^o_a / dt) was essentially the same (~1,525 mg/L hr) for all of the systems evaluated. This occurred because the equilibrium PCE solubility ($C^o_{a,sat}$) increased with increasing EtOH concentration. Thus, as the concentration of EtOH increased, the corresponding increase in the driving force (i.e., $C^o_{a,sat} - C^o_a$) was compensated for by a reduction in the mass transfer coefficient ($K_{e,int}$). It should also be noted that the observed values of $K_{e,int}$ were approximately one order-of-magnitude less those obtained under steady-state conditions ($K_{e,ss}$) at a Darcy velocity of 4.0 cm/hr. This difference in mass transfer rates can be attributed to the fact that under no flow conditions the system is purely diffusion driven, and thus, the boundary layer thickness extends to the pore scale.

3.3 2-D Box Experiments

The effects of rate-limited solubilization, subsurface layering and flushing solution density on PCE recovery were evaluated in two separate box studies (Box A and Box B). Box A was flushed with 4% Tween 80 alone to serve as the control case, while Box B was flushed with 4% Tween 80 + 5% EtOH to evaluate the effects of cosolvent addition on PCE solubilization, cumulative PCE recovery, and surfactant delivery. Each box was packed with 20-30 mesh Ottawa sand as the background porous medium, with one rectangular layer of F-70 Ottawa sand above two side-by-side rectangular layers of Wurtsmith aquifer material (Figure 5). The overall PCE saturation in Box A (s_{PCE} = 3.45% or 38 mL) was slightly higher than that obtained in Box B (s_{PCE} = 2.65% or 28.9 mL). Regions of high PCE saturation or "pools" formed above each fine layer, and PCE flowed horizontally around the layers.

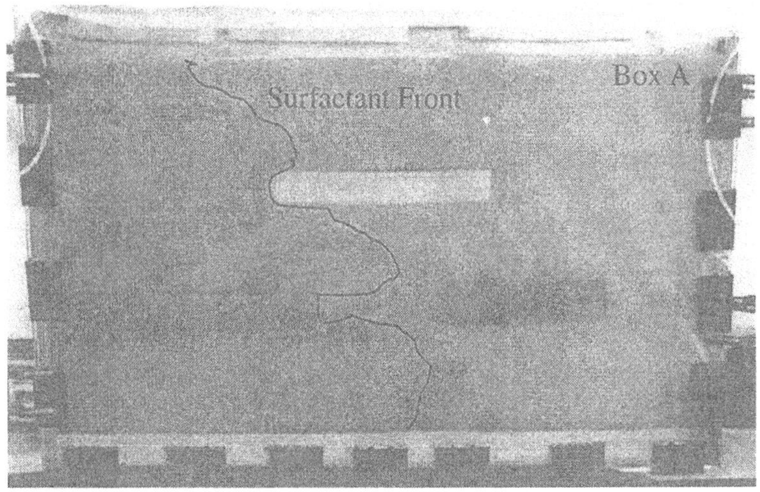

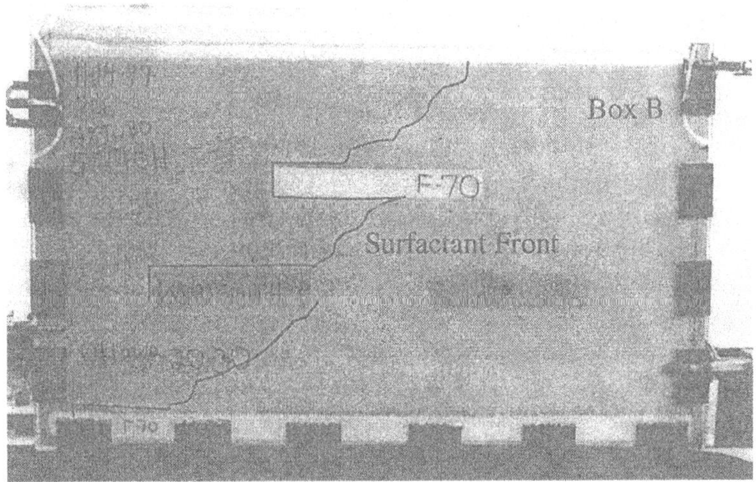

Figure 5. Digital images of surfactant fronts for a) Box A flushed with 4% Tween at a Darcy velocity of 4.7 cm/hr (0.4 pore volumes) and b) Box B flushed with 4% Tween 80 + 5% EtOH at a Darcy velocity of 4.8 cm/hr (0.5 pore volumes).

The resulting PCE distribution within each box consisted of relatively narrow vertical zones of residual PCE and horizontal regions of high PCE saturation above the fine layers. Experimental conditions and relevant physical properties of each box experiment are summarized in Table 4.

Table 4. Summary of experimental conditions and physical properties of the 2-D flushed with 4% Tween 80 (Box A) and 4% Tween 80 plus 5% EtOH (Box B).

Parameter	Box A 4% Tween 80	Box B 4% Tween 80 + 5% EtOH
q (cm/hr)	4.70	4.80
ρ_b 20-30 mesh Ottawa (g/cm^3)	1.70	1.69
ρ_b F-70 Ottawa (g/cm^3)	1.75	1.78
ρ_b Wurtsmith (g/cm^3)	1.89	1.69
Total pore volume (mL)	1104	1,090
Total Trapping Number (N_T)	2.48 e^{-6}	2.79 e^{-6}
Total PCE added (mL)	38.05	28.9
Overall PCE saturation (%)	3.45	2.65
Volume of surfactant flushed (mL), (pore volumes)	8,717 (7.9)	8,317 (7.6)
PCE recovered during flush (mL), (% total)	26.52 (69.7%)	20.28 (70.2%)
Total PCE recovered (mL), (% total)	36.45 (95.8%)	27.41 (94.8%)

$N_T = \sqrt{N_{Ca}^2 + N_B^2}$ (Pennell et al., 1996)

Box A was flushed with 4% Tween 80 at a Darcy velocity of approximately 4.7 cm/hr. The density of 4% Tween 80 was slightly greater than the density of the resident pore fluid (Nanopure water with 0.5 g/L $CaCl_2$ + 1 g/L NaN_3) at the onset of surfactant flushing ($\rho_{4\%Tween}$ = 1.002 g/cm^3 versus ρ_{H20} = 0.998 g/cm^3). As a result of this solution density contrast and the fact that the end chambers were fully screened, the surfactant solution flowed preferentially along the bottom of Box A, as illustrated in Figure 5a. This density effect becomes more pronounced as flushing velocity is further decreased (Taylor, 1999). For contamination scenarios involving zones of high DNAPL saturation above a confining layer, the use surfactant solutions that possess a slightly greater density than the resident aqueous phase may be advantageous due to preferential flow of the more dense surfactant solution through the most contaminated region of the aquifer. If a vertical surfactant flushing profile (i.e., piston-displacement) is desired, the flow rate must be sufficient to minimize density effects. Alternatively, an alcohol cosolvent could be added to the surfactant formulation to equalize the density of flushing solution and the resident aqueous phase.

To evaluate the effects of cosolvent on surfactant delivery and PCE recovery, Box B was flushed with 4% Tween 80 + 5% EtOH at a Darcy velocity of 4.8 cm/hr. The surfactant/cosolvent mixture, which had a density of 0.994 g/cm^3, was also representative of a neutral buoyancy flood solution (Shook *et al.*, 1998). It is important to recognize that "neutral buoyancy" refers to density of flushing fluid after solubilization of the DNAPL. Thus, the initial density of the surfactant formulation must be less than that of the resident aqueous phase. Figure 5b shows the location and shape of the 4% Tween 80 + 5% EtOH front after flushing Box B with 0.5 pore volumes of solution. The lower density of the 4% Tween 80 + 5% EtOH solution (0.994 g/cm^3) relative to the density of resident pore water (0.998 g/cm^3) caused the injected solution to flow preferentially along the top of Box B (Figure 5b). This effect can become severe at low flow rates (Taylor, 1999). The observed flow behavior, commonly referred to as density over ride, could lead to substantially lower DNAPL recovery rates. Under these conditions, lower regions of the formation, which are likely to contain DNAPL, may never become completely saturated with the flushing solution or may be flushed with a reduced volume of solution.

In Boxes A and B, micellar solubilization was the primary PCE recovery mechanism during surfactant flushing. No visible mobilization of PCE, existing either as residual droplets or in pools above fine layers, was observed in either box. The lack of PCE displacement as free product was consistent with values of the total trapping number (N_T) calculated for each box (Table 4). The values of N_T were well below the critical value of 2.0 X 10^{-5} required to displace residual PCE from a sandy aquifer material (Pennell *et al.*, 1996). In Box A, approximately 70% of the PCE (26.5 mL) was recovered via solubilization after flushing with 8 pore volumes (8,717 mL) of 4% Tween 80. In box B, 70% of the PCE (20.3 mL) was recovered after flushing with 7.6 pore volumes (8,300 mL) of the 4% Tween 80 + 5% EtOH formulation. PCE remaining in Boxes A and B after surfactant flushing existed primarily as thin pools above the lower confining layer and above fine lenses. During excavation and extraction of each box, no neat PCE was found within the fine layers, indicating that the entry pressure had not been exceeded even after the interfacial tension between PCE and the aqueous phase was reduced to ~5 dyne/cm during surfactant and surfactant+EtOH flushing.

Aqueous phase concentrations of PCE measured in extraction well effluent of Box A and Box B are presented in Figures 6a and 6b, respectively. These data are expressed as the concentration of PCE (mg/L) versus cumulative volume of effluent solution (mL) collected during each experiment. A sharp rise in effluent PCE concentration occurred after approximately 1 pore volume of surfactant solution had been injected into each box, which coincided with surfactant breakthrough. In Box A, the initial PCE concentrations reached a maximum value of approximately 7,200 mg/L. This value was similar to initial steady-state effluent concentration of PCE measured in column 1 (7,900 mg/L), flushed at a similar Darcy velocity (4.9 cm/hr).

The greater overall PCE mass loading in the 1-D columns (s_{PCE} = 10.6-12.6%), resulted in prolonged periods of steady-state effluent concentrations compared to the 2-D systems (s_{PCE} = 2.7-3.5%). For Box A, effluent PCE concentrations began to decline after flushing with less than 2 pore volumes of 4% Tween 80. In Box B, the initial concentration of PCE in the column effluent prior to the first flow interruption reached a value of approximately 7,500 mg/L. The initial PCE concentration was similar to that observed in Box A (~7,200 mg/L), even though the equilibrium solubility of PCE in the 4% Tween 80 + 5% EtOH was substantially greater than solubility of PCE in 4% Tween 80 alone (26,900 versus 33,100 mg/L). The failure of the EtOH amended system to achieve greater PCE recovery can be attributed, in large part, to density over ride of the flushing solution, which reduced contact with high PCE saturation zones located in the lower regions of Box B.

Following the initial rise in PCE concentration and a brief period of steady-state conditions observed in Box A and Box B, effluent PCE concentrations declined sharply as more surfactant solution was displaced through each box. This behavior is characteristic of transient NAPL dissolution or solubilization, in which the interfacial area of NAPL available for mass transfer is reduced over time (e.g., Powers *et al.*, 1992; Pennell *et al.*, 1994). In both 2-D box studies, PCE existing at residual saturation (small droplets or ganglia) was rapidly removed from the domain. Regions of high PCE saturation (i.e., pools) above fine layers gradually became smaller as the flushing process continued. However, these high saturation zones, which possessed relatively small specific interfacial area, persisted throughout the course of box experiments.

The effect of flow interruptions (q = 0 cm/hr), ranging in duration from 12 to 17 hours, on effluent PCE concentrations was also evaluated in the box experiments. The steep rise in effluent PCE concentrations after flow interruption was indicative of rate-limited micellar solubilization. Although these data cannot be directly compared to those obtained in the column studies, the magnitude of the PCE concentration increases following flow interruption were substantially less than those observed in the more uniform 1-D soil columns. Taken in concert, the results of the 2-D box studies demonstrate that rate-limited mass transfer influenced PCE recovery, and that subsurface layering and density over ride effects were key factors contributing to reduced PCE recovery in heterogeneous, 2-D systems.

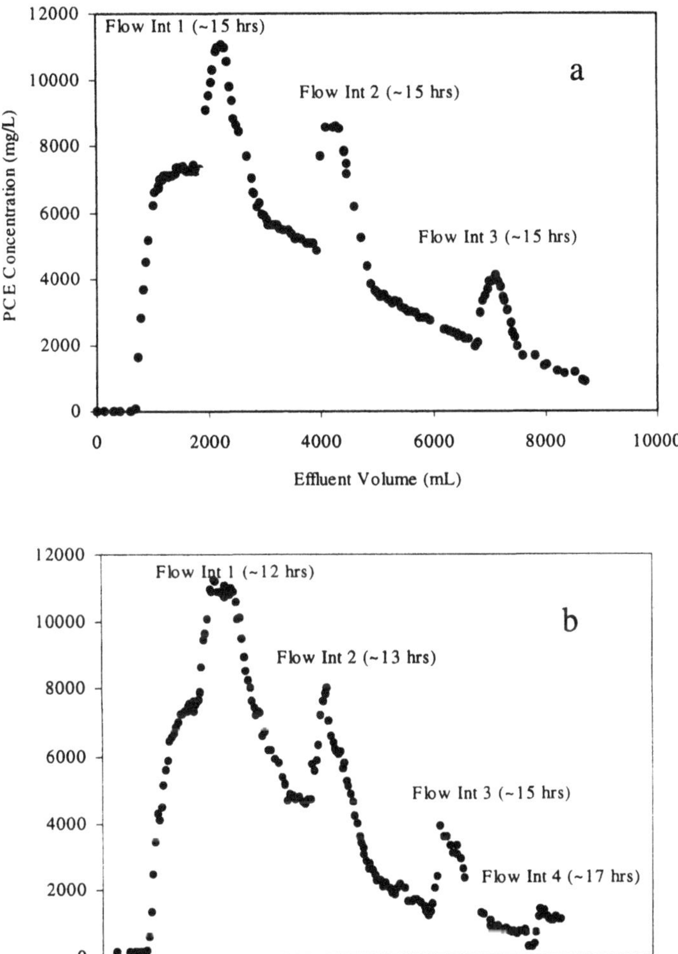

Figure 6. Measured effluent concentrations of PCE versus column effluent volume for a) Box A (4% Tween 80) and b) Box B (4% Tween 80 + 5% EtOH).

4. CONCLUSIONS

In this study, the effects of cosolvent (EtOH) addition on the solubilization and recovery of PCE by a nonionic surfactant (Tween 80) was evaluated using a combination of batch, column and 2-D box studies. Batch results demonstrated that the addition of 5% and 10% EtOH increased the solubilization capacity of Tween 80 from 0.69 g PCE/g surfactant to 1.09 g PCE/g surfactant. For a 4% Tween 80 solution, this translates into a solubility enhancement of more than 50%, from 26,900 mg/L to 42,300 mg/L. When the surfactant formulations were flushed through soil columns containing residual PCE, effluent concentration data clearly showed that PCE solubilization was rate-limited, regardless of the EtOH concentration. Using analytical solutions to the 1-D ADR equation, effective mass transfer coefficients (K_e) were obtained from the effluent concentration data for both steady-state ($K_{e,ss}$) and no flow conditions ($K_{e,int}$). The addition of EtOH had no discernable effect on steady-state mass transfer coefficients ($K_{e,ss}$), which ranged from 0.13 to 0.14 1/hr. Under no flow conditions, the effective mass transfer coefficient ($K_{e,int}$) decreased from 0.026 to 0.015 1/hr when the EtOH concentration was increased from 0 to 10%. The observed reduction in $K_{e,int}$ was compensated for by a corresponding increase in the driving force ($C^{o}_{a,sat} - C^{o}_{a}$) due to an increase in the equilibrium solubility of PCE in the presence of EtOH. Thus, actual changes in aqueous phase PCE concentrations with flow interruption time were identical for all systems investigated. In 2-D box studies, the addition of 5% EtOH to 4% Tween 80 did not result in improved PCE recovery despite the fact that the equilibrium solubility of PCE was approximately 6,000 mg/L (23%) greater than in 4% Tween 80 alone. This finding was attributed to the lower density of the 4% Tween 80 + 5% EtOH solution (0.994 g/cm^3) relative to the resident pore fluid (0.998 g/cm^3), which resulted in preferential flow of the surfactant formulation through the upper third of the box. In contrast, the 4% Tween 80 solution had a density of 1.002 g/cm^3, which gave rise to preferential flow through the heavily contaminated regions of the box near the lower confining layer. These results demonstrate the importance of accounting for relatively small differences in solution density, which can dramatically influence surfactant flushing profiles and subsequent recovery of target contaminants. In practice, relatively high flushing rates and judicious positioning or screening of injection wells may be utilized to reduce or mitigate density contrast effects. Regardless of the specific surfactant formulation employed, this work demonstrates that alteration of flushing solution density must be accounted for during surfactant delivery to the DNAPL source zone and during recovery of the solubilized-DNAPL plume.

ACKNOWLEDGEMENTS

Funding for this research was provided by the U.S. Environmental Protection Agency, Great Lakes and Mid-Atlantic Center for Hazardous Substance Research (GLMAC-HSRC) under Grant No. R-825540 and by the U.S. Environmental Protection Agency, Office of Research and Development under Grant No. R-825409. Additional matching funds were provided to the GLMAC-HSRC by the Michigan Department of Environmental Quality. The content of this publication does not necessarily represent the views of either agency and has not been subject to agency review.

BIBLIOGRAPHY

Advanced Applied Technology Demonstration Facility (AATDF) Program (1997). "AATDF Technology Practices Manual for Surfactants and Cosolvents." Rice University, Energy and Environmental Systems Institute.

Edwards, D.A., Adeel, Z., and Luthy, R.G. (1994). "Distribution of nonionic surfactant and phenanthrene in a sediment/aqueous system." *Environ. Sci. Technol*, 28, 1550-1560.

Fountain, J.C. (1997). "The role of field trials in development and feasibility assessment of surfactant-enhanced aquifer remediation." *Water Environ. Res.*, 69, 188-195.

Fountain, J.C., Starr, R.C., Middleton, T., Beikirch, M., Taylor, C., and Hodge, D. (1996). "A controlled field test of surfactant-enhanced aquifer remediation." *Ground Water*, 34, 910-916.

Fountain, J.C. Waddell-Sheets, C., Lagowski, A., Taylor, C., Frazier, D., and Byrne, M. (1995). "Enhanced Removal of Dense Nonaqueous Phase Liquids using Surfactants – Capabilities and Limitations from Field Trials." In: *Surfactant Enhanced Subsurface Remediation, ACS Symposium Series Emerging Technologies 594*, Sabatini, D.A., Knox, R.D., Harwell, J.H., (ed.), American Chemical Society, 177-190.

Kostarelos, K., Pope, G.A., Rouse, B.A. and Shook, G.M. (1998). A new concept: The use of neutrally buoyant microemulsions for DNAPL remediation." *J. Contam. Hydrol.*, 34, 383-397.

Martel R. and Gelinas, P.J. (1996). "Surfactant solutions developed for NAPL recovery in contaminated aquifers." Ground Water, 34, 143-154.

Martel, R., Gelinas, P.J., Desnoyers, J.E., and Masson, A. (1993). "Phase diagrams to optimize surfactant solutions for oil and DNAPL recovery in aquifers." *Ground Water*, 31, 789-800.

Pennell, K.D., Adinolfi, A.M., Abriola, L.M., and Diallo, M.S. (1997). Solubilization of dodecane, tetrachloroethylene, and 1,2-dichlorobenzene in micellar solutions of ethoxylated nonionic surfactants." *Environ. Sci. Technol.*, 31, 1382-1389.

Pennell, K.D., Pope, G.A., and Abriola, L.M. (1996). "Influence of viscous and buoyancy forces on the mobilization of residual tetrachloroethylene during surfactant flushing." *Environ. Sci. Technol*, 30, 1328-1335.

Pennell, K.D., Jin, M., Abriola, L.M., and Pope, G.A. (1994). "Surfactant enhanced remediation of soil columns contaminated by residual tetrachloroethylene." *J. Contam. Hydrol.*, 16, 35-53.

Pennell, K.D., Abriola, L.M., and Weber Jr., W.J. (1993). "Surfactant-enhanced solubilization of residual dodecane in soil columns. 1. Experimental investigation." *Environ. Sci. Technol.*, 27, 2332-2340.

Powers, S.E., Abriola, L.M., and Weber Jr., W.J. (1992). "An experimental investigation of nonaqueous phase liquid dissolution in saturated subsurface systems: Steady state mass transfer rates." *Water Resour. Res.*, 28, 2691-2705.

Powers, S.E., Loureiro, C.O., Abriola, L.M., and Weber Jr., W.J. (1991). "Theoretical study of the significance of nonequilibrium dissolution of nonaqueous phase liquids in subsurface systems." *Water Resour. Res.*, 27, 463-477.

Rouse, J.D., Sabatini, D.A., Deeds, N.E., and Brown, R.E. (1995). "Micellar solubilization of unsaturated hydrocarbon concentrations as evaluated by semiequilibrium dialysis." *Environ. Sci. Technol*, 29, 2484-2489.

Shiau, B.J., Sabatini, D.A., and Harwell, J.H. (1994). "Solubilization and microemulsification of chlorinated solvents using direct food additive (edible) surfactants." *Ground Water*, 32, 561-569.

Shook G.M., Pope, G.A., and Kostarelos, K. (1998). "Prediction and minimization of vertical migration of DNAPLs using surfactant enhanced aquifer remediation at neutral buoyancy." *J. Contam. Hydrol.*, 34, 363-382.

Shook, G.M., Kostarelos, K., and Pope, G.A. (1997). "Minimization of vertical migration of DNAPLs using surfactant enhanced aquifer remediation at neutral buoyancy." Presented at SPE Annual Tech. Conf., *Soc. Petrol. Eng.*, 39294, 1-8.

Smith, J.A., Sahoo, D., McLellan, H.M., and Imbrigiotta, T.E. (1997). "Surfactant-enhanced remediation of trichloroethene-contaminated aquifer. 1. Transport of Triton X-100." *Environ. Sci. Technol.*, 31, 3565-3572.

Taylor, T.P. (1999). "Characterization and Surfactant Enhanced Remediation of Organic Contaminants in Saturated Porous Media." Doctoral Dissertation, Georgia Institute of Technolgy, Atlanta, GA.

van Genuchten, M.Th. (1981). "Non-equilibrium Transport Parameters from Miscible Displacement Experiments." Research Report No. 119, U.S. Dept. of Agriculture, U.S. Salinity Lab, Riverside, CA.

Verschueren, K. (1983). "Handbook of Environmental Data on Organic Chemicals, 2nd Ed." Van Nostrand Reinhold Co., New York, NY.

Waddill, D.W. and Parker, J.C. (1997). "Recovery of light, non-aqueous phase liquid from porous media: Laboratory experiments and model validation." *J. Contam. Hydrol.*, 27, 127-155.

Chapter Fourteen

Unsaturated-Zone Airflow: Implications for Natural Remediation of Groundwater by Contaminant Transport through the Subsurface

FRED D. TILLMAN, JR.[1], JEE-WON CHOI[1], WHITNEY KATCHMARK[2], JAMES A. SMITH[1], and HOUSTON G. WOOD, III[3]

[1] *Department of Civil Engineering, University of Virginia, Charlottesville, VA 22903*

[2] *Naval Engineering Command, Norfolk, VA*

[3] *Department of Mechanical, Aerospace and Nuclear Engineering, University of Virginia, Charlottesville, VA 22903*

Key words: unsaturated zone; barometric pumping; natural remediation

Abstract: Both diurnal and weather-system-induced atmospheric pressure changes can cause air pressure gradients to form in the unsaturated zone. As the subsurface pressure re-establishes equilibrium with the changing surface pressure, gas is "breathed" in and out of the unsaturated zone. This movement of gas and subsequent advection of organic vapors present above a contaminated aquifer may provide a significant mechanism of ground-water contaminant mass loss through the unsaturated zone to the atmosphere. Previous research into the nature and effects of barometric pumping are presented. Studies involving field measurements and modeling of subsurface airflow and vapor fluxes at land surface are reviewed. This research indicates that, under certain circumstances, gas-phase contaminant transport in the unsaturated zone is influenced by atmospheric pressure changes at land surface. New data and modeling results involving unsaturated-zone pressure gradients and airflow are also presented. Air pressure, moisture content, and water table elevation were measured as functions of depth and time in the unsaturated zone at Picatinny Arsenal, New Jersey during dry periods in August and October, 1996. Significant air-pressure gradients between the subsurface and the atmosphere were observed, while little variation in air pressure was measured at depths between 0.5 and 1.7 m. Changing subsurface air pressures in response to varying atmospheric pressure were successfully simulated using a simplified, 1-dimensional, finite-difference model. Further model results indicated non-zero subsurface air velocity during most of the simulation period. Based on these results, it was concluded that airflow is occurring in the unsaturated zone and that this airflow can be explained by one-dimensional (vertical) pressure gradients driven by atmospheric-pressure variations.

1. INTRODUCTION

Intrinsic remediation (or "natural attenuation") refers to the reduction in mass, mobility or toxicity of contaminants in soils, sediments, or ground water by naturally occurring physical, chemical, and/or biological processes. It may include biodegradation, dilution, dispersion, sorption, volatilization, chemical or biochemical stabilization, or a combination of these processes (Swett and Rapaport 1998). Many physical, biochemical, and inter-phase processes contribute to the fate and transport of dissolved organic pollutants in a contaminated ground-water plume. Physical processes such as advection with the bulk fluid phase (and subsequent longitudinal dispersion) and molecular diffusion, biochemical processes such as biodegradation and hydrolysis, and inter-phase processes such as sorption and volatilization are all important mechanisms determining the transport and fate of ground-water pollutants. Of these different mechanisms, natural volatilization of pollutants from ground water into the unsaturated zone soil gas and eventual transport to the atmosphere has received comparatively little attention in the scientific literature, probably due to the general belief that it is not a significant transport process relative to other processes. The concentration gradient that drives mass transfer from ground water to overlying soil gas is relatively small because of transport limitations in both the saturated and unsaturated zone. However, given the large surface area of many contaminated aquifers and the high Henry's Law constants of many common organic ground-water contaminants, volatilization into unsaturated-zone soil gas and subsequent transport to land surface may represent a significant ground-water contaminant mass loss. The determination of a contaminated aquifer's potential for natural remediation should therefore include an investigation of dissolved contaminant mass loss due to volatilization and subsequent transport through the unsaturated zone.

Most intrinsic remediation literature has been devoted to research involving *in-situ* bioremediation processes. Additionally, significant research has been conducted on engineered remediation solutions such as soil-vapor extraction and ground-water sparging that seek to utilize and optimize naturally occurring processes. Several researchers have quantified the presence and concentration of volatile organic compounds in unsaturated-zone soil gas above VOC contaminated aquifers (Batterman et al. 1992; Cho et al. 1993; Kerfoot 1987; Marrin and Kerfoot 1988; Smith et al. 1990) and have correlated these soil-gas concentrations with the concentration of the contaminant in the shallow ground water. However, little published information is available concerning the effect of natural atmospheric pressure variations on the volatilization of dissolved organic compounds in underlying ground water and their subsequent transport to the atmosphere through unsaturated-zone soil gas. This chapter presents a review of research into the effects of barometric pressure changes on organic vapor transport in the unsaturated zone and presents results of a field study on the flow of unsaturated-zone air in response to atmospheric pressure changes.

2. PREVIOUS RESEARCH ON BAROMETRIC PUMPING

Atmospheric pressure near land surface is constantly changing, due both to short-term diurnal temperature and gravitational fluctuations as well as longer-term cycles due to the passage of high-and-low pressure weather systems (Figure 1). Analysis of data taken by the National Oceanographic and Atmospheric Administration (NOAA) in 1998 at the Newark International Airport in Newark, NJ show pressure fluctuations that vary from the mean by 250 N/m^2 (2.5 mbar) during 73% of the observations, with variations from the mean greater than 1000 N/m^2 (10 mbar) occurring during nearly 20% of the hourly measurements (Figure 2)(NOAA 1999). Rarely will spatial measurement of atmospheric pressure vary more than 300 N/m^2 (3 mbar) per 100 km and the rate of pressure change with distance is usually much less than 150 N/m^2 (1.5 mbar) per 100 km (Miller and Thompson 1975). Figure 3 shows atmospheric pressure readings for the first 6 months of 1998 for the Newark International Airport along with readings from Philadelphia International Airport, some 114 km (70.8 mi) away. Differences greater than 300 N/m^2 (3 mbar) occur during only 6.2% of the hourly measurements. Pressure gradients may therefore be considered a one-dimensional phenomenon at the field-site scale. Weeks (1978) observed barometric pressure fluctuations at depths up to hundreds of feet in deep unsaturated zones. As the subsurface pressure reestablishes equilibrium with the atmospheric pressure, air is "breathed" in and out of the unsaturated zone. This phenomenon is know as barometric pumping and can affect the transport of volatile organic compounds (VOC) in the unsaturated zone both directly and indirectly. At first thought, it may appear that this "ratcheting" of contaminated soil gas up and down during a barometric pressure cycle would result in a net zero effect on VOC transport. As the decreasing surface pressure that moved contaminated soil gas upward begins the increasing portion of its cycle, the organic vapors would return to their original position in the unsaturated zone. However, at land surface organic vapors that exit the unsaturated zone during decreasing atmospheric pressure are almost instantaneously diluted to nondetectable concentrations. By contrast, because the diffusion coefficient for most VOCs in water is approximately 4 orders of magnitude less than in air, organic vapor transport is limited across the capillary fringe. Therefore, the effect of cyclic atmospheric pressure would be a net flux of vapors out of the system and into the atmosphere. Periods of decreasing atmospheric pressure cause a release of near-surface soil gas to the atmosphere, thus contaminants escape from the shallow unsaturated zone faster than they would by pure diffusion alone. Periods of increasing atmospheric pressure will force uncontaminated surface air into the shallow unsaturated zone. This in turn increases the rate of diffusion by lowering the surface where the contaminant concentration is zero below the true ground level (Auer et al. 1996). Massman and Farrier (1992) simulated gas transport in the unsaturated zone resulting from atmospheric pressure variations using the single-component advection-dispersion equation. They found that "fresh" air can migrate several meters into the subsurface during a typical barometric pressure cycle. These effects may be compounded by the presence of fractures, partially penetrating cracks or boreholes open to the unsaturated zone (Holford et al. 1993; Schery et al. 1988).

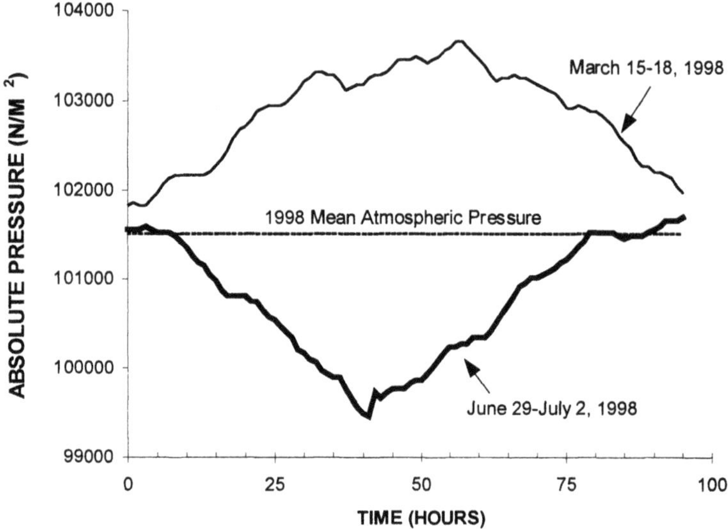

Figure 1. Atmospheric pressure variations between March 15-18, 1998 and June 29-July2, 1998 at Newark International Airport, NJ.

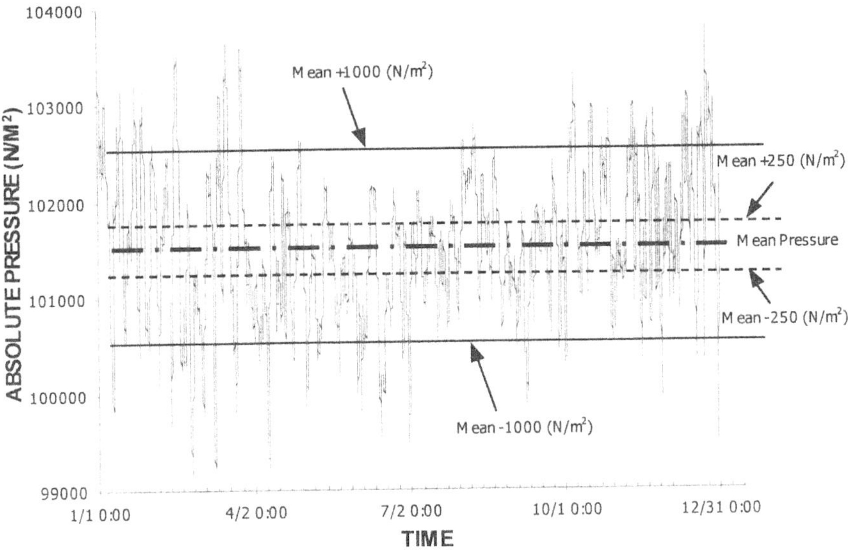

Figure 2. Atmospheric pressure for 1998 at Newark International Airport, NJ.

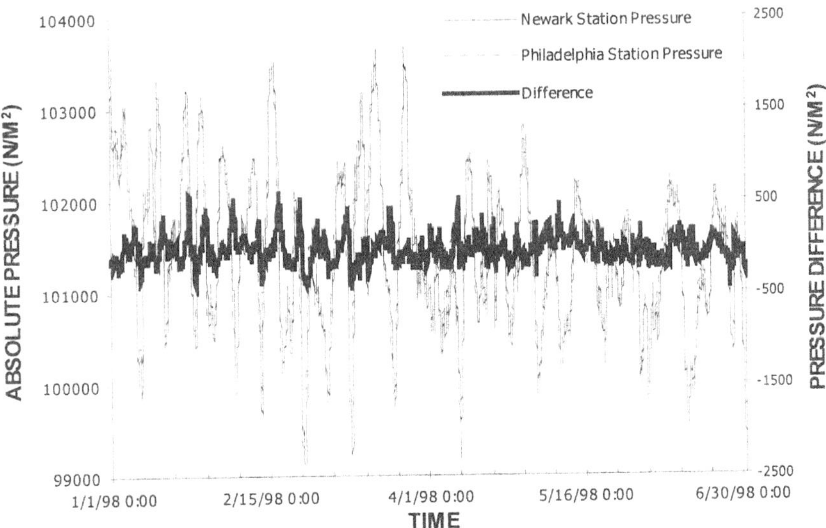

Figure 3. Atmospheric pressure at Newark International Airport, NJ, Philadelphia International Airport, PA and the difference between them.

In an early study by Schery and Gaeddert (1982), an "accumulator" device was used to measure the effect of atmospheric pressure variations on the flux of Radon (^{222}Rn), an inert radioactive element with a half-life of 3.8 days, from the soil. Fluxes measured by the accumulator were compared with predictions for flow-free diffusion from a model developed by Clements and Wilkening (1974), which applies Fick's law. A mean ^{222}Rn-flux enhancement of about 10%, with a high value of 20%, due to cyclic atmospheric pressure variations was observed. However, the device's effectiveness was limited by back diffusion from the accumulator to the subsurface, leading the authors to view the flux values as semi-quantitative.

In an effort to simulate flux of the fumigant 1,3-dichloropropene (1,3-D) to the atmosphere, Chen et al. (1995) found that existing models for chemical transport and flux in soils considered advection and dispersion of solute in the liquid phase with organic vapor diffusion being the only transport mechanism for the air phase. Organic chemicals with high vapor pressures were typically modeled by increasing the effective vapor-phase diffusion coefficient in order to simulate measured experimental fluxes. The researchers modified an existing pesticide version of the simulation model LEACHP by adding explicit simulation of advective transport of the vapor phase driven by barometric pressure change at the soil surface. The resulting model, LEACHV, was then used to simulate 1,3-D fluxes from a Salinas Valley, CA field site which were compared to field fluxes determined by the aerodynamic method. The model was able to simulate the same general pattern as field measurements of flux in response to changes in barometric pressure (Figure 4). Model results showed that barometric pressure changes over the measurement period drove advective vapor movement through unsaturated zone soil and ultimately to land surface. However, these researchers did not collect subsurface pressure measurements with which to calibrate and verify the airflow part of the model.

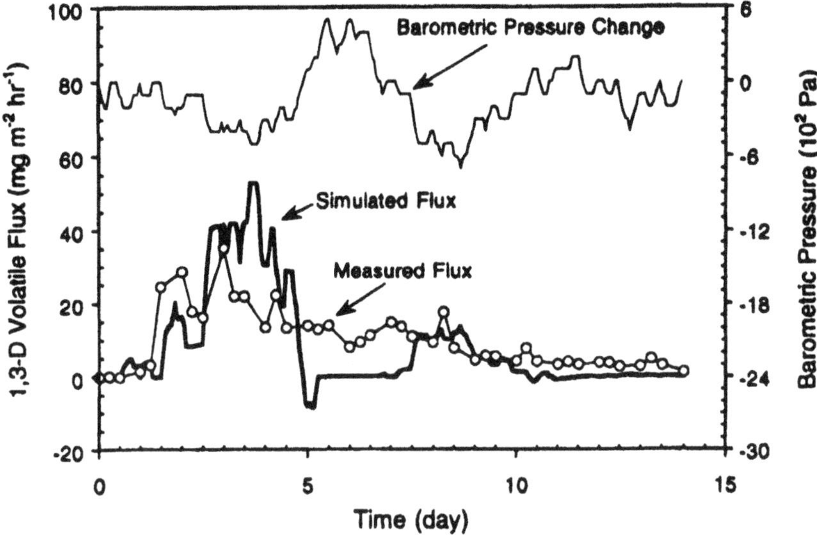

Figure 4. Simulated and measured 1,3-dichloropropene (1,3-D) volatilization flux in response to atmospheric pressure changes at Salinas Valley. Reprinted with permission from Chen et al. (1995). Copyright 1995, American Chemical Society

Other researchers have proposed that the direct affects of barometric pumping are negligible with respect to the indirect effects. In a theoretical modeling study by Auer et al. (1996), the authors looked at two effects of barometric pumping on organic vapor transport in the unsaturated zone: the direct effect of the evacuation of the top layer of unsaturated zone soil gas with decreasing atmospheric pressure and the replacement of this layer with "clean" surface air during increasing atmospheric pressure and the indirect effect of increasing velocity-dependent dispersion. The researchers argue that the direct pumping mechanism of atmospheric pressure changes has little or no effect on the overall rate of contaminant transport. While the effect near land surface is to lower the effective surface (where the concentration is zero) to below true ground level, thus decreasing the distance that organic vapors must travel before evacuation, the pumping simply moves the volatilized compound up and down at depth, not affecting the chemical gradient or diffusion rate. The study found that, for permeabilities greater than 10^{-8} cm^2 and/or surface-pressure periods greater than 1 day, the surface-pressure perturbation propagates with some delay to even the deepest depths of moderately thick systems (60^+ m). While the pressure perturbation is felt over a large vertical length of the unsaturated zone, the air itself moves only a short distance during an atmospheric pressure cycle. Numerical markers placed at depths during model simulations of a 60-m unsaturated zone with air permeability of 10^{-11} m^2

show that subsurface air deeper than ~1.75 m never reach land surface (Figure 5). The volume of air removed by barometric pumping was shown to be independent of the period of the pressure cycle, but proportional to the depth of the lower boundary.

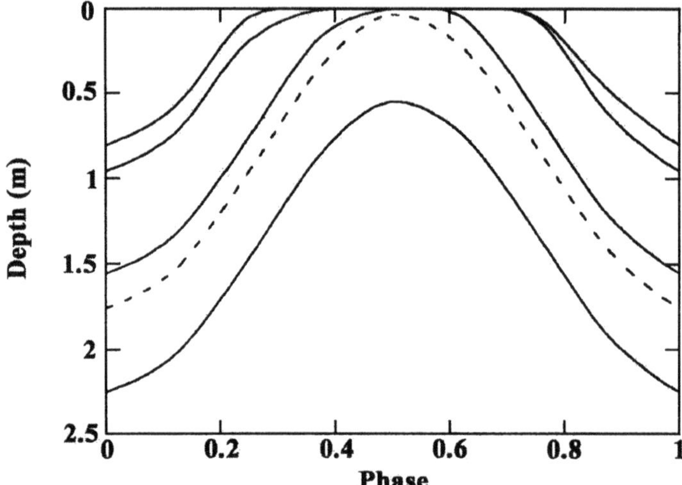

Figure 5. Position of subsurface numerical markers during an atmospheric pressure cycle. Simulation is of a 60-m unsaturated zone with air permeability of 10^{-11} m^2. Note that soil gas initially deeper than ~1.75 m never reaches the surface while soil gas at shallower depths would be flushed from the system. Reprinted from Auer et al. (1996), Copyright 1996, pg. 152, with permission from Elsevier Science.

The indirect effects of barometric pumping are derived from an analysis of the one-dimensional air-phase transport equation:

$$\frac{\partial C}{\partial t} = -V\frac{\partial C}{\partial z} + \frac{\partial}{\partial z}\left(D\frac{\partial C}{\partial z}\right) \tag{1}$$

where C is the concentration per unit volume air; V is the pore-scale velocity; and D is the coefficient of hydrodynamic dispersion. The hydrodynamic dispersion term can be written as:

$$D = D_{mol} + \alpha|V| \tag{2}$$

where D_{mol} is the intrinsic molecular diffusion in the porous medium; α is the dispersivity; and $|V|$ is a positive measure of the air velocity.

Dispersion arises from the fact that, even in a relatively homogenous porous medium, small-scale heterogeneities exist which cause airflow to proceed along various channels at different rates. Barometric pumping causes a significant increase in the coefficient of hydrodynamic dispersion over a pure diffusion-based transport model, thus increasing the overall transport rate.

The authors presented three simulations to underscore the relative contributions of barometric pumping on unsaturated zone transport: a baseline case of pure molecular diffusion transport alone, a case of direct effect of barometric pumping of the soil gas, and a final simulation incorporating both the direct effects of soil-gas pumping and indirect effects of increased mechanical dispersion. It was found that the surface boundary condition altered by barometric pumping in the second simulation did not significantly affect the rate of diffusion transport. However, when the complete effects of barometric pumping were included in the model, as little as 13% of the original contaminant mass remained after 50 years, compared with 49% for diffusion only (Figure 6).

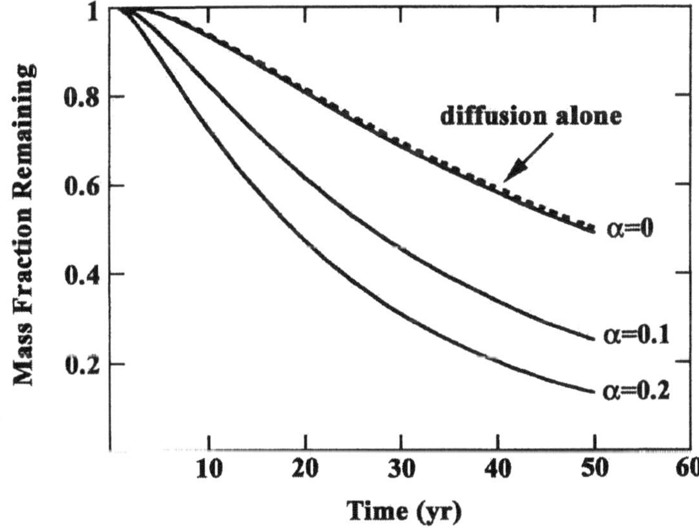

Figure 6. Mass fraction of theoretical contaminant mass remaining vs. time for various values of the hydrodynamic dispersivity, α. Atmospheric pressure changes at land surface cause advective subsurface air flow, increasing dispersivity which significantly affects the rate at which volatile contaminants escape from a layer. Reprinted from Auer et al. (1996), Copyright 1996, pg. 157, with permission from Elsevier Science.

In a recent study, Elberling et al. (1998) looked at the effect of barometric pumping on the transport of atmospheric oxygen within the unsaturated zone. A field study was carried out in a glacial till aquifer with a 10-12-m thick sand unsaturated zone. The sand deposit is overlain by up to 20 m of very low permeability clayey till, except for a small "geological window" where the clay cap has been eroded away, exposing a 30m by 20m outcrop area of relatively high permeability sand. The authors proposed that atmospheric pressure variations are propagated throughout the sandy unsaturated zone through this window, causing both vertical and horizontal pressure gradients and resulting in 2-dimensional airflow and transport of oxygen in the subsurface. Vadose zone pressure variations and oxygen concentrations in response to changing atmospheric pressure were also modeled using the numerical groundwater model SUTRA. The authors achieved good matches for both pressure and oxygen concentration modeling (Figures 7 and 8). They found a clear relationship between variations in atmospheric pressure and changes in the subsurface soil gas pressure. During a barometric pressure cycle, atmospheric oxygen was transported more than 10 md^{-1} horizontally in the clay-capped unsaturated zone.

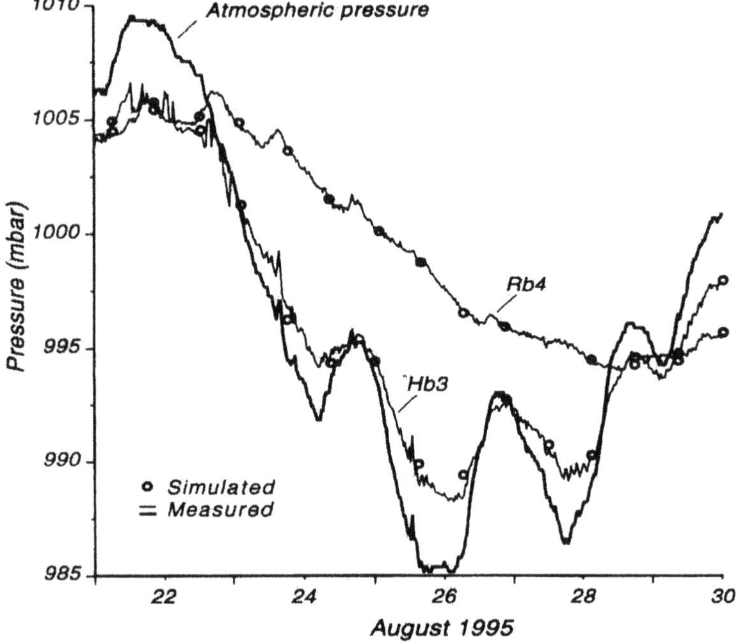

Figure 7. Changing subsurface pressures in response to falling atmospheric pressure at two boreholes in August 1995. Boreholes Hb3 and Rb4 are located 50 and 300 m from the geological window, respectively. The study by Elberling et al. assumes two-dimensional subsurface pressure gradients and air flow. Reprinted with permission from Elberling et al. (1998). Copyright 1998, American Geophysical Union.

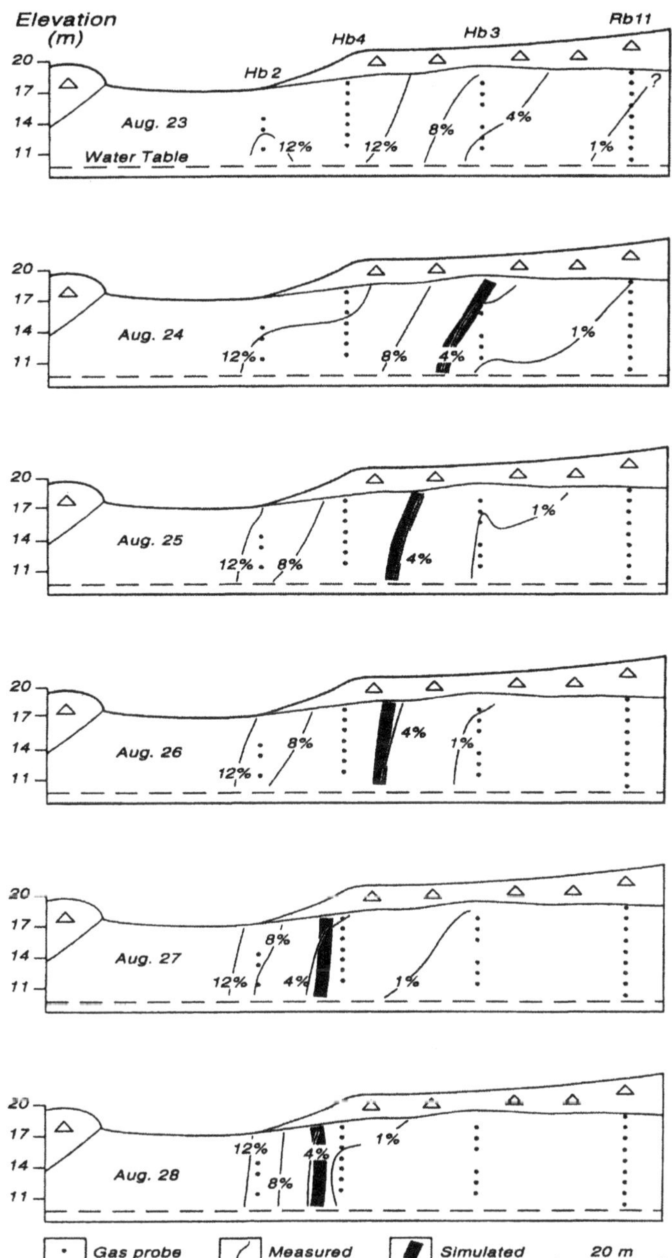

Figure 8. Measured and modeled results for O_2 volume % within the unsaturated zone, showing oxygen transport in response to barometric pressure changes. The simulated 4% vol O2 is shown as a solid band. Reprinted with permission from Elberling et al. (1998). Copyright 1998, American Geophysical Union.

Other researchers have indicated the possible significance of barometric-pumping-induced advective transport based on the discovery of non-linear vertical concentration profiles in the unsaturated zone. In an extensive year-long field study, Smith et al. (1996) measured total and diffusive vertical fluxes of trichloroethylene (TCE) from the unsaturated zone to the atmosphere above a contaminated water-table aquifer at Picatinny Arsenal, NJ. Total TCE flux measurements were made with a vertical flux chamber (VFC) while diffusive fluxes were estimated with Fick's law. Unsaturated-zone vapor-phase concentration profiles necessary for calculations using Fick's law were obtained by soil-gas sampling using nested vapor probes and adsorbent traps. Of the 29 different vertical profiles measured at 3 different locations during the study, a majority increased approximately linearly with depth. However, distinctly nonlinear concentration gradients were also observed. These researchers found that of the 15 attempts made to simultaneously measure diffusive and total fluxes, 7 were unsuccessful because of non-linear unsaturated-zone soil-gas concentration profiles. The remaining eight pairs of measurements were compared and it was found that the total and diffusive fluxes were of similar magnitude only three times. In most cases, the total flux was 1-4 orders of magnitude greater than the diffusive flux predicted by Fick's law. Using a spatial averaging technique, the researchers calculated that the average annual flux from the subsurface to the atmosphere using VFC measurements was approximately 50 kg/yr, which is comparable in magnitude to the TCE removal rate of approximately 70 kg/yr from a pump-and-treat system operated at the site over the same time period (Imbrigiotta et al. 1995; Smith et al. 1999). The total flux was significantly larger than the 0.05 kg/yr diffusive flux predicted from vapor probe measurements and Fick's law. These researchers concluded that the comparison of total and diffusive fluxes suggested that processes other than steady-state diffusion were occurring in the unsaturated zone and contributing to the transport of TCE vapors to land surface. One possibility cited by the researchers was the advection of TCE vapors in response to changes in atmospheric pressure.

Numerous studies have indicated the potential importance of atmospherically-induced subsurface pressure gradients on unsaturated-zone airflow and contaminant transport. While modeling studies have proven effective in gaining insight into the mechanisms of barometric pumping in the unsaturated zone (Auer et al. 1996; Elberling et al. 1998; Massman and Farrier 1992), actual field data on atmospheric and subsurface air pressures are necessary to calibrate and validate vadose-zone airflow models. Except for the specific study of 2-dimensional subsurface airflow caused by a "geological window" performed by Elberling et al. (1998), none of the research presented attempts to match modeling results with subsurface data collected during natural atmospheric pressure changes. Also, before conclusions can be drawn about the ability of flux-measuring technologies to accurately and precisely quantify solute transport from the unsaturated zone to land surface, rigorous laboratory testing of these instruments must be conducted.

3. MEASUREMENT AND SIMULATION OF SUBSURFACE AIR PRESSURE AT PICATINNY ARSENAL, NEW JERSEY

3.1 Description of Field Site

The field site for the air pressure studies was Picatinny Arsenal in Morris County, New Jersey, a U.S. Geological Survey Toxic Substances Hydrology Program National Research Site. The site is located in a glaciated valley on top of 50-65 m of stratified and unstratified drift, which overlies a weathered bedrock surface. The unconsolidated sediments form three major hydrogeologic units: a 15-21-m thick unconfined sand-and-gravel aquifer, an 8-21-m thick confining layer composed of fine sand, silt, and clay, and an 8-35-m thick confined sand-and-gravel aquifer (Smith et al. 1992). From 1960 to 1981, wastewater from metal-plating and degreasing operations was discharged into two unlined wastewater lagoons and an unlined overflow dry well adjacent to building 24 (Figure 9). As a result, several chlorinated organic compounds, including trichloroethene (TCE), have contaminated the unconfined sand and gravel aquifer that underlies the site. The unsaturated zone is 2-4 m thick over most of the study area (Smith et al. 1992) and TCE has been detected in the vadose-zone water and soil gas at the site (Smith et al. 1990)). This site was chosen for study because of extensive previous characterization of the unsaturated zone (Smith et al. 1990; Smith et al. 1996) and the presence of primarily a single organic contaminant. A location above the TCE-contaminated ground-water plume at Picatinny Arsenal was chosen (Figure 9) with an unsaturated zone composed of medium-to-course sand with low organic-carbon content ranging from 0.08% (Choi 1999) to 1.02% (Sahoo and Smith 1997) and a thickness of 2.2-2.3 m.

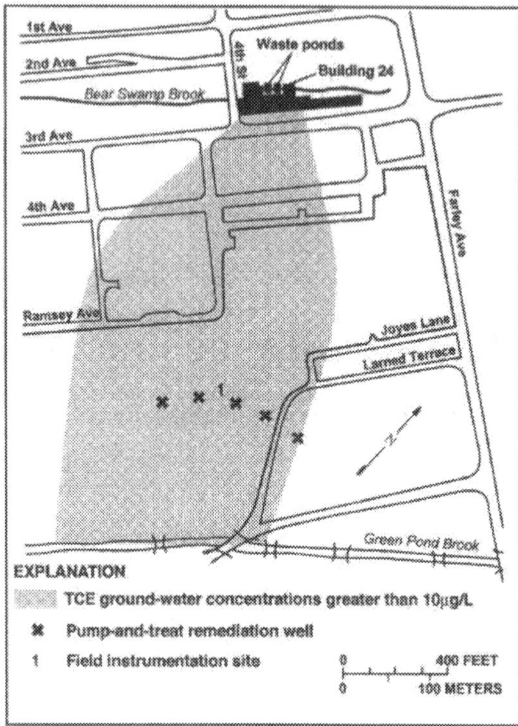

Figure 9. Map of field site at Picatinny Arsenal, NJ showing the location of 10 μg/L trichloroethene (TCE) plume boundary, subsurface air pressure and moisture content sampling site, and withdrawal wells for the pump-and-treat remediation system.

3.2 Air Pressure

Unsaturated-zone air pressure and temperature were measured using four unvented Geokon, Inc. model 4500LP vibrating-wire pressure transducers with a range of 0-10 psi (0-70,000 N/m^2). All the transducers are accurate to 0.1% of full scale, corresponding to 0.01 psi for the model 4500LP. Measurements were collected at several depths ranging from 0.33 m to 1.68 m (Table 1). The transducers were attached to standard AW-threaded drill rods and installed at the bottom of separate 7-cm diameter boreholes. The holes were filled with native soil around the transducer and a mixture of soil and bentonite from a distance of 30 cm above the transducer to land surface. A Campbell Scientific CR10 datalogger with an AM416 Relay Multiplexer was interfaced with the 4 pressure transducers to record temperature and pressure every 5 minutes. Atmospheric pressure at land surface was taken from the National Oceanic and Atmospheric Administration's National Data Center [http://www.nndc.noaa.gov/cgi-bin/nndc/buyOL-002.cgi]. This website contains an archive of surface climatological data from airports around the country. Surface data from Newark International Airport, 43.23 km (26.8 mi) away, was used.

Table 1. Air pressure measurement events.

Sampling Date	Field Location	Sampling Frequency (min)	Sampling Depths (m)
August 19-22, 1996	1	5.0	0.00, 0.49, 0.73, 1.01, 1.25, 1.68
October 5-8, 1996	1	5.0	0.00, 0.33, 0.65, 0.98, 1.28, 1.68

3.3 Moisture Content

Time domain reflectometry (TDR) was used to measure volumetric moisture content in the subsurface (Smith et al. 1996). The probes consist of an aluminum block stacked on top of a Delrin® block (6.99 cm by 3.81 cm by 1.27 cm thick) with three parallel, stainless steel prongs attached to the bottom of the Delrin® block. The prongs are 22.2 cm long and 6 mm in diameter and spaced 7.6 cm apart. Separate boreholes were hand augered and the three-prong probes were pushed into undisturbed soil at depths given in Table 2. The 7-cm diameter boreholes were filled with native soil. A Tektronix 1502B Metallic TDR cable tester was attached to the probes by 50-ohm coaxial cable and used to quantify their apparent length, which is used to calculate the apparent dielectic constant of the soil surrounding the probes. The apparent dielectric constant of the soil measured by the TDR probes was converted to volumetric moisture content by the empirical relationship given by Topp et al. (1980):

$$\theta_v = -5.3\times10^{-2} + 2.92\times10^{-2} K_a - 5.5\times10^{-4} K_a^{\,2} + 4.3\times10^{-6} K_a^{\,3} \quad (3)$$

where θ_v is volumetric moisture content and K_a is the apparent dielectric constant of the soil as read by the cable tester.

Table 2. Moisture content measurement events.

Sampling Date	Field Location	Sampling Frequency (hrs)	Sampling Depths (m)
August 19-22, 1996	1	4*	0.30, 0.61, 1.07, 1.52
October 5-8, 1996	1	4*	0.00, 0.30, 0.61, 1.07, 1.52

* TDR probes read by hand and irregularly over the sampling event.

3.4 Water Level

The water-table elevation was measured at the field-sampling site throughout the investigation. Measurements were collected with an electric water level meter at a monitoring well at location 1 (Figure 9) which was installed during previous work at Picatinny Arsenal. Depth to water table readings were measure from the top of the well casing and were taken approximately every 4 hours.

4. MODELING APPROACH

For geologic systems with air permeabilities greater than 10^{-10} cm^2, the air velocity, v, can be estimated by Darcy's Law (Massmann 1989):

$$v = \frac{-k}{\mu\theta_g}\frac{\partial P}{\partial z} \tag{4}$$

where v = air velocity (m s^{-1}), k = intrinsic air permeability (m^2), μ = air viscosity (kg m^{-1} s^{-1}), θ_g = volumetric gas content, P = air pressure (N m^{-2}) and z = depth (m). This formulation assumes that slip flow is negligible relative to viscous flow, which is a valid approximation for the pore sizes associated with the sandy porous media at Picatinny Arsenal.

The pressure gradient in equation (4) is obtained by solving the following linearized flow equation (Massman and Farrier 1992; Massmann 1989):

$$\theta_g \frac{\partial P}{\partial t} = \frac{P_o}{\mu}\frac{\partial}{\partial z}\left(k\frac{\partial P}{\partial z}\right) \tag{5}$$

where t = time (s), P_o = mean atmospheric pressure (N m^{-2}). If the soil gas pressure varies by less than 10,000 N m^{-2} throughout the flow field, then the error associated with using equation (5) rather than the nonlinear flow equation is less than 1% (Massman and Farrier 1992; Massmann 1989).

Gravity effects are not considered in the above equations as they are written in terms of pressure instead of potentiometric head. For a homogenous system with constant fluid density, (5) can be written as:

$$\frac{\partial P}{\partial t} = \beta \frac{\partial^2 P}{\partial z^2} \tag{6}$$

where $\beta = (kP_o)/(\theta_g \mu)$.

Solution of equation (6) requires two boundary conditions and an initial condition. The upper boundary condition at land surface is specified by the actual atmospheric pressure measurements at the field site. For the analytical solution presented in the following section, a sinusoidal pressure cycle at land surface was used as the upper boundary condition:

$$P(0,t) = \Delta P \sin\left(\frac{2\pi t}{T}\right) \tag{7}$$

The lower boundary condition is located at the top of the capillary fringe, which was approximated as the location of the water table. For a constant-position water table, the velocity of subsurface air at this boundary is necessarily zero. With consideration of equation (4), this gives rise to the following lower boundary condition:

$$\frac{\partial P}{\partial z}(L,t) = 0 \tag{8}$$

The initial conditions were typically a mean pressure throughout the depth of the unsaturated zone, but any type of initial pressure profile could be chosen as the initial condition.

A finite difference model was written to simulate the subsurface air pressure data collected at the field site in Picatinny Arsenal, NJ. The model first uses the linearized flow equation (6) to calculate the changes in subsurface air pressure with respect to time and depth for a given atmospheric pressure change. Having solved for air pressure throughout the system, the model then calculates soil-gas velocities according to Darcy's law using equation (4). Both equations are one-dimensional with a mobile gas phase and immobile aqueous and soil phases and were modeled in FORTRAN using the finite difference scheme described by Stone and Brian (1963).

5. MODEL VERIFICATION

An analytical solution was derived and used to calculate subsurface air pressure (equation 6) for a sinusoidal pressure fluctuation at the soil surface (equation 7), with the unsaturated zone consisting of homogenous soil. The initial condition was the mean air pressure throughout the unsaturated zone. Verification results are presented in the following section and the full derivation is presented in Appendix A.

The analytical solution is given by:

$$P(z,t) = \sum_{n=1}^{\infty} b_n(t)\phi_n(z) + \Delta P \sin(\omega t) \tag{9}$$

where $\omega = 2\pi/T$ and

$$b_n(t) = e^{-\lambda_n kt}\left(\frac{2P_o}{\left(n-\frac{1}{2}\right)\pi}\right) + \frac{2k\Delta P\sqrt{\lambda_n}}{L}\left(\frac{\frac{\sin(\omega t)}{a} - \frac{\omega}{a^2}\cos(\omega t) - \frac{\omega}{a^2}e^{-\lambda_n kt}}{1+\frac{\omega^2}{a^2}}\right)$$

$$-\frac{2\Delta P\omega}{\left(n-\frac{1}{2}\right)\pi}\left(\frac{\frac{\cos(\omega t)}{a} + \frac{\omega}{a^2}\sin(\omega t) - \frac{1}{a}e^{-\lambda_n kt}}{1+\frac{\omega^2}{a^2}}\right) \tag{10}$$

and

$$\phi_n(z) = \sin\left(\frac{\left(n-\frac{1}{2}\right)\pi z}{L}\right) \qquad n = 1,2,3...$$

$$\lambda_n = \left(\frac{\left(n - \frac{1}{2}\right)\pi}{L} \right)^2 \qquad n = 1,2,3...$$

6. RESULTS AND DISCUSSION

6.1 Field Data Collection

Air pressure, moisture content, and water-table elevation were monitored in August and October 1996 at Picatinny Arsenal, New Jersey. Figure 10 shows land-surface and unsaturated-zone air pressures (at three depths) as a function of time for the August and October 1996 sampling events. Based on the surface pressure data, it is apparent that no sudden, large pressure changes occurred during the August 1996 sampling event. This is consistent with the observation that no major regional storms entered or left the area during this time. During the October 1996 measurements, a low pressure front entered the area, dropping the surface pressure approximately 2700 N/m^2 over three days. The measured air pressures in the unsaturated zone for both sampling events are relatively constant with depth for any given time. The small differences measured between subsurface pressure transducers are, for the most part, less than the accuracy of the transducer (70 N/m^2), suggesting that any real pressure differences between these sampling locations cannot be reliably quantified by the pressure transducers used in this study. However, during most of the sampling periods, measurable pressure differences existed between the subsurface and the atmosphere. Both positive and negative pressure gradients were observed.

Moisture content (Figure 11) remained relatively constant in time for any given depth for both the August and October sampling events, which is consistent with the observation that there was no precipitation during the

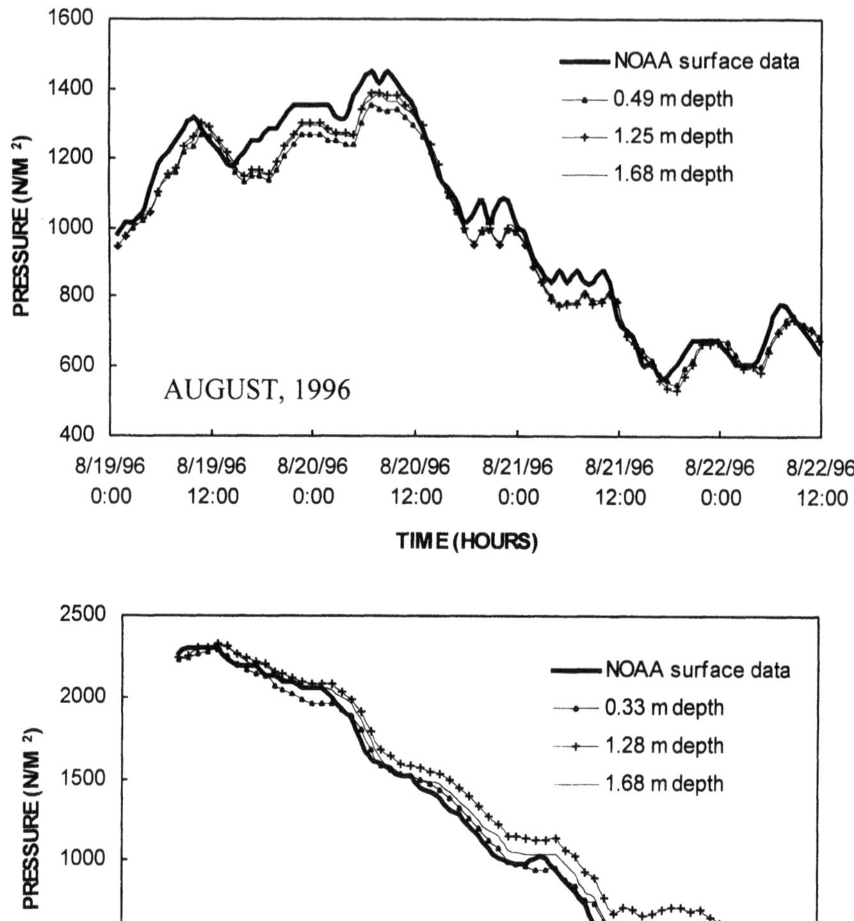

Figure 10. Air pressure at land surface and three depths in the unsaturated zone as a function of time for the August and October 1996 sampling events. Symbols are only used to help distinguish between lines and do not necessarily represent actual data points. Actual data points were collected at 5-minute intervals for all subsurface depths and hourly for surface pressure.

data collection periods. The highest moisture contents were observed within the top 0.30 m of the unsaturated zone. This is probably due to the finer-grained soil in this region compared to deeper soil. Smith et al. (1990) have reported that surficial soil from the site (0.0 to 0.01-m depth) has a higher silt and clay content than deeper unsaturated-zone soil (1.5 to 2.0-m depth).

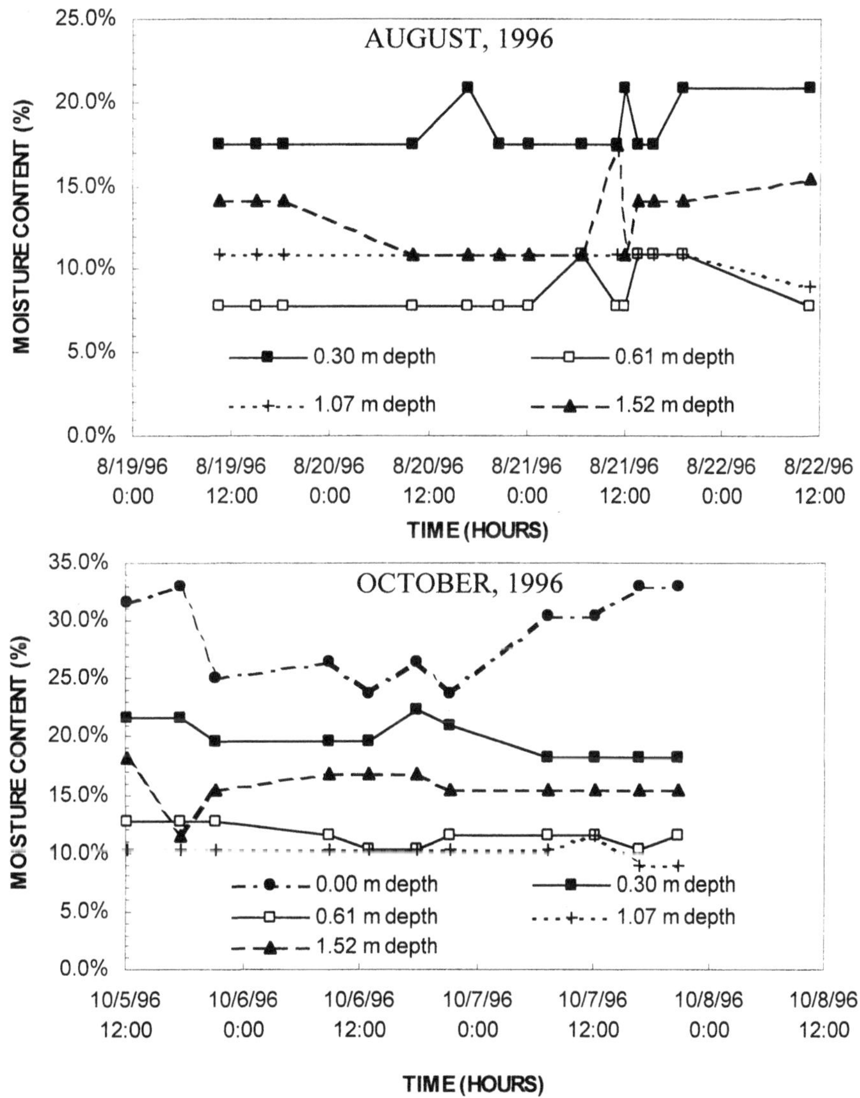

Figure 11. Moisture content as a function of time for the August and October 1996 sampling events.

The unsaturated zone airflow model described previously assumes that the water table is at a constant position throughout the simulation. Under certain circumstances it is possible that soil gas pressure gradients may result from a fluctuating water table. If the water table rises significantly during a short period of time, unsaturated zone soil gas just above the capillary fringe will be temporarily under higher pressure than the rest of the soil profile. This higher pressure will propagate with a time lag to the surface, thus "breathing out" soil gas. Surface air may be introduced into the subsurface during a similarly falling water table. However, during the August and October 1996 soil air pressure sampling events, the depth to the water table (Figure 12) averaged 2.26 m and 2.25 m respectively, with standard deviations of 0.02 m for both sampling events, thus validating the constant-water-table-position assumption.

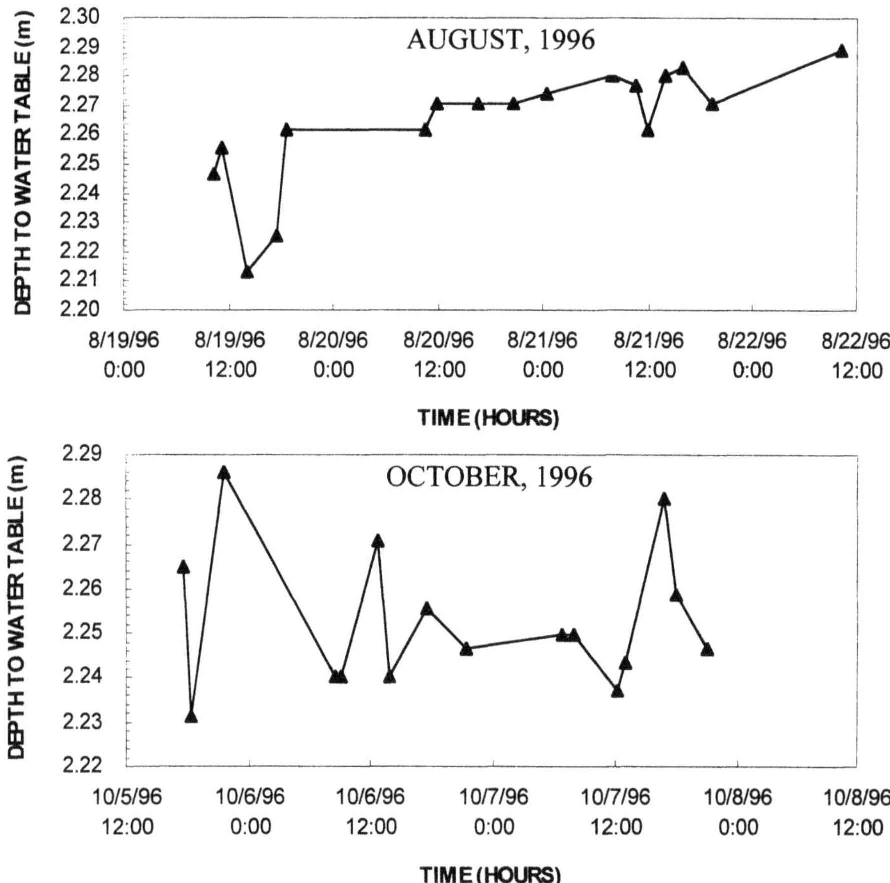

Figure 12. Water-table position during the August and October 1996 sampling events.

6.2 Modeling Results and Discussion

The finite difference air pressure model was compared with the analytical solution for depths of 1.0 and 2.0 m (Figure 13). As can be seen, the model matches the analytical solution to subsurface air pressure very well, thus verifying the accuracy of the finite difference model.

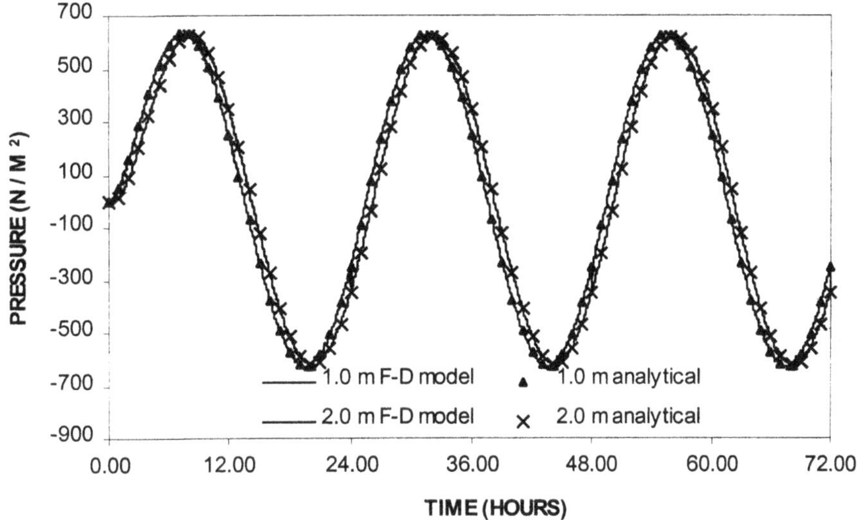

Figure 13 Simulations of air pressure at depths of 1.0 m and 2.0 m for the analytical solution and the finite difference model.

The subsurface air pressure measurements collected during the August 19-22 and October 5-8, 1996 were simulated using the finite difference model described previously. The surface pressure measured by NOAA at Newark International Airport, NJ during the sampling events was used as the upper boundary condition for these simulations. For both events, atmospheric pressure measured at the start of the simulation was assigned throughout the unsaturated zone as the initial condition. The volumetric gas content, θ_g, was based on the moisture content measured by the TDR probes assuming a constant porosity of 30%. Air permeability, k, was used as a calibration parameter using Powell's direction set method for parameter optimization (Press et al. 1986). This technique minimizes the squared normalized differences between the observed and modeled concentrations. The merit function, or error, is defined as the sum of the square of the normalized deviation and is mathematically given as:

$$Merit\ function = \sum \left[\frac{(data - simulation\ result)}{data} \right]^2 \quad (11)$$

This merit function ensures that the percentage error at each comparison point is given equal weight.

Figure 14 shows field data and model results of subsurface air pressure for two depths for the August and October 1996 sampling events. Surface air pressure is also presented in the figure. An optimized fit of the August subsurface air pressure was achieved with a permeability value of 2.73×10^{-14} m^2. For the October data, an optimized permeability value of 7.02×10^{-15} m^2 was used for the simulation. The lower air permeability for the October simulation as compared with the August simulation agrees with a measured increase in moisture content during the latter period (Figure 12). Increased moisture content in the soil profile indicates unsaturated-zone void space being filled by water, thus reducing its permeability to air. As evidenced by the simulation results for both sampling events, the 1-dimensional linear model was able to accurately describe the measured subsurface air pressure using atmospheric pressure as the upper boundary condition. This result implies that the water table movement measured during the sampling events had no effect on subsurface pressure.

Modeling results of subsurface pressure gradients were used to simulate subsurface soil gas velocity throughout the unsaturated zone profile. Figure 15 shows vertical profiles of unsaturated-zone air velocities for 12-hr time periods for August and October 1996. Results show that subsurface airflow is almost never zero, as is assumed in a diffusion-only transport model. Air-phase solute transport models based solely on diffusion would therefore not be able to accurately predict contaminant flux from the subsurface.

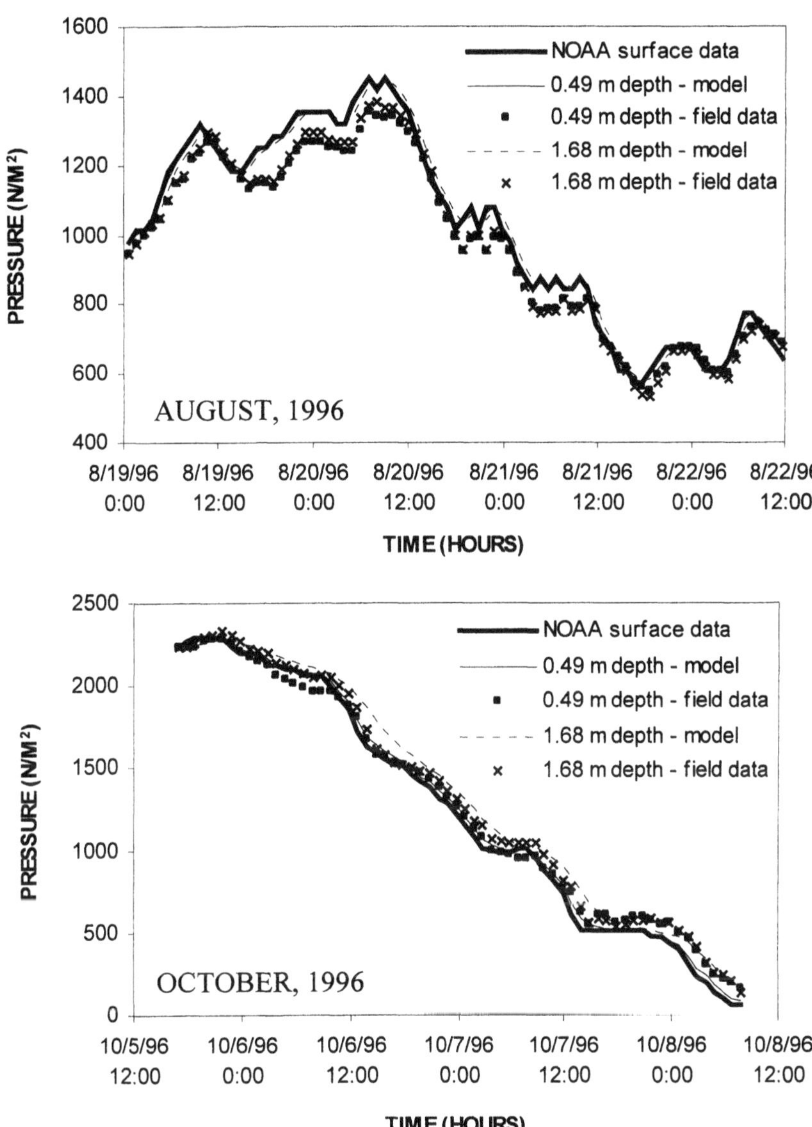

Figure 14. Measured and model-simulated air pressures as a function of time for two depths for the August and October 1996 sampling events. A single air permeability was used for the simulation. Symbols are only used to help distinguish between lines and do not necessarily represent actual data points. Actual data points were collected at 5-minute intervals for subsurface pressure and hourly for surface pressure.

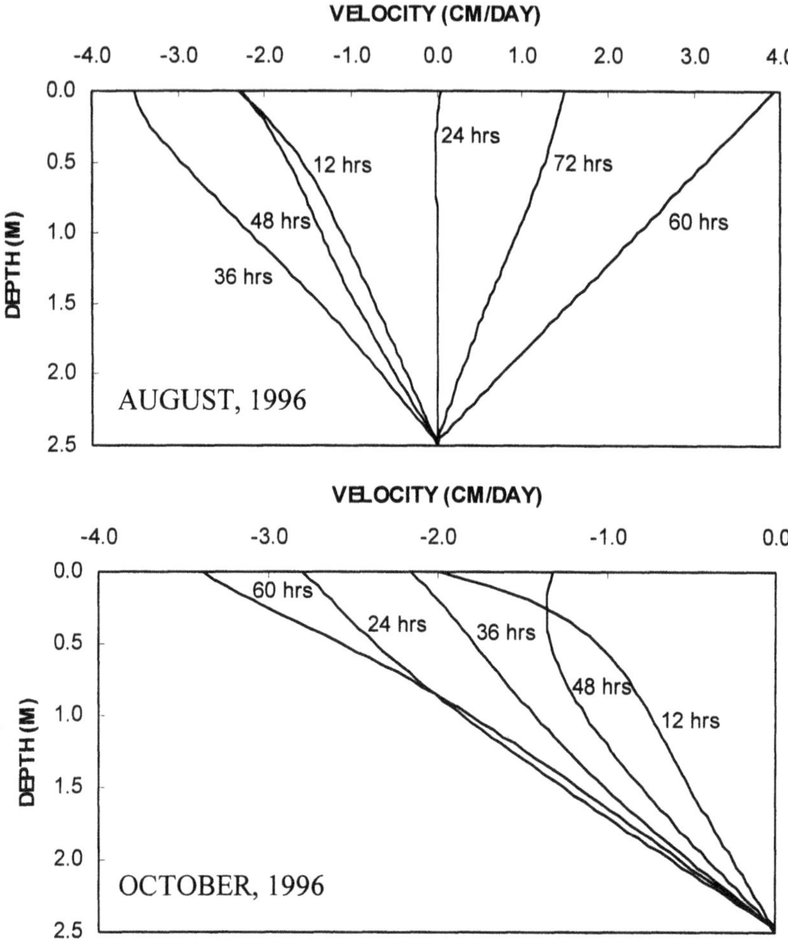

Figure 15. Simulated subsurface air velocities for August and October 1996 sampling events. Hours indicate time from start of data collection (8/19/96 0:00 and 10/5/96 17:50 for August and October sampling periods, respectively). Air flow from the subsurface toward land surface is presented as negative velocities, with downward air flow indicated by positive values. All velocities during the October 1996 time period are in the upward direction because of a falling surface pressure throughout the simulation.

The unsaturated zone airflow model is capable of addressing other issues occurring in a natural subsurface environment including heterogeneity and infiltrating rainwater. The model allows the subsurface to be divided into layers containing homogenous properties in order to simulate the effect heterogeneity plays in unsaturated-zone airflow. Each layer is assigned a single permeability value, then the entire subsurface is modeled as before. Infiltrating rainwater serves to increase the unsaturated-zone moisture content resulting in a decrease in volumetric gas content. As shown by equation 4, this decrease in gas content would produce an increase in subsurface air velocity for the same pressure gradient. It has been shown, however, that infiltrating rainwater may cause partitioning of the gas-phase VOC into the water phase, which is subsequently carried down to the water table (Cho et al. 1993). Therefore, increases in subsurface air velocities caused by infiltrating rainwater may not result in an increase in VOC flux from the subsurface to the atmosphere.

Failing to incorporate soil-gas advection induced by barometric pumping into gas-phase subsurface transport models may, under certain conditions, under predict contaminant flux to the atmosphere. As previously described, Smith et al. (1996) compared TCE vapor fluxes measured with a chamber device to TCE in groundwater being removed by a pump-and-treat system and discharge into a surface-water receiving body at the same site. These researchers found VOC removal rates by flux to the atmosphere comparable in magnitude to both of the other attenuation pathways.

7. CONCLUSIONS

The effect of atmospheric pressure changes on subsurface air velocity and contaminant transport has received relatively little attention in scientific literature, probably due to its perceived lack of importance in relation to other transport mechanisms. Research was presented that indicates that, under certain circumstances, air-phase contaminant transport in the subsurface is influenced by barometric pressure changes at land surface. Combined with the large surficial area of many contaminated ground-water plumes and relatively high Henry's law constants of many common ground-water pollutants, this natural atmospheric-pressure-fluctuation-enhanced transport of unsaturated-zone vapor-phase contamination may present a significant path of contaminant mass loss to the atmosphere. New data was presented that show pressure gradients resulting from natural pressure fluctuations exist in the unsaturated zone. A mathematical model involving simplifying assumptions was presented to describe these pressure gradients. The model was validated against an analytical solution that was developed. The linear model fit the subsurface pressure field data well using a 1-

dimensional formulation with atmospheric pressure as the only driving force for pressure change. It was therefore inferred that water-table movement had little or no influence on air pressure in the unsaturated zone over the simulation period. Further model results indicated nonzero subsurface air velocities, with implications for advective transport of air-phase TCE at the site.

BIBLIOGRAPHY

Auer, L. H., Rosenberg, N. D., Birdsell, K. H., and Whitney, E. M. (1996). "The effects of barometric pumping on contaminant transport." *Journal of Contaminant Hydrology*, 24, 145-166.

Batterman, S. A., McQuown, B. C., Murthy, P. N., and McFarland, A. R. (1992). "Design and evaluation of a long-term soil gas flux sampler." *Environmental Science and Technology*, 26(4), 709-714.

Chen, C., Green, R. E., Thomas, D. M., and Knuteson, J. A. (1995). "Modeling 1,3-dichloropropene fumigant volatilization with vapor-phase advection in the soil profile." *Environmental Science and Technology*, 29(7), 1816-1821.

Cho, H. J., Jaffé, P. R., and Smith, J. A. (1993). "Simulating the volatilization of solvents in unsaturated soils during laboratory and field infiltration experiments." *Water Resources Research*, 29, 3329-3342.

Choi, J.-W. (1999). "Effect of TX-100 on TCE Degradation," Master of Science, University of Virginia, Charlottesville.

Clements, W. E., and Wilkening, M. H. (1974). "Atmospheric Pressure Effects on ^{222}Rn Transport Across the Earth-Air Interface." *Journal of Geophysical Research*, 79(33), 5025-5029.

Elberling, B., Larsen, F., Christensen, S., and Postma, D. (1998). "Gas transport in a confined unsaturated zone during atmospheric pressure cycles." *Water Resources Research*, 34(11), 2855-2862.

Holford, D. J., Schery, S. D., Wilson, J. L., and Phillips, F. M. (1993). "Modeling radon transport in dry, cracked soil." *Journal of Geophysical Research*, 98(B1), 567-580.

Imbrigiotta, T. E., Ehlke, T. A., Martin, M., Koller, D., and Smith, J. A. (1995). "Chemical and biological processes affecting the fate and transport of trichloroethylene in the subsurface at Picatinny Arsenal, New Jersey." *Hydrological Science and Technology*, 11(1-4), 26-50.

Kerfoot, H. B. (1987). "Soil-gas measurement for detection of groundwater contamination by volatile organic compounds." *Environmental Science and Technology*, 21, 1022-1024.

Marrin, D. L., and Kerfoot, H. B. (1988). "Soil-gas surveying techniques." *Environmental Science and Technology*, 22, 740-745.

Massman, J., and Farrier, D. F. (1992). "Effects of atmospheric pressures on gas transport in the vadose zone." *Water Resources Research*, 28(3), 777-791.

Massmann, J. W. (1989). "Applying Groundwater Flow Models in Vapor Extraction System Design." *Journal of Environmental Engineering*, 115(1), 129-149.

Miller, A., and Thompson, J. C. (1975). *Elements of Meterology*, Charles E. Merrill Publishing Co., Columbus, Ohio.

NOAA. (1999). "National Oceanic and Atmospheric Administration National Data Center." http://www.nndc.noaa.gov/cgi-bin/nndc/buyOL-002.cgi.

Press, W. H., Flannery, B. P., Teukolsky, S. A., and Vetterling, W. T. (1986). *Numerical Recipes*, Cambridge University Press, Cambridge.

Sahoo, D., and Smith, J. A. (1997). "Enhanced trichloroethene desorption from long-term contaminated soil using Triton X-100 and pH increases." *Environmental Science and Technology*, 31(7), 1910-1915.

Schery, S. D., and Gaeddert, D. H. (1982). "Measurements of the effect of cyclic atmospheric pressure variation on the flux of ^{222}RN from the soil." *Geophysical Research Letters*, 9(8), 835-838.

Schery, S. D., Holford, D. J., Wilson, J. L., and Phillips, F. M. (1988). "The flow and diffusion of radon isotopes in fractured porous media: Part 2, Semi-infinite media." *Radiation Protection Dosimetry*, 24(1/4), 191-197.

Smith, J. A., Chiou, C. T., Kammer, J. A., and Kile, D. E. (1990). "Effect of soil moisture on the sorption of trichloroethene vapor to vadose-zone soil at Picatinny Arsenal, New Jersey." *Environmental Science and Technology*, 24(5), 676-683.

Smith, J. A., Cho, H. J., MacLeod, C. L., and Koehnlein, S. A. (1992). "Sampling unsaturated-zone water for trichloroethene at Picatinny Arsenal, New Jersey." *Journal of Environmental Quality*, 21(2), 264-271.

Smith, J. A., Katchmark, W., Choi, J.-W., and Fred D Tillman, Jr. "Unsaturated Zone Air Flow at Picatinny Arsenal, New Jersey: Implications for Natural Remediation of the Trichloroethylene-Contaminated Aquifer." *U.S. Geological Survery Toxic Substances Hydrology Program - Proceedings of the Technical Meeting*, Charleston, South Carolina, 625-634.

Smith, J. A., Tisdale, A. K., and Cho, H. J. (1996). "Quantification of natural vapor fluxes of trichloroethene in the unsaturated zone at Picatinny Arsenal, New Jersey." *Environmental Science and Technology*, 30(7), 2243-2250.

Stone, H. L., and Brian, P. L. T. (1963). "Numerical solution of convective transport problems." *American Institute of Chemical Engineering Journal*, 9(5), 681-688.

Swett, G. H., and Rapaport, D. (1998). "Natural Attenuation: Is the Fit Right?" *Chemical Engineering Progress*, 94(6), 37-43.

Topp, G. C., Davis, J. L., and Annan, A. P. (1980). "Electromagnetic determination of soil water content: Measurements in coaxial transmission lines." *Water Resources Research*, 16(3), 574-582.

Weeks, E. P. (1978). "Field Determination of Vertical Permeability to Air in the Unsaturated Zone." *U.S. Geological Survey Professional Paper*, 1051, 1-41.

APPENDIX

ANALYTICAL SOLUTION FOR SUBSURFACE AIR PRESSURE

This appendix presents derivation of the analytical solution to the linearized, transient, one-dimensional unsaturated zone air-pressure equation with a no-flow lower boundary condition and sinusoidal pressure upper boundary condition for a homogenous porous medium.

The partial differential equation is equal to:

$$v(z,t) = \sum_{n=1}^{\infty} b_n(t)\phi_n(z); P = P(z,t) \quad \text{(A1)}$$

where k = constant. The boundary conditions are:

$$P(0,t) = \Delta P \sin\left(\frac{2\pi t}{T}\right) \quad \text{(A2)}$$

$$\frac{\partial P}{\partial z}(L,t) = 0 \quad \text{(A3)}$$

The initial condition is:

$$P(z,0) = P_o \quad \text{(A4)}$$

A reference pressure distribution that satisfies the given non-homogenous boundary conditions (A2) and (A3) is:

$$r(0,t) = A(t) = \Delta P \sin(\omega t) \quad \text{(A5)}$$

$$\frac{\partial r}{\partial z}(L,t) = 0 \quad \text{(A6)}$$

where $\omega = 2\pi/T$

The simplest function is $r(z,t) = A(t)$ (A7)
Define: $v(z,t) = P(z,t) - r(z,t)$ (A8)
Then v satisfies the homogenous boundary conditions:

$$v(z,t) = 0 \quad \text{(A9)}$$

$$\frac{\partial v}{\partial z}(L,t) = 0 \quad \text{(A10)}$$

Write: $P(z,t) = v(z,t) + r(z,t) = v(z,t) + A(t)$ (A11)

and substitute (A11) into the partial differential equation (A1).

$$\frac{\partial v}{\partial t} = k\frac{\partial^2 v}{\partial z^2} - A'(t) \tag{A12}$$

Initial condition: $v(z,0) = P(0,t) - A(0) = P_o$ (A13)

Solve the homogenous problem related to equations (A9), (A10), and (A12) to find the eigenvalues:

$$\lambda_n = \left(\frac{\left[n - \frac{1}{2}\right]\pi}{L}\right)^2 \quad n = 1,2,3... \tag{A14}$$

and the eigenfunction:

$$\phi_n(z) = \sin\left(\frac{\left[n - \frac{1}{2}\right]\pi z}{L}\right) \quad n = 1,2,3... \tag{A15}$$

Find a solution to equation (A12) in the form:

$$v(z,t) = \sum_{n=1}^{\infty} b_n(t)\phi_n(z) \tag{A16}$$

Then:

$$\frac{\partial v}{\partial t} = \sum_{n=1}^{\infty} \frac{\partial b_n(t)}{\partial t}\phi_n(z) \tag{A17}$$

$$\sum_{n=1}^{\infty} \frac{\partial b_n}{\partial t}\phi_n(z) = k\frac{\partial^2 v}{\partial z^2} - A'(t) \tag{A18}$$

$$\frac{\partial b_n}{\partial t} = \frac{\int_0^L \left(k\frac{\partial^2 v}{\partial z^2} - A'(t) \right) \phi_n(z)\partial z}{\int_0^L \phi_n^2(z)\partial z} \tag{A19}$$

Integrate equation (A19):

$$\frac{\partial b_n}{\partial t} = \frac{2k}{L}\sqrt{\lambda_n}\Delta P \sin \omega t - \lambda_n k b_n - \frac{2\Delta P}{\left(n-\frac{1}{2}\right)\pi}\omega \cos \omega t \tag{A20}$$

$$\frac{\partial b_n}{\partial t} + \lambda_n k b_n = 2\Delta P \left(\frac{k\sqrt{\lambda_n}}{L}\sin \omega t - \frac{\omega}{\left(n-\frac{1}{2}\right)\pi}\cos \omega t \right) \tag{A21}$$

Use an integrating factor $e^{-\lambda_n kt}$:

$$b_n(t) = e^{-\lambda_n kt} b_n(0) + \frac{2k\Delta P\sqrt{\lambda_n}}{L}\int_0^t e^{\lambda_n k\tau} \sin \omega\tau \partial\tau$$

$$- \frac{2\Delta P\omega}{\left(n-\frac{1}{2}\right)\pi} e^{-\lambda_n kt} \int_0^t e^{\lambda_n k\tau} \cos \omega\tau \partial\tau \tag{A22}$$

Let $a = \lambda_n k$ and solve equation (A22) using (A16) and initial condition in (A13) to solve for $b_n(0)$:

$$b_n(t) = e^{-\lambda_n kt}\left(\left(\frac{2P_o}{\left(n-\frac{1}{2}\right)\pi}\right) + \frac{2k\Delta P\sqrt{\lambda_n}}{L}\left(\frac{\frac{\sin(\omega t)}{a} - \frac{\omega}{a^2}\cos(\omega t) - \frac{\omega}{a^2}e^{-\lambda_n kt}}{1+\frac{\omega^2}{a^2}}\right)\right.$$

$$\left. - \frac{2\Delta P\omega}{\left(n-\frac{1}{2}\right)\pi} + \left(\frac{\frac{\cos(\omega t)}{a} + \frac{\omega}{a^2}\sin(\omega t) - \frac{1}{a}e^{-\lambda_n kt}}{1+\frac{\omega^2}{a^2}}\right)\right) \qquad \text{(A23)}$$

Using equations (A11), (A15), (A16), and (A23), the solution is equal to:

$$P(z,t) = \sum_{n=1}^{\infty} b_n(t)\phi_n(z) + A(t) \qquad \text{(A24)}$$

Chapter Fifteen

High-Vacuum Soil Vapor Extraction in a Silty-Clay Vadose Zone

Richard E. Meixner[1], Richard Heibel[2], and James E. Sadler[3]

[1] *Daniel B. Stephens & Associates, Inc., Albuquerque, NM*
[2] *New Mexico State Highway and Transportation Department, Santa Fe, NM*
[3] *AcuVac Remediation Inc., Houston, TX*

Key words: soil vapor extraction, clay, remediation, bioventing, air sparging

Abstract: A high-vacuum soil vapor extraction (SVE) system was successfully used as part of a sequence of corrective action technologies at an underground storage tank site at which diesel and gasoline had leaked to the silty-clay vadose zone soils and underlying perched aquifer. The efficiency of the system was continually assessed, and operational modifications were made to optimize system performance. The use of and adjustments to the high-vacuum SVE system removed much of the large mass of phase-separated hydrocarbon that had collected on the perched groundwater surface, successfully preparing the site for bioventing associated with air sparging. Decreasing concentrations of the regulated contaminants, especially benzene, attest to the effectiveness of this step in the sequence of corrective action technologies used at this site.

1. INTRODUCTION

This chapter presents the history of remediation activities at an underground storage tank (UST) site where high-vacuum soil vapor extraction (SVE) was used to remove a large mass of phase-separated hydrocarbon (PSH) and improve observed site conditions. The successful operation of the high-vacuum SVE system to meet New Mexico air quality discharge requirements prepared the subsurface for bioventing, the next in the sequence of overlapping remedial technologies that have been used at this site. This success also suggests that high-vacuum SVE may be a useful option at other sites where fine-grained sediments are contaminated with petroleum hydrocarbons or other suitably volatile compounds, such as some chlorinated solvents.

The vadose zone soils at the New Mexico State Highway and Transportation Department (NMSHTD) District 2 Maintenance Patrol Yard in Artesia, New Mexico (Artesia Yard) consist primarily of massive to poorly stratified silty clay, clay, or clayey silt. Soils of such low permeability ($<10^{-10}$ square centimeters) are generally not amenable to the use of SVE (U.S. EPA 1995). However, as this case history shows, operation of a high-vacuum SVE system, with periodic monitoring and adjustment, can be very successful in removing sizeable secondary sources of petroleum hydrocarbons (both PSH and residual soil contamination) from the subsurface.

2. BACKGROUND

The town of Artesia is located in the southeast corner of New Mexico (Fig. 1). The Artesia Yard, which is used for a variety of road repair and maintenance activities, is located along the west side of town on flat terrain that slopes from west to east toward the Pecos River.

On May 17, 1993, two underground storage tanks (USTs) at the Artesia Yard failed tank-tightness tests. A representative of the New Mexico Environment Department (NMED) confirmed the release on August 2, 1993, when the tanks were removed (Fig. 2). In addition to removing the tanks and associated piping, which were the primary sources of contamination, the excavation contractor removed approximately 100 yards of hydrocarbon-contaminated soil from the former tankhold.

The initial consultant for the NMSHTD conducted an investigation in the fall of 1993, installing seven soil borings to determine the extent of contamination (INTERA/BAI 1994a). The supervising geologist described the sediments as massive to poorly stratified, silty clay/clayey silt resting on a thick, hard clay that occurred between about 85 and 96 feet below ground surface (bgs). The borings intersected a zone of perched groundwater extending from about 60 feet to about 80 feet bgs, and three of the borings were converted to perched-zone monitor wells. The supervising geologist also observed occasional caliche layers between the surface and the perched groundwater table.

Figure 1. Site location

After the primary source (the underground tanks and piping) and accessible contaminated soil were removed, recovery of the PSH that had collected in the perched-zone wells was initiated. Due to the fortuitous presence of the perched aquifer (the regional groundwater occurs at about 150 feet bgs), the PSH pool occurred between 55 and 60 feet bgs in the alluvial sediments. Between November 1993 and February 1994, NMSHTD personnel removed PSH by hand bailing the three monitor wells, recovering about 400 gallons of PSH from the perched groundwater surface.

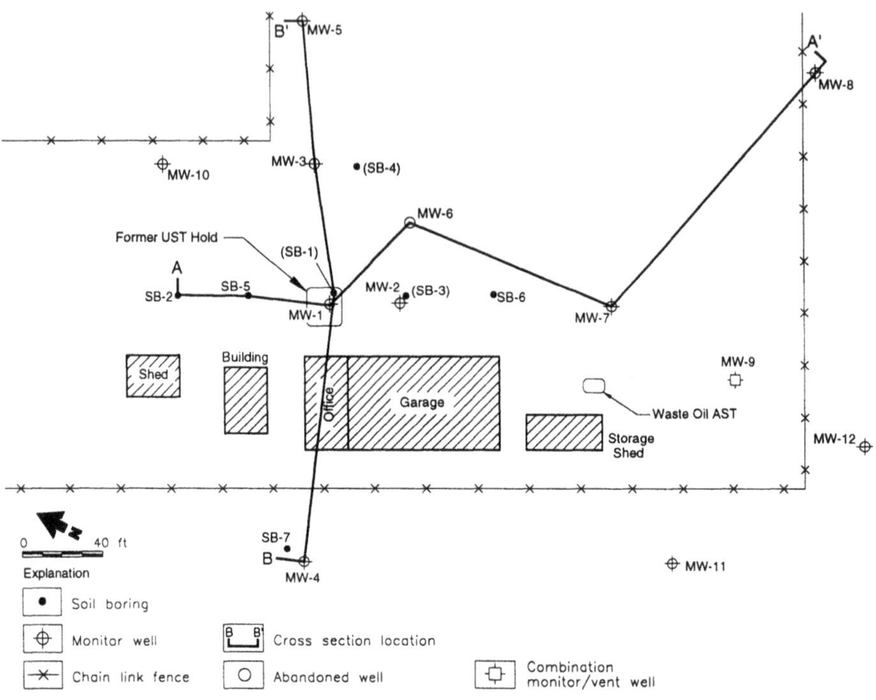

Figure 2. Site layout

In January 1994, personnel from INTERA/BAI installed five additional monitor wells to further define the extent of soil and perched groundwater contamination (one well installed below the hard clay perching layer remained dry and was abandoned). They discovered that the petroleum hydrocarbons had penetrated the vadose zone down to the perched water table, where the PSH then spread laterally (Figs. 3 and 4). The results of four slug tests suggested that the hydraulic conductivity ranged from about 2.7×10^{-5} to 1.8×10^{-6} centimeters per second. The actual values are probably lower, given the fine-grained nature of the alluvial deposit, as described by the field geologist.

In late April 1994, one additional monitor well (MW-9) was installed in an attempt to define the southern boundary of the contaminant plume. The consultant also conducted a pilot test to determine the design parameters for an SVE system (INTERA/BAI 1994b). The pilot test was done using 50 inches water column (wc) vacuum, resulting in a well flow rate of 3 cubic feet per minute (cfm) and extracted vapor concentrations of less than 1,000 parts per million volume (ppmv). The radius of influence (ROI) suggested by the test results ranged from about 6.8 to 8.6 feet, which would result in an extraction well spacing of 10 to 15 feet.

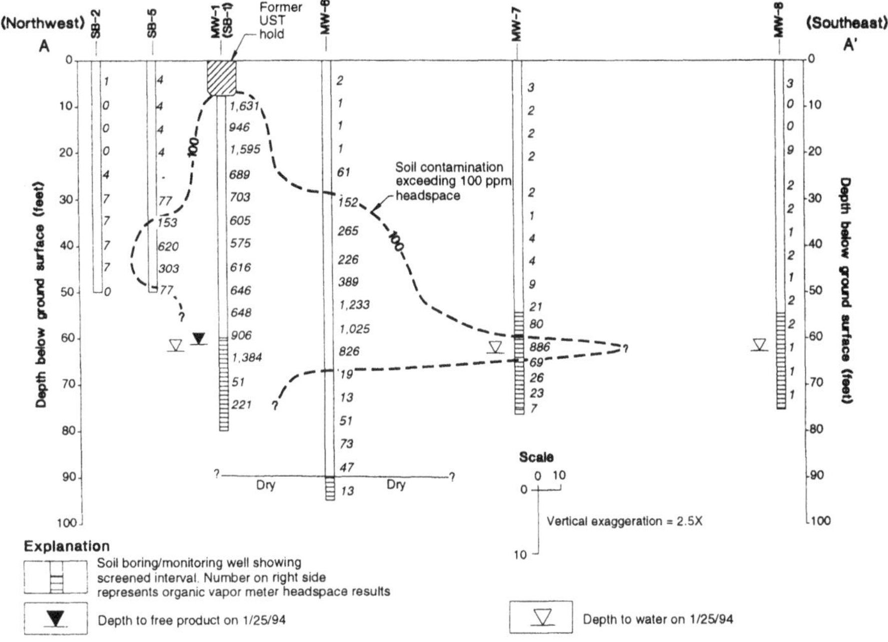

Figure 3. Cross section A-A'

3. HIGH-VACUUM SOIL VAPOR EXTRACTION SYSTEM

In September 1994, the NMSHTD retained a second consultant, Daniel B. Stephens & Associates, Inc. (DBS&A) to design and implement a corrective action system at the Artesia Patrol Yard. Sections 3.1 and 3.2 outline DBS&A's rationale for and design of the high-vacuum SVE system for the Artesia Yard. The techniques used to enhance the recovery of petroleum vapors from the vadose zone are described in Sections 3.3 and 3.4. These techniques extended the operation of the high-vacuum SVE system and allowed a more effective cleanup of the vadose zone in preparation for the current phase of bioventing.

3.1 Evaluation for High-Vacuum Soil Vapor Extraction

Although the Artesia Yard was not an ideal site for the use of SVE technology, it was the most reasonable alternative, given the depth of contamination and the estimated volume of product that had been released. Excavation to the contamination depth of 55 or 60 feet bgs was not attractive from an operational or economic standpoint. Direct recovery of PSH was a very slow process (hand bailing was tried for a period of time, with limited success), especially for the conservatively estimated 50,000 gallons of gasoline that had been released over the years at this site.

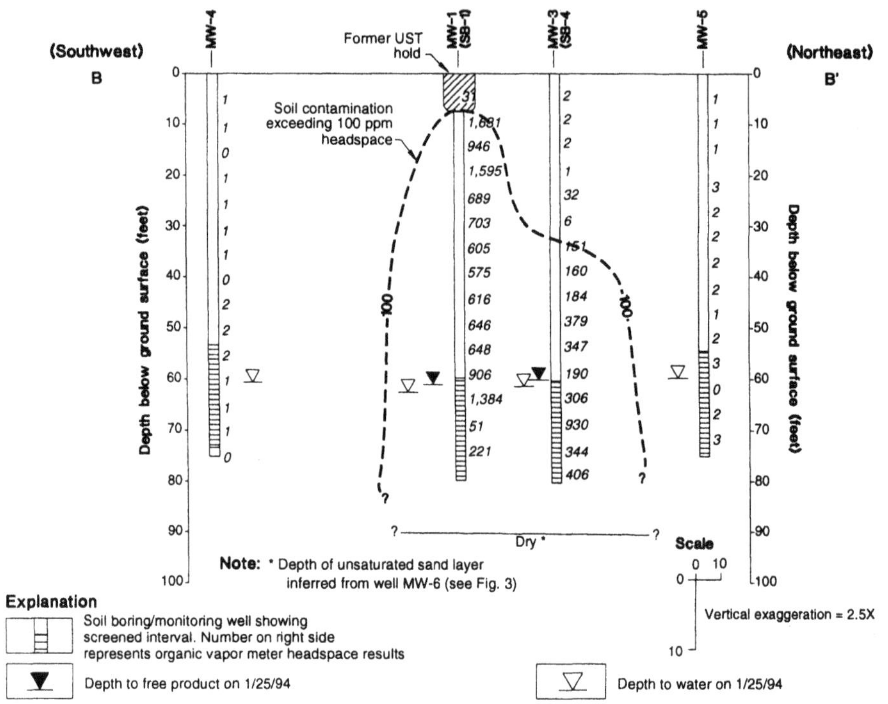

Figure 4. Cross section B-B'

Accordingly, in preparation for designing the corrective action system, DBS&A carefully evaluated the existing SVE pilot test data. If the design parameters suggested by this test had been used, an inordinately large number of SVE wells (more than 100) would have been required for the site. Therefore, DBS&A elected to conduct a high-vacuum (75 to 200 inches wc) SVE pilot test prior to developing a design for the corrective action system.

DBS&A also defined the southern extent of perched groundwater with the installation of a dry borehole south of the site. Field measurements in the existing monitor wells showed that an extensive layer (up to 4 feet thick in the well bores) of PSH remained on the perched water table (Fig. 5).

AcuVac Remediation Inc. of Houston, Texas (AcuVac) conducted the high-vacuum (up to 205 inches wc) pilot tests. These tests yielded well flows of 12 to 18 cfm and extracted vapor concentrations in the tens of thousands of ppmv. The ROI was calculated using a rearranged equation from Johnson et al. (1990, Eq. 22) which is based on the vacuums measured at the extraction and the observation wells and the distance between them. The ROI can also be approximated using data from all observations wells. In this case, at a given imposed vacuum (and flow rate), the vacuum measurements (inches wc) in observation wells were plotted versus distance from the extraction well. The distance at which the imposed vacuum fell to between 0.25 and 0.30 inch wc was considered to be the ROI. This distance is larger for the higher vacuums used in these tests and resulted in an effective ROI in excess of 25 feet. These results suggested that a high-vacuum SVE system could successfully remove volatile contaminants from the fine-grained soils at the site. Using a conservative (but still larger than the original 8.6 feet) ROI of 20 feet, DBS&A prepared a design with 21 SVE wells to address the contamination at the site.

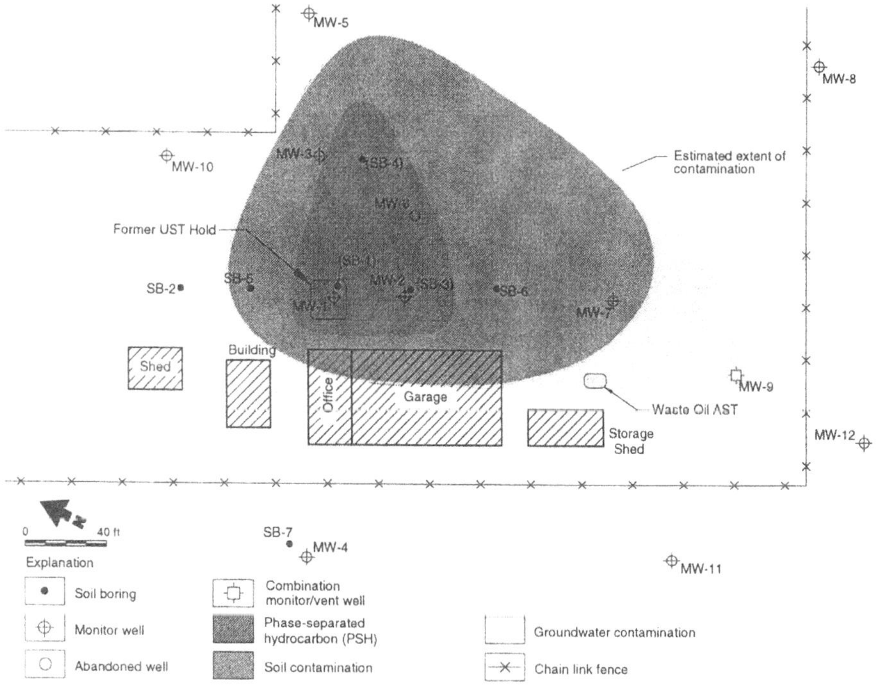

Figure 5. Estimated extent of contamination, December 1994

The practicality of using very high vacuums in the field is limited by the upwelling that can occur and eventually close off the screened interval that spans the contaminated soil. In addition, while a higher vacuum generally increases the ROI, it also increases the chances for short-circuiting to the surface, especially as the top of the SVE screen gets closer to the ground surface.

3.2 High-Vacuum SVE System Design and Installation

In December 1994, DBS&A submitted a corrective action system design that included 21 combination SVE/air sparge (AS) wells in three circuits of seven wells each (Fig. 6). The dual completion wells were laid out using the 20-foot ROI determined from the high-vacuum SVE pilot test results (Fig. 6). At each location, an SVE well with a screened interval ranging in length from 15 to 45 feet was installed, along with a separate 1.5-foot AS point installed between 10 and 15 feet below the water table (Fig 7). In addition to a control valve on each of the three SVE circuits at the main manifold in the equipment compound, each SVE/AS well was outfitted with control (ball) valves for both the SVE and AS components. The system also included a vapor treatment system consisting of an internal combustion (IC) engine fitted with a supplemental fuel source to destroy the vapors collected by the SVE wells.

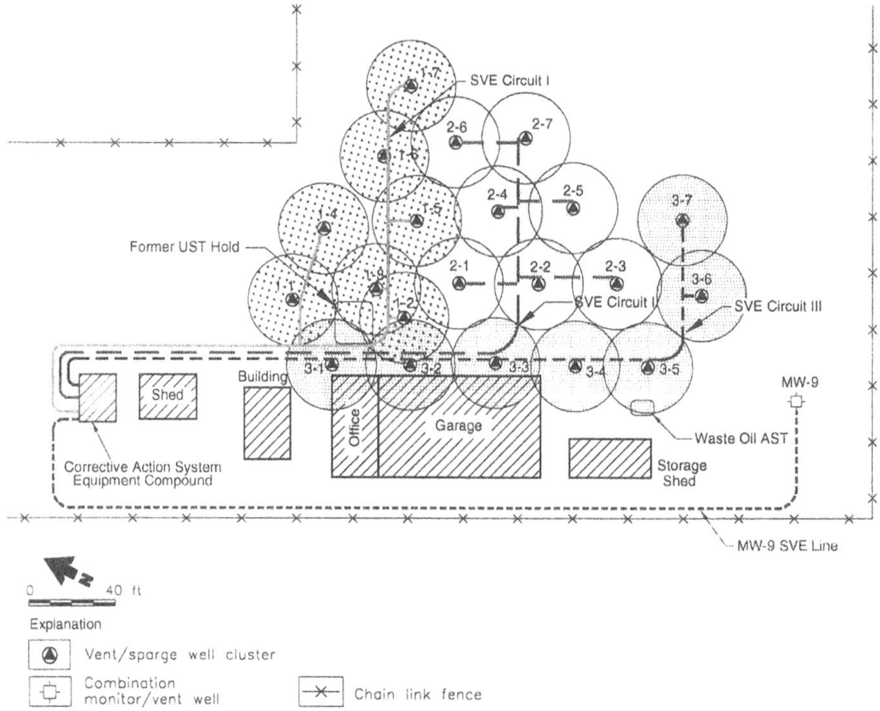

Figure 6. SVE/sparge well and conveyance system layout

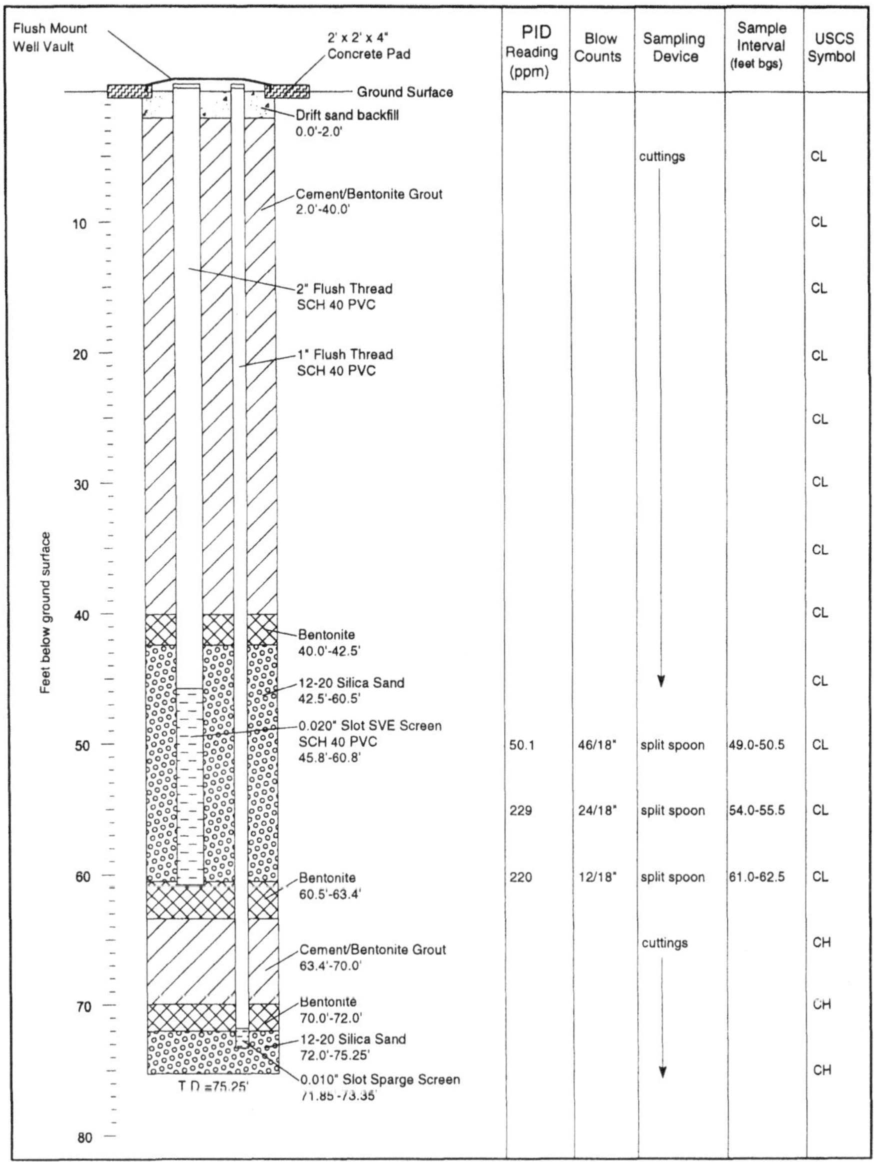

Hydrologists: Jordan
Driller: Pool Environmental

Drilling method: Hollow stem auger, CME 75
Bit diameter: 6.25 in. I.D., 10.5 in. O.D.
Date completed: 6/23/95

Figure 7. Example combination SVE/AS well

After approvals from the NMSHTD and the NMED, DBS&A constructed the system during the summer of 1995. Details of the original system installation are included in the installation report provided to the

NMED (DBS&A 1995). DBS&A's careful evaluation of site conditions resulted in cost savings to NMSHTD in excess of $200,000 on well construction alone. DBS&A further reduced construction costs by using a portable shed to house the IC engine and blowers and the assistance of NMSHTD personnel during construction.

The NMED approved an operation and maintenance plan in November 1995, and DBS&A started operation of the SVE system in early December 1995. On April 29, 1996, monitor well MW-9 was incorporated into the SVE network to provide further control of the groundwater plume (Fig. 6).

3.3 Routine High-Vacuum SVE Operations

The SVE system required no significant adjustments during the first six months of operation. Every 2 to 3 days, a local technician performed routine maintenance on the IC engine, removed condensed water from the SVE piping runs, and recorded the volume of supplemental natural gas that had been consumed.

DBS&A engineers prepared an equation to estimate the rate and volume of gasoline-equivalent hydrocarbon recovery. This equation is based on the BTU requirements of the IC engine operating at a given rate (measured in revolutions per minute). Using a standard BTU value for the natural gas and an estimated average BTU content for the recovered vapors, DBS&A engineers tabulated the rate and volume of gasoline-equivalent hydrocarbons that had been recovered (Table 1). During the initial two months of operation, these rates generally averaged more than 1 gallon per hour (gal/hr). Over the first three months of operation, more than 2,000 gallons of gasoline-equivalent hydrocarbons were recovered and consumed during approximately 1,750 hours of IC engine operation (1.15 gal/hr; see Table 1). Recovery rates declined during the second quarter of operations, but the total recovery still exceeded 1,600 gallons (Table 1, Fig. 8).

3.4 Modified High-Vacuum SVE Operations

In June 1996, the operating data exhibited a continuing decline in vapor recovery rates, as indicated by the corresponding increase in supplemental fuel use. During the first quarter of operation the gasoline-equivalent hydrocarbon recovery rate decreased from about 2 gallons per hour (gal/hr) to about 1 gal/hr. This decrease continued during the second quarter from about 1 gal/hr to about 0.6 gal/hr. In response, DBS&A and AcuVac began to modify (fine-tune) the SVE system using a variety of means. These included using well-specific information to change and adjust the operating

SVE well configuration and enhancing vapor flows by injecting air, first into the PSH and then into the perched groundwater.

Table 1. SVE System Gasoline-Equivalent Hydrocarbon Recovery

Quarter	Dates	Cumulative Months	Average Recovery Rate (gal/hr)	Recovery (gal) Quarter	Recovery (gal) Cumulative
1	12/95-02/96	3	1.15	2,003	2,003
2	02/96-05/96	6	0.78	1,609	3,612
3	05/96-08/96	9	0.72	1,528	5,140
4	08/96-11/96	12	0.40	694	5,834
5	12/96-02/97	15	0.54	945	6,779
6	03/97-05/97	17[a]	0.16	185	6,964
7	06/97-08/97	20	0.43	746	7,714
8	09/97-12/97	23[b]	0.29	217	7,931
9	12/97-01/98	25[b,c]	0.25	89	8,019

[a] The SVE system was shut down between May 3 and June 10, 1997.
[b] Full SVE system ran every fourth day.
[c] SVE system operation ended on January 30, 1998.

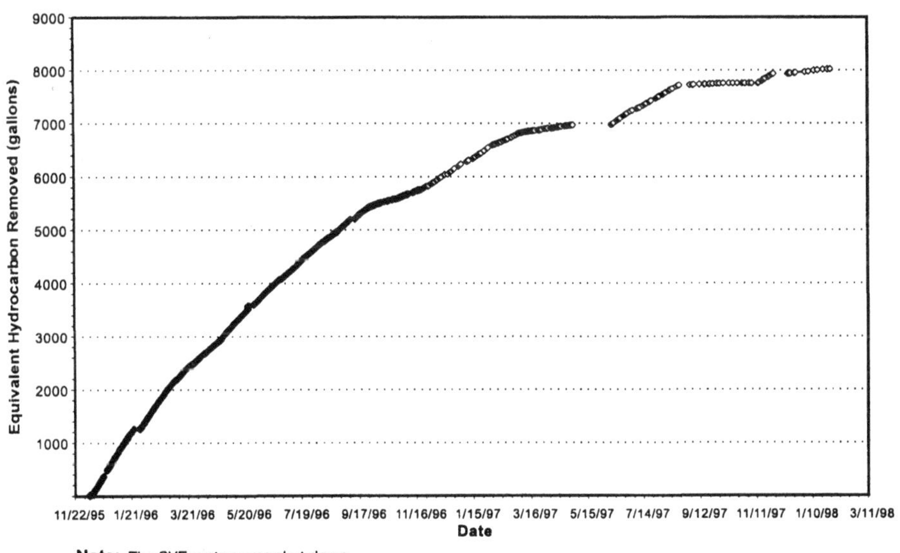

Figure 8. Cumulative gasoline-equivalent hydrocarbon removal

3.4.1 Initial SVE Mini-Pilot Tests

The primary tool used to better optimize the SVE system was a short-term testing procedure to obtain well-specific information. DBS&A and

AcuVac conducted the first of these short-term SVE mini-pilot tests in late June 1996 and completed nine tests over the next 18 months. The periodic mini-pilot tests provided information about well vacuum, vapor flow rates, and contaminant concentrations. DBS&A and AcuVac used these operational data, along with PSH and water level measurements, to balance the operation in the SVE circuits, adjusting SVE operations as necessary to optimize contaminant removal.

The mini-pilot tests consisted of measuring well vacuum and airflow in each SVE well using AcuVac's portable IC unit. During the initial tests in June 1996, stepped vacuum increases were also applied to further characterize each well's response. In addition, water levels and/or PSH were measured, and individual influent vapor samples were collected and analyzed for total hydrocarbon and carbon dioxide (CO_2) concentrations using a portable HORIBA gas analyzer.

The June mini-pilot tests showed that the flow from individual wells varied from less than 1 to 16 cfm at a vacuum of approximately 70 inches of water. The hydrocarbon vapor concentrations ranged from 188 to 4,630 ppmv (Table 2). Based on the June vapor flows and hydrocarbon concentrations, 13 wells were selected for continued SVE operation to maintain hydraulic control. These included wells 1-2, 1-5, and 1-7 from circuit 1, wells 2-1 and 2-6 from circuit 2, all wells except well 3-3 in circuit 3, and combination monitor/SVE well MW-9 (Fig. 6).

Table 2. Mini-Pilot Test Data Summary

Quarter of Remedial System Operations	Mini-Pilot Test Number	Date of Test	Maximum PSH Thickness in Monitor Wells (ft)	Concentration Range		
				Hydro-carbons (ppmv)	CO_2 (%)	O_2 (%)
3	1	6/26-27/96	~3.5 (MW-1)	188-4,630	0.60-3.08	---
3	2	8/24/96	1.21 (MW-1)	34-5,360	0.64-2.54	---
4	3	9/27-28/96	1.13 (MW-1)	40-2,040	0.56-3.28	---
4	4	11/19/96	1.30 (MW-1)	18-1,944	0.22-1.94	---
5	5	2/4/97	0.53 (MW-3)	114-7,960	0.26-2.18	---
6	6	5/1-2/97	0.36 (MW-3)	14-842	0.04-2.04	18.9-20.8
7	7	6/9-10/97	0.22 (MW-3)	22-1,672	0.08-3.36	17.2-20.2
7	8	8/23-24/97	0.24 (MW-3)	18-480	0.08-0.60	19.2-20.8
8	9	12/3/97	None	30-474	0.20-1.44	19.2-20.8
10	10	6/12/98	None	20-270	0.04-0.42	19.9-20.8

PSH = Phase-separated hydrocarbon --- = Not measured
CO_2 = Carbon dioxide O_2 = Oxygen

Satisfactory results, as indicated by relatively steady natural gas consumption and slowly increasing vacuum, were recorded until early August 1996, when flows increased and the system vacuum began to drop. A second set of mini-pilot tests was completed on August 24, 1996. The results of the August 1996 pilot tests showed that SVE well 2-6 was short-circuiting the system by providing high air flows at low hydrocarbon concentrations (12 cfm and 136 ppmv). Apparently, fresh air had found a preferential pathway into the screened interval (41 to 61 feet bgs; sand pack at 38 to 61 feet bgs), thereby leaving stagnant air in the lower section of the well (this well also contained 2.73 feet of PSH).

Based on the August vapor flow and hydrocarbon concentration results, six SVE wells were selected for continued operation: wells 1-3, 1-5, and 1-7 from circuit 1, wells 2-5 and 2-7 from circuit 2, and for downgradient hydraulic control, well MW-9. All of the SVE wells in circuit 3 were turned off due to the relatively high groundwater elevations, the high vacuums, and low air flows measured in these SVE wells during the August 1996 test. In addition, the SVE wells in circuit 3 generally had very low hydrocarbon concentrations, ranging from 34 to 249 ppmv. During this third quarter of SVE operation, the IC engine consumed another 1,500 gallons of gasoline-equivalent hydrocarbons, for a total of more than 5,000 gallons since the start of SVE system operation (Table 2; Fig. 8).

In general, air flow rates greater than 10 cfm resulted in low (500 ppm) hydrocarbon concentrations, but some moderate flow rates (5 to 10 cfm) resulted in quite high (2,000 ppm) hydrocarbon concentrations. Numerous low flow (0.0 to 3 cfm) wells also had low concentrations, while a few of these had hydrocarbon concentrations greater than 2,000 ppm. Generally, the system vacuum varied between 50 and 80 inches wc.

3.4.2 SVE with Air Bubbling into PSH

In late September 1996, the persistence of PSH was addressed by bubbling air at a low flow directly into two SVE wells (1-5 and 1-6) that contained the thickest (about 1-foot) layers of PSH. Overall, a positive vacuum was maintained on the SVE system. Although the bubbling probably diluted the overall concentration of hydrocarbon vapors recovered by the system (see Table 1, recovery rate for the fourth quarter), continued SVE operation along with the bubbling was successful in removing the PSH from these two wells.

After the adjustments made in late August 1996, the gasoline-equivalent hydrocarbon recovery rate increased from about 0.55 gal/hr to between 0.65 and 0.70 gal/hr. AcuVac conducted mini-pilot tests in the SVE wells on September 27-28, 1996. These tests showed that the flow from individual wells varied from less than 1 to 16 cfm at a vacuum of approximately 90 inches of water. Hydrocarbon vapor contents ranged from 40 to 2,040 ppmv. Based on the observed vapor flows and hydrocarbon concentrations, and with the objective of maintaining hydraulic control by applying suction to raise water levels, 10 wells were selected for continued operation. These included wells 1-5, 1-6, and 1-7 from circuit 1, wells 2-1 and 2-5 from circuit 2, wells 3-2, 3-4, 3-5, and 3-6 from circuit 3, and combination monitor/SVE well MW-9. In addition, air bubblers were installed into the PSH layers in SVE wells 1-5 and 1-6 to help remove the product that persisted in those two wells.

Mixed results, as indicated by the decreasing gasoline-equivalent hydrocarbon burn rate and increasing natural gas consumption, were observed after these adjustments to the system. The gasoline-equivalent hydrocarbon burn rate decreased from between 0.55 and 0.60 gal/hr to between 0.25 and 0.30 gal/hr. However, by the next set of mini-pilot tests, completed on November 19, 1996, the bubblers were successful in removing not only the PSH in these two wells, but also a significant portion of the PSH present in monitor wells in the vicinity.

3.4.3 SVE with Air Sparging in Groundwater

After the experience with air bubbling in SVE wells 1-5 and 1-6, operation of the AS system was initiated at two adjacent SVE wells (1-3 and 1-4) where PSH was then localized. The air flow at these two sparge points was set to a low volume (3 cfm) in an attempt to continue the PSH removal that resulted from the air bubbling in wells 1-5 and 1-6. The pressure associated with this low flow rate was about 55 pounds per square inch.

Hydrocarbon recoveries were enhanced during this (fourth) quarter of SVE system operation, and PSH thicknesses in both the monitor and the SVE wells decreased. DBS&A continued operating both the SVE and AS systems using a variety of configurations to focus on areas with the highest hydrocarbon vapor concentrations or to maintain hydraulic control. Because the systems had been constructed with separate controls located at the equipment compound and at the wellhead, the SVE and AS systems at each location could be operated independently of each other.

Similar adjustments were made during the next three quarters (5, 6, and 7) with mixed recoveries (945, 185, and 746 gallons). With SVE recovery rates declining, the schedule of operation was reduced for the eighth and ninth quarters to run the SVE system every fourth day in rotation with various AS cycles. An additional 217 gallons were recovered and destroyed during the eighth quarter and 89 gallons during the ninth quarter.

During this period of combined operation, PSH was successfully removed from all of the monitor and SVE wells (including SVE well 2-6) sometime after August 1997, and no PSH was observed for a number of months between August 1997 and December 1998. When hydrocarbon recoveries declined to approximately one tenth the rate allowed in the air permit (2 pounds per hour), the SVE system was shut down. After two years of operation, the SVE system had logged more than 13,500 hours and removed and treated more than 8,000 gallons of hydrocarbon product. The AS operation continued, but the focus of attention switched from vapor recovery and destruction to in situ biodegradation.

4. AIR-SPARGE OPERATIONS

Except for a temporary shutdown (due to a lapse in the NMSHTD contract) between July and December 1998, the AS system has operated separately since the end of January 1998. The AS system provides air flow to sweep volatile contaminants from the groundwater and promotes bioventing in the vadose zone.

One additional mini-pilot test was conducted in June 1998 to provide additional data for evaluating the success of the AS operation and the potential need to restart the SVE system. The results of this test showed that hydrocarbon concentrations were generally less than 100 ppm and oxygen concentrations were above 20 percent in all wells except one, which had an oxygen content of 19.9 percent (DBS&A 1998). These data showed that continued operation of the SVE system would not be effective.

During 1998 groundwater samples were collected for analysis of input parameters for the EPA natural attenuation model BIOSCREEN (Newell et al. 1996). These parameters included dissolved oxygen, nitrate, sulfate, ferrous iron, and dissolved methane. Although the site conditions are not ideal for the simplified transport assumptions in the BIOSCREEN model, the biogeochemical results of the modeling conducted suggest that sulfate reduction accounts for about 90 percent of natural biodegradation in the perched groundwater (Table 3).

Table 3. Biodegradation Capacity, NMSHTD Artesia Maintenance Patrol Yard

Biological Activity Indicator	Artesia Value (mg/L)	Utilization Factor[a]	Biodegradation Capacity	
			mg/L	%
Dissolved oxygen difference	4.9	3.14	1.561	1
Nitrate difference	0.75	4.9	0.153	0
Ferrous iron	0.38	21.8	0.017	0
Sulfate difference	560	4.7	119.1	90
Methane	9.5	0.78	12.18	9
		Total capacity	133.06	100

[a] Utilization factor = mg indicator/mg hydrocarbon (taken from Newell et al., 1996).

After temporarily shutting down the AS system, a thin (0.1-foot) layer of PSH reappeared in monitor well MW-2. This reappearance was associated with an extended drought during which water levels declined up to 4 feet in the perched aquifer. After several months of renewed air sparging, the PSH in this well again disappeared.

5. DISCUSSION AND CONCLUSIONS

One of the technical challenges originally encountered during remediation activities at the Artesia Yard was the persistence of PSH at the perched water table. In part, this persistence was probably associated with the fine-grained soils into which the hydrocarbon has moved over the course of a number of years. However, the NMSHTD and its consultants removed significant volumes of petroleum hydrocarbons from the low-permeability soils using a sequence of standard remediation techniques. High-vacuum SVE was the primary technique used to remove the bulk of the PSH and prepare the subsurface for ongoing bioventing.

The extension of SVE techniques to low-permeability soils was based on monitoring the vapor recovery rate and using the results of mini-pilot tests to adjust SVE system operation. The periodic mini-pilot tests provided information from individual SVE wells, including air flow, well vacuum, and hydrocarbon concentration in the extracted vapors, that was used to balance the flows. Wells with low hydrocarbon concentrations were shut off to focus remedial efforts on the most contaminated locations, and during some periods, wells with high flows were shut off to allow a more balanced flow from low flow wells or to provide hydraulic control along the periphery of the perched groundwater contaminant plume.

One of the potential shortcomings associated with the use of SVE is the development of preferential flow paths due to soil drying and cracking. The recovery of condensed water in the system attests to the effectiveness of the SVE system in removing water from the vadose zone soils. Based on results of a mini-pilot test, SVE well 2-6 was identified as a location with clear signs of this kind of cracking and, ultimately, short-circuiting. The data from the mini-pilot test indicated that this well had a relatively high air flow with very low hydrocarbon concentrations, suggesting that a pathway had developed through which fresh air from the atmosphere could enter the vadose zone. Therefore, this well was removed from the operating configuration, despite the fact that the well was located downgradient from the former tank hold and contained a significant layer of PSH. Instead, the remediation effort was switched to adjacent locations where a pathway to fresh air had not developed. This adjustment was successful in removing PSH from this location.

Direct bubbling into the PSH and air sparging were also used to enhance SVE recovery rates. Again, the hydrocarbon recovery rates (and supplemental fuel use) and the results of the periodic mini-pilot tests provided the basis on which to determine the SVE and AS operating configurations.

The circumstances frequently changed, and scientific planning and frustration often went hand-in-hand. Although a variety of strategies continued to be used to maintain the hydrocarbon recovery rate, at one point (May 1997) the entire remediation system was simply shut off to allow the subsurface to equilibrate. By this time nearly 7,000 gallons of gasoline-equivalent hydrocarbons had been recovered. The results of the seventh mini-pilot test were then used to make adjustments to the operating configuration. The recovery rate temporarily improved in the following quarter, during which the last PSH observed in the wells disappeared for a number of months.

Quarterly monitoring of groundwater chemistry indicates that the ongoing air sparging in the perched aquifer and resulting bioventing in the vadose zone is continuing to be successful, as evidenced by sharp decreases in the concentrations of benzene and total BTEX. Table 4 shows the concentrations of these constituents in three representative wells: MW-2, which is located in the PSH area, and MW-7 and MW-9, which are outside and downgradient of the main (PSH-affected) contaminated soil area. The decreases in benzene and BTEX are also shown in Figs. 9 and 10, respectively. Although the concentration of benzene is still far above the regulatory standard, continued bioventing and degradation should continue to remove this compound from both the groundwater and the vadose zone beneath the Artesia Yard.

Table 4. Selected Groundwater Data

Well Number	Date Sampled	PSH	Concentration (μg/L) Benzene	Toluene	Ethyl-benzene	Total Xylenes	Total BTEX
MW-2	01/25/94	Y	NS	NS	NS	NS	NS
	12/06/95	Y	23,000	22,000	440	6,400	51,840
	03/06/96	Y	NS	NS	NS	NS	NS
	06/13/96	Y	28,000	29,000	3,800	9,200	70,000
	02/16/98	N	520	100	98	630	1,348
	05/21/98	N	190	<20	27	380	597
	09/15/99	N	9.5	<1.0	2.2	10.9	23
MW-7	01/25/94	N	880	4,500	810	4,100	10,290
	12/05/95	N	1,800	2,500	1,200	3,800	9,300
	03/06/96	N	270	520	170	430	1,390
	06/13/96	N	460	1,300	530	1,700	3,990
	09/04/96	N	91	92	81	130	394
	12/03/96	N	400	62	240	120	822
	02/26/97	N	430	510	190	1,900	3,030
	05/07/97	N	270	130	210	300	910
	08/13/97	N	200	<2.5	73	310	583
	11/20/97	N	10	19	24	120	173
	02/16/98	N	<5.0	25	74	450	549
	05/21/98	N	0.7	1.2	6.6	63	72
	12/04/98	N	2.1	<1.0	4.7	19.3	26
	03/19/99	N	<1.0	<1.0	4.3	29	33
	06/25/99	N	<1.0	<1.0	5.9	90	96
	09/15/99	N	<5.0	<5.0	33	335	368
MW-9	04/29/94	N	96	180	45	190	511
	12/05/95	N	4,400	14,000	2,600	9,000	30,000
	03/06/96	Y	NS	NS	NS	NS	NS
	06/13/96	N	3,600	17,000	3,500	12,000	36,100
	09/04/96	N	1,200	5,700	1,700	6,700	15,300
	12/03/96	N	690	6,100	2,100	7,900	16,790
	02/26/97	N	350	2,200	1,300	6,000	9,850
	05/07/97	N	240	840	410	1,400	2,890
	08/13/97	N	90	1,100	1,400	4,900	7,490
	11/20/97	N	160	220	660	2,500	3,540
	02/16/98	N	170	140	600	700	1,610
	05/21/98	N	120	120	780	1,300	2,320
	12/04/98	N	43	52	590	810	1,495
	03/19/99	N	50	18	400	380	848
	06/25/99	N	190	<10	27	151	368
	09/15/99	N	200	5.1	570	131	906

PSH = Phase-separated hydrocarbon (Y=yes; N=no) μg/L = Micrograms per liter
BTEX = Benzene, toluene, ethylbenzene, and xylenes NS = Not sampled

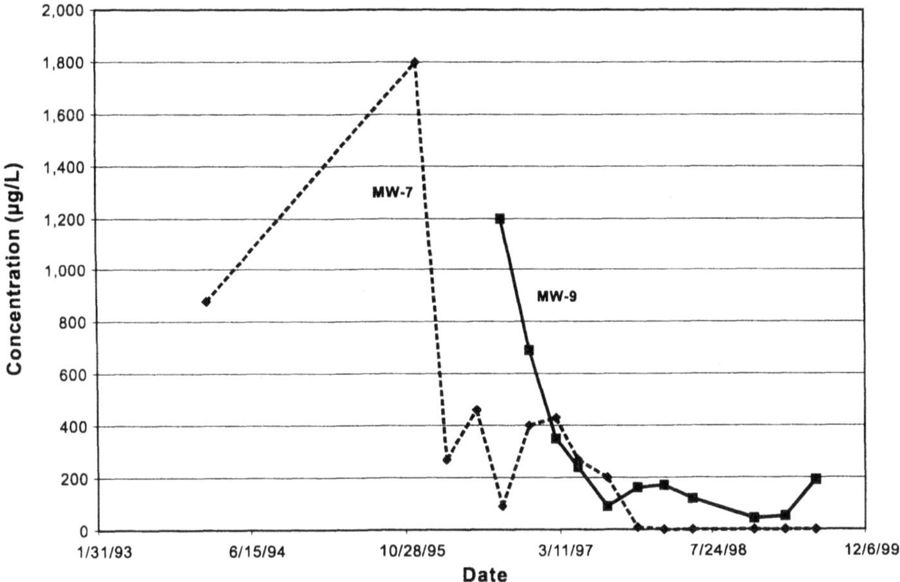

Figure 9. Benzene concentration in groundwater

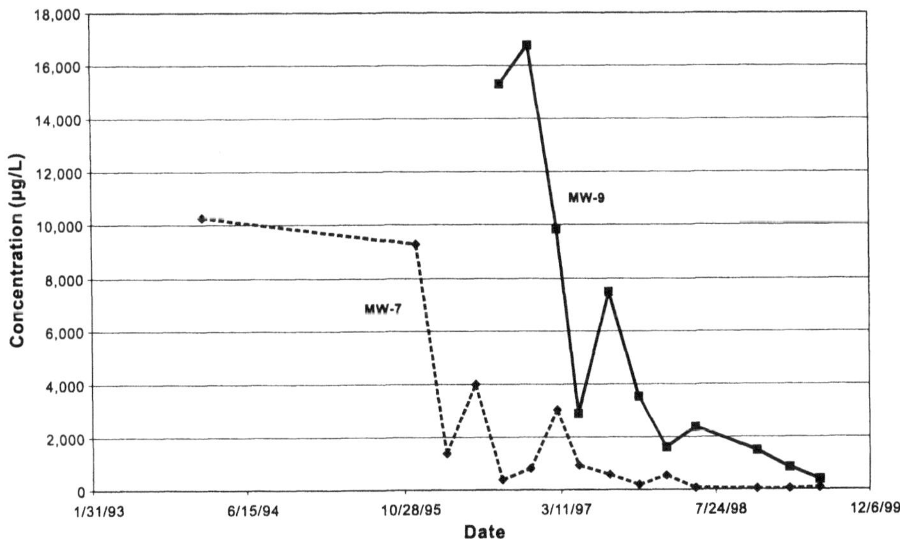

Figure 10. BTEX concentration in groundwater

BIBLIOGRAPHY

Daniel B. Stephens & Associates, Inc. (DBS&A). (1995). "Ground-Water and Soil Reclamation System Installation Summary, Artesia Maintenance Patrol Yard." Prepared for New Mexico State Highway and Transportation Department, District 2 Office, Roswell, New Mexico, November 28, 1995.

———. (1998). "Quarterly monitoring report, March 1, 1998 through June 12, 1998, NMSHTD Artesia Maintenance Patrol Yard." June 18, 1998.

INTERA, Inc. and Billings and Associates, Inc. (BAI). (1994a). "Hydrogeologic Investigation Report, USTR 1210, NMSHTD District 2 Patrol Yard, Artesia, New Mexico." Prepared for New Mexico State Highway and Transportation Department and New Mexico Environment Department Underground Storage Tank Bureau, March 25, 1994.

———. (1994b). "Monitor Well Installation and SVE Pilot Test, NMSHTD District 2 Patrol Yard, Artesia, New Mexico." Prepared for the New Mexico State Highway and Transportation Department and the New Mexico Environment Department, Underground Storage Tank Bureau, July 7, 1994.

Johnson, P.C., Kemblowski, M.W., and Colthart, J.D. (1990) "Quantitative analysis for the cleanup of hydrocarbon-contaminated soils by in-situ soil venting." *Ground Water,* 28(3), 413-429.

Newell, C.J., McLeod, R.K., and Gonzales, J.R. (1996). BIOSCREEN Natural Attenuation Decision Support System, User's Manual, Version 1.3. EPA/600/R-96/087. Office of Research and Development, Washington D.C. 63 p.

U.S. Environmental Protection Agency (EPA). (1995). "How to Evaluate Alternative Cleanup Technologies for Underground Storage Tank Sites: A Guide for Corrective Action Plan Reviewers." EPA 510-B-95-007. Office of Solid Waste and Emergency Response, Washington D.C., May 1995.

Contributors

David P. Ahlfeld
University of Massachusetts
Amherst, MA

Aysegul Aksoy
University of Virginia
Charlottesville, VA

Souhail R. Al-Abed
US EPA
Cincinnati, OH

Akram N. Alshawabkeh
Northeastern University
Boston, MA

Robert S. Bowman
New Mexico Institute of Mining & Technology
Socorro, NM

Todd Burt
New Mexico Institute of Mining & Technology
Socorro, NM

Elizabeth R. Carraway
Clemson Institute of Environmental Toxicology
Pendleton, SC

Jiann-Long Chen
United States Environmental Protection Agency
Cincinnati, OH

Hefa Cheng
University of Oklahoma
Norman, OK

Jeffrey D. Childs
University of Oklahoma
Norman, OK

Jee-Won Choi
University of Virginia
Charlottesville, VA

Teresa B. Culver
University of Virginia
Charlottesville, VA

James J. Deitsch
University of Kentucky
Lexington, KY

Baolin Deng
New Mexico Institute of Mining and Technology
Socorro, NM

Kim F. Hayes
University of Michigan
Ann Arbor, MI

Richard Heibel
New Mexico State Highway and Transportation Department
Santa Fe, NM

Amy Chan Hilton
Florida State University
Tallahassee, FL

Shaodong Hu
New Mexico Institute of Mining and Technology
Socorro, NM

Richard L. Johnson
Oregon Graduate Institute of Science and Technology
Beaverton, OR

Timothy L. Johnson
Oregon Graduate Institute of Science and Technology
Beaverton, OR

Whitney Katchmark
Naval Engineering Command
Norfolk, VA

Ashutosh Khandelwal
S.S. Papadopulos & Associates
Bethesda, MD

Tohren G. Kibbey
University of Oklahoma
Norman, OK

Seok-Oh Ko
Daewoo Institute of Construction Technology
Seoul, Korea

Zhaohui Li
New Mexico Institute of Mining and Technology
Socorro, NM

Krishnanand Maillacheruvu
Bradley University
Peoria, IL

Richard E. Meixner
Daniel B. Stephens & Associates, Inc.
Albuquerque, NM

Ann E. Mulligan
Woods Hole Oceanographic Institution
Woods Hole, MA

Naoko Munakata
Stanford University
Stanford, CA

Kurt D. Pennell
Georgia Institute of Technology
Atlanta, GA

Alan J. Rabideau
University at Buffalo
Buffalo, NY

C. Andrew Ramsburg
Georgia Institute of Technology
Atlanta, GA

Martin Reinhard
Stanford University
Stanford, CA

Craig R. Repp
West Valley Nuclear Services LLC
West Valley, NY

Elizabeth J. Rockaway
University of Kentucky
Lexington, KY

Stephen J. Roy
New Mexico Institute of Mining and Technology
Socorro, NM

David A. Sabatini
University of Oklahoma
Norman, OK

James E. Sadler
AcuVac Remediation, Inc.
Houston, TX

Mark A. Schlautman
Clemson University and Clemson Institute of Environmental Toxicology
Clemson, SC and Pendleton, SC

James A. Smith
University of Virginia
Charlottesville, VA

Tammy P. Taylor
Los Alamos National Laboratory
Los Alamos, NM

Fred D. Tillman, Jr.
University of Virginia
Charlottesville, VA

John van Benschoten
University at Buffalo
Buffalo, NY

Houston G. Wood, III
University of Virginia
Charlottesville, VA

Bin Wu
University of Oklahoma
Norman, OK

Index

www.ingramcontent.com/pod-product-compliance
Ingram Content Group UK Ltd.
Pitfield, Milton Keynes, MK11 3LW, UK
UKHW051127260726
13967UKWH00010B/2908

* 9 7 8 1 4 7 5 7 8 7 6 5 8 *